Glacial Geologic Processes

David Drewry

Scott Polar Research Institute, University of Cambridge

Edward Arnold

© David Drewry 1986

First published in Great Britain 1986 by
Edward Arnold (Publishers) Ltd, 41 Bedford Square, London WC1B 3DQ

Edward Arnold (Australia) Pty Ltd, 80 Waverley Road, Caulfield East,
Victoria 3145, Australia

Edward Arnold, 3 East Read Street, Baltimore, Maryland 21202, USA.

British Library Cataloguing in Publication Data

Drewry, David
   Glacial geologic processes.
   1. Glaciers
   I. Title
   551.3′1   QE576
   ISBN 0 7131 6485–9
   ISBN 0–7131–6390–9 Pbk

Text set in 10/12pt Times Compugraphic
by Colset Private Ltd., Singapore
Printed and bound by The Bath Press, Bath

Contents

Contents

Contents

Preface

It is often said that if the Royal Albert Hall in Kensington, London were filled with chimpanzees, each with a typewriter tapping away at random, it would take several hundreds of thousands of years for the complete works of Shakespeare to be produced. Whilst this book can claim no comparison with Shakespeare in other respects I have often been struck by the appropriateness of the analogy. However, not having as much time as the chimpanzees I finally decided to abandon a first draft of this book (GGP as it has become known to long-suffering friends and helpers) to the publishers in late 1983.

I am acutely aware of the considerable shortcomings of GGP; of the sections that deserve more considered development, those that have given inadequate reviews of previous and current research, and sections that will turn out to be misleading or plainly wrong. In all these matters responsibility is entirely my own.

GGP has evolved over the last few years out of a series of lectures I have given in Cambridge. The substance of the text is founded on my belief that it is only by scientifically establishing a background understanding of primary physical processes that the geological role of ice can be adequately understood. The interpretation and correlation of complex stratigraphic glacial sequences, for instance, and palaeo-environmental reconstructions therefrom require an initial familiarization with basic glaciological concepts in order to establish and develop genetic explanations. There is little merit, in my opinion, attached to the simple description of glacial landforms.

GGP was written, therefore, as an attempt to bridge a gap between purely glaciological texts (such as Paterson's excellent general introduction or Hutter's indigestible but rigorous mathematical treatise) and the community of geologists and geomorphologists who deal with the products of ice activity at the surface of the Earth. Notwithstanding this aim I have during the writing of this book been constantly impressed by how little is really known about physical process in the glacial environment. If for no other reason than to point out these deficiencies and encourage others to undertake future investigations I shall feel the effort in preparing this book has been worthwhile.

Numerous people have aided me (willingly and under duress!) in the production of GGP. In the first instance I thank those innumerable and unsuspecting Part II students upon whom these ideas fell. David Stoddart and D. Pasquale between them suborned me with numerous glasses of good Frascati and Barolo into considering that I should indeed write this text. Stoddart introduced me, fatefully, to my editor at Edward Arnold whose enthusiasm and sheer bonhomie has seen this book to completion. I owe him especial thanks for his encouragement, understanding and friendship. I would not have completed GGP without the assistance of Jane Torrance who organized my manuscript on our word processor, sorted diagrams, arranged much of the bibliography and countless other tasks. I am sorry she escaped before the task was completed! Sylvia Peglar and Sue Jordan and Mark Welford spent many hours drawing clear, intelligible diagrams from my sketches. I am indebted to them for their patience and understanding. Paul Cooper assisted with aspects of computer processing of some data and Eva Novotny, Liz Morris, Norris Riley, Julian Dowdeswell, Robert Delmas, Gerry Lock, and Bernard Hallet kindly read through some or all parts of the MS. Other persons who helped in various ways were Neil McIntyre, Martin Horkley, Jenny Walsby and Sue Pilkington.

I am particularly grateful to the Scott Polar Research Institute and Gordon Robin its former Director, for excellent working facilities. Harry King and later Valerie Galpin and their library staff helped with many questions and queries.

I wish to finally acknowledge and thank my wife, Gill, who has endured countless bleak and wretched weekends and evenings when work on 'the book' took precedence.

David Drewry,
Cambridge, 1985

Acknowledgements

The author and publishers wish to thank the following for permission to reproduce copyright material. Full citations can be found by consulting the captions and references:

Academic Press for Figs. 5.3 and 5.8; the American Concrete Institute for Figs. 5.5 and 5.9; the American Society of Civil Engineers for Fig. 3.3; *Annual Review of Earth and Planetary Sciences* for Fig. 1.4, © Annual Reviews Inc;, vol. 11; *Arctic and Alpine Research* and the University of Colorado for Fig. 11.7; the Arctic Institute of North America, Arlington (Virginia) for Figs. 14.12 and 14.13; the Arctic Institute of North America, the University of Calgary for Fig. 14.10; the Australian Academy of Science for Fig. 13.10; Balkema (Rotterdam) for Fig. 11.16; Bell & Hyman, Publishers for Fig. 5.6; H. Björnsson for Fig. 2.8; Blackwell Scientific Publications Ltd for Fig. 7.13; G.S. Boulton for Figs. 4.9; 8.1 and 8.3A; the British Hydromechanics Research Association for Fig. 5.4; Butterworth Scientific Ltd for Fig. 8.11; *Canadian Journal of Earth Sciences* for Fig. 11.13; *Canadian Journal of Fisheries and Aquatic Sciences* for Figs. 12.3 and 12.4; the Canadian Society of Petroleum Geologists for Fig. 10.14; C-Core for Fig. 14.5; Chapman & Hall Ltd for Fig. 3.2; Cold Regions Research and Engineering Laboratory (CRREL) for Figs. 7.3, 7.5, 7.6, 7.7, 8.15A–D and 14.14; Department of Geography, University of Quelph for Figs. 4.11, 4.12, 9.3, 9.4 and 11.5A and C; Dover Publications Inc. for Fig. 11.8; Lon D. Drake for Fig. 8.13; *Earth Surface Processes* for Fig. 5.11; Elsevier Scientific Publications Ltd for Figs. 10.11, 14.15 and 14.16; Eros Data Center for Fig. 13.1A; Forschungsinstitut Senkenberg for Fig. 12.11; Geo Books for Figs. 4.11, 4.12, 9.3 and 9.4; *Geografiska Annaler* for Fig. 10.15; *Geological Society of America Bulletin* for Figs. 8.13 and 13.12; The Geological Survey of Canada for Fig. 10.7; R.P. Goldthwait for Figs. 8.10 and 9.5; B. Hallet for Fig. 6.1; F.G. Hammett for Fig. 5.7; S.M. Hodge for Fig. 2.1; R. Hooke for Fig. 1.5; Institute of Hydrology for Fig. 7.15; Institute of Polar Studies for Fig. 8.14; The Institution of Mining and Metallurgy for Fig. 4.4; the International Association of Hydrological Sciences for Fig. 10.2; The International Glaciological Society for Figs. 1.5, 2.1, 2.4, 2.5, 2.9, 3.11, 4.7, 4.9, 4.10, 5.13, 5.14, 6.1, 7.9, 8.3, 9.6, 9.9, 9.10, 9.13, 10.3B, 12.12, 13.5, 13.7 and 13.8; *Jökull* for Figs. 2.8 and 2.11; *Journal of Sedimentary Petrology* for Fig. 8.15E; K. Kitzaki for Fig. 8.3B; D. Lawson for Fig. 9.11; Macmillan for Figs. 4.6 and 11.8; L.W. Morland for Fig. 3.8; Museum National d'Histoire Naturelle for Fig. 14.4; the National Book Service, Washington for Fig. 7.12; Norges Vassdrags- og Elektrisitetsvesen — Vassdragsdirektoratet Hydrologist Avdeling for Fig. 2.10; *Norsk Geografisk Tidsskrift*, Universitets Forlaget for Fig. 12.10; Norsk Polarinstitutt for Figs. 1.12B, 1.13, 12.6 and 12.9; Ohio State University Press for Figs. 8.10 and 9.5; G. Østrem for Fig. 10.4; Pergamon Press for Fig. 1.2; Plenum Publishing Corpn for Fig. 14.4; *Polar Record* for Fig. 13.4; R. Powell for Fig. 12.8; the Royal Geographical Society for Fig. 14.1B; The Royal Society, London for Fig. 3.8; *Science* for Fig. 1.12A; Society of Economic Palaeontologists and Mineralogists, Tulsa, Okla. for Figs. 10.3A, 10.6B, 10.12, 10.13, 11.11, 11.12 and 11.15B; P. Tchernia for Fig. 14.4B; R. Thomas for Fig. 13.5; G.P. Tilly for Fig. 5.3; J. Weertman for Figs. 2.5, 2.6 and 7.9; John Wiley & Sons Ltd for Fig. 8.7; R. Wright for Fig. 13.12 and *Zeitschrift für Gletscherkunde und Glazialgeologie* for Figs. 2.3, 2.7B, 3.6, 3.7 and 5.12.

1 Physical properties of ice, glacier dynamics and thermodynamics

1.1 Introduction

The study of the processes of glacial erosion and glacial sedimentation must commence with familiarization with some basic concepts of physical glaciology. The processes that control and characterize the ice–rock and ice–water interfaces are particularly important. The glaciological concepts required fall into five principal categories: physical properties of ice, large-scale dynamics, **thermodynamics**, hydrological conditions and time-variant behaviour. All of these aspects have been dealt with in specialized texts (Hobbs, 1974; Paterson, 1981; Colbeck, 1980; Hutter, 1983). The intention in this chapter is to provide a short summary of fundamental ice and glacier physics relevant to the development of ideas in later sections of this book.

1.2 Physical properties: structure and composition of ice

Naturally occurring inorganic substances can be classified in several different ways: as fluids or solids, as metals and non-metals, or by chemical composition in general or mineralogical habit in particular. Ice is a *mineral* as it possesses a definite chemical composition and atomic structure and would come, therefore, within the field of study of geology or earth science. Nevertheless it is also possible to view ice as a material substance and one that is of interest to engineers. Of the four basic groups of materials (metals and alloys, **polymers, composites and ceramics**) ice falls into the latter category by being non-metallic, inorganic and crystalline.

Naturally occurring ice (Ih) is polycrystalline, composed of hydrogen-bonded oxygen atoms, the latter forming the core of regular tetrahedra. There are four bonds with adjacent oxygen atoms located at the corners: two positive, two negative. The molecular spacing in ice is 2.76 Å and the weight of a molecule is 2.992×10^{-26} kg.

Important **isotopes** of hydrogen and oxygen (nuclides with same atomic number but varying in atomic mass because of a difference in number of neutrons in the nucleus), are Deuterium (D), Tritium (^{3}H) ^{16}O and ^{18}O. The composition of most natural water is given in Table 1.1.

Table 1.1 Isotopic composition of natural water

	%
$^1H_2\ ^{16}O$	99.73
$^1H_2\ ^{18}O$	0.20
$^1H_2\ ^{17}O$	0.04
$^1HD^{16}O$	0.03

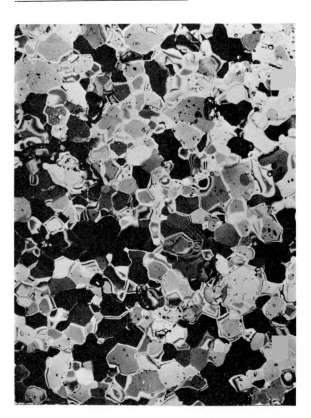

Figure 1.1 Photo-micrograph of thin section of ice from McCall Glacier, Alaska viewed in polarized light. Ice sample from a depth of 10 m from the centre of the glacier (photo G. Wakahama courtesy International Glaciological Society).

The structure of ice can readily be observed in thin section on a universal microscopic stage using plane or **polarized light** (Figure 1.1) in a manner similar to optical analysis of rocks and minerals. A mosaic of interlocking crystals is evident. Growth of crystals by low stress metamorphism (compaction and **sintering**) results in a predominantly random orientation of grains. However, because of the effects of **stress** and subsequent deformation induced by glacier flow, ice may exhibit a preferred orientation of crystals.

Some important properties of ice are summarized in Table 1.2. The most important of the properties for general glaciological considerations are those relevant to ice flow.

1.3 Deformational behaviour: creep

To investigate deformational behaviour of ice, experiments may be conducted either on single crystals and **polycrystalline ice** grown in the laboratory or on natural glacier ice. While the results from single crystal experiments especially have assisted our understanding of the role of several important physical effects (e.g. defects and **dislocations**), it is the behaviour of naturally occurring polycrystalline ice that is of concern in a study of mechanisms important for erosion and sedimentation.

Ice deformational behaviour is confined to an elastic response and plastic (viscous) **creep** which can be understood by reference to a typical stress–strain curve (Figure 1.2). With only small stresses (σ), there is complete recovery of the original form: i.e. an elastic response occurs. If, however, the stress exceeds a critical amount plastic deformation takes place involving incomplete recovery of the original form if the stress is removed. Irreversible changes occur to the structure of the ice when the **elastic limit** or **yield strength** (σ_y) is reached. Such deformation under constant stress is termed creep and is due to the breaking of the bonds between atoms. If ice crystals were perfect the strength of the ice would be given by **Young's modulus** (E) and the interatomic potential – see Equation (3.10) ($E/15 = 6.1 \times 10^8$ N m^{-2})(Ashby and Jones, 1980, p. 87). The actual strength of ice is much lower (of the order of 8.5×10^7 N m^{-2}) due to presence of defects or dislocations within the crystal lattice. A dislocation is a line of atoms with slightly different coordination than normal.

Table 1.2 Selected mechanical and thermal properties of ice

	Property	Units	Symbol	Quantity
1	*Mechanical*			
	Density	kg m^{-3}	ρ_i	920
	Young's modulus	GN m^{-2}	E	9.10
	Yield strength	MN m^{-2}	σ_y	85.0
	Fracture toughness	MN m$^{-3/2}$	K_c	0.2
	Toughness	kJ m^{-2}	G_c	0.003
	Creep activation energy	J mol^{-1}	Q	6.07×10^4
	Flow law constant	Pa^{-3} s^{-1}	B	8.75×10^{-13}
	Pressure melting coefficient	Deg Pa^{-1}		-0.7×10^{-7}
2	*Thermal*			
	Melting temperature	K		273.1
	Thermal conductivity	W m^{-1} Deg^{-1}	K_i	2.51
	Thermal diffusivity	m^2 s^{-1}	k_i	1.33×10^{-6}
	Heat capacity	J mol^{-1} Deg^{-1}		37.7
	Latent heat (fusion)	kJ kg^{-1}		
	0°C			334
	-10°C			285
	-20°C			241

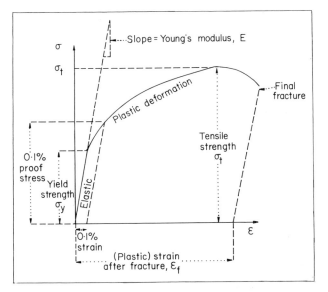

Figure 1.2 A typical stress–strain curve for ice as described in the text. σ_y is the yield strength and defines the stress at which plastic creep begins. σ_t is the tensile strength at which specimen necking commences in extensional tests. 0.1% proof stress is the load required to induce 0.1% permanent strain in the material; it is a particularly useful parameter in substances like ice which show no distinctive yield point and deform gradually (from Ashby and Jones, 1980).

1.3.1 Single crystals

Single crystals readily undergo deformation by slip along the basal crystallographic plane (that perpendicular to the 'c' axis of hexagonal symmetry, with **Miller Index** 0001). Slip on other systems is possible but requires much higher stresses. The basal plane possesses dislocations and these provide an energetically favourable means of allowing planes of hydrogen atoms to move over each other (i.e. to glide) and hence deform even under very small applied stresses. The principal dislocations found in ice are of the screw variety.

Unlike the case for most metals work softening rather than **work hardening** occurs during basal gliding; that is, the strain-rate increases with time. Glen (1955) has explained this phenomenon by suggesting that, during initial straining, there are few mobile dislocations and deformation is restricted to a few locations with relatively low speeds. The density and velocity of dislocations increase slowly. Only if normal basal gliding is prevented is work hardening observed.

1.3.2 Polycrystalline ice

Randomly oriented polycrystalline ice is less readily deformable than monocrystals by as much as two orders of magnitude. This is due primarily to the fact that few crystals are initially oriented for **gliding** in the direction of an applied stress. Figure 1.3 illustrates the creep behaviour of polycrystalline ice at various stresses but at constant temperature.

Loading results in instantaneous creep (elastic strain). Then the strain rate decreases with time (transient creep). There is, therefore, work hardening unlike the case for single crystals and termed transient creep. At higher stresses (more than several tens of kPa) a constant rate of creep sets in (secondary creep). This is succeeded by a tertiary stage of accelerating creep, caused by

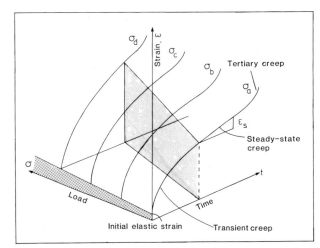

Figure 1.3 Creep curves for polycrystalline ice indicating general behaviour at various applied loads (σ_a to σ_d). Note the initial elastic strain followed successively by transient (or primary), steady-state (secondary) and tertiary creep. At low stresses, simulating conditions in a glacier, it may take several years to achieve even steady-state creep.

growth of strain-free crystals (**recrystallization**) with preferred orientation, many of the grains having 'c' axes nearly orthogonal to the maximum shear stress (which eases gliding). The creep rate is then intermediate between that of randomly oriented polycrystalline ice and single crystals oriented for basal gliding. Finally there may be fracture. From numerous laboratory experiments at different stresses (σ) it is possible to determine the relationship between minimum strain rate in steady-state secondary creep ($\dot{\epsilon}_s$) and applied stress (Figure 1.4). The form of the curve in Figure 1.4 indicates that the creep relationship is a simple power relationship termed the 'Glen Flow Law':

$$\dot{\epsilon}_s = B\sigma^n \qquad (1.1)$$

where B = temperature-dependent ice hardness parameter

   n = creep exponent (a visco-elastic parameter)

The exponent n is given by the slope of the line through experimentally determined points of a plot of log $\dot{\epsilon}_s$ verses log σ.

Two limiting cases can be easily appreciated: n = 1 (linear or Newtonian flow) and n = ∞ (perfectly plastic creep). The latter is often used as an approximation to more realistic behaviour. Glen (1955) found the value of n for ice to vary between 1.9 and 4.5 with a mean value of 3, in reasonable agreement with field determinations of n for the intermediate range of stress 0.1–1.0 MPa (Nye, 1953; Budd, 1969; Thomas, 1973a). It is clear that the creep rate is relatively unaffected by **cryostatic pressures** of up to 30 MPa and provides the physical basis for rejecting older ideas of **'extrusion' flow** in glaciers.

Polycrystalline ice 'softens' remarkably at temperatures above about – 10°C so that the steady-state creep rate may also be expressed in terms of temperatures, usually by an Arrhenius-type equation:

$$\dot{\epsilon}_s = B_o \exp\left(-Q/R_g T_a\right)\sigma^n \qquad (1.2)$$

where B_o = constant independent of temperature

   Q = **activation energy**

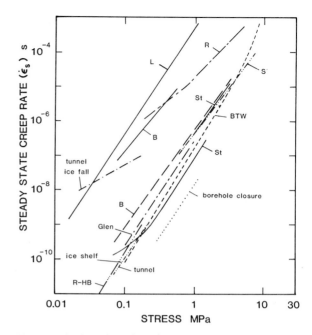

Figure 1.4 Log–log plot of the steady-state creep rate against stress (after Weertman, 1983). The creep rate has been normalized to – 10°C, with Q = 134 kJ mole^{-1} for polycrystalline ice above – 10°C and 60 kJ mole^{-1} below this temperature with Q set to that given by the experimenter. Data from numerous sources are given in Weertman, 1983.

   R_g = gas constant (8.314 × 10^3 JK^{-1} kmol^{-1})

   T_a = absolute temperature

Equation (1.2) suggests that under conditions of constant stress the logarithm of the strain-rate shows a linear dependence on absolute temperature.

Figure 1.4 plots the temperature-normalized secondary creep rate against stress as measured in a variety of laboratory and environmental experiments.

1.3.3 Creep of natural ice

Significant differences exist between laboratory polycrystalline ice and ice naturally occurring in glaciers and ice sheets. These are principally differences in grain size, crystal fabric, stress and temperature regime and content of water, air and impurities. In general laboratory ice has smaller crystals with less variation in size than ice found in glaciers where crystals as large as several tens of cm^2 may occur towards the base of large ice sheets or in slowly moving ice. Laboratory ice may be bubble-free whereas glacier ice commonly contains air trapped during the sealing-off of pores at the **firn**–ice transition. These bubbles occur in concentrations that are typically in the order of 100 ppm (weight) and become smaller in size under increasing cryostatic pressure. When the pressure exceeds 10 MPa the air molecules may diffuse into the ice. The orientation of crystals in glacier ice often shows very pronounced, tight single pole 'c' axis fabrics, which have been impossible to reproduce in the laboratory even under the highest strains.

The low stress region commonly encountered in glaciers at temperatures close to the **melting point** presents particular problems for simulation by short-period laboratory studies. Considerable time (years) is necessary to achieve **steady-state** creep and often the environmental conditions are not easy to produce or maintain.

Ice at the melting point may contain a significant amount of water located at grain boundaries, **triple junctions**, grain corners and around air bubbles. In addition the presence of soluble (chemicals) and insoluble impurities (debris) is very common in natural ice and further complicates deformation behaviour.

It becomes clear from such factors that creep measurements on specimens of glacier ice (either brought to the laboratory or *in situ*) are necessary in order to fully explore and evaluate the flow of ice in glaciers and ice sheets. Some examples are discussed below to show field-related studies can assist in formulating realistic understanding of ice deformation.

1   The first example is provided by the work of Colbeck and Evans (1973), who performed creep experiments on ice samples taken from the Blue Glacier, Washington within the stress range 6–10 kPa and with temperatures at the pressure melting point. Their results suggest a polynomial flow law:

$$\dot{\epsilon}_s = (2.1 \text{ MPa})\, \sigma + (1.4 \text{ MPa}^3)\sigma^2 + (5.5 \text{ MPa}^5)\sigma^5 \tag{1.3}$$

At low stresses a more linear viscous creep is evident while intercrystalline gliding is the principal mechanism in deformation at the higher stresses.

2   A second technique for determining creep rate is to examine the flow of ice under its own weight in an **ice shelf** resting upon a frictionless bed (i.e. sea water). This method has been applied by Dorrer (1971) and more generally by Thomas (1973a) to determine B and n in the flow law from data for the Amery, Ross, Brunt and Maudheim Ice Shelves in Antarctica and the Ward Hunt Ice Shelf in NWT, Canada. Thomas's results suggest that Equation (1.1) is applicable with n $\approx$ 3, for the stress range 1 MPa–0.04 MPa and maybe to stresses as low as 0.01 MPa. They do indicate, however, that natural ice appears to be 'stiffer' than laboratory samples. It is uncertain whether the relatively good correspondence of laboratory and field determinations observed by Thomas applies at temperatures higher than $-6°C$ to $-10°C$.

3   A final field method is reported by Paterson (1977, 1981) namely the measurement of borehole or tunnel closure rates, at temperatures between $-16°C$ and $-28°C$ and with shear stresses of between 15 kPa and 100 kPa. Paterson's results for steady-state secondary creep are given in Figure 1.4 and yield a value of n = 2.7 by regression, quite close to 3. The borehole strain rates are again significantly lower than those for laboratory polycrystalline ice and confirm Thomas's idea that natural ice may be stiffer.

1.3.4 Discussion

The logarithm of strain rate when plotted against stress for laboratory and field experimental data covering a wide range of temperatures and stresses may show an increase with increasing stress and, at low stresses (those of importance in most glaciers), the relationship may be close to $n \approx 1$ (Newtonian flow). Such results suggest that a power flow law with constant n may be inapplicable. To accommodate this evidence a two-termed creep equation can be used:

$$\dot{\epsilon}_s = k_1\sigma + k_2\sigma^n \qquad (1.4)$$

The first term on the left-hand side accounts for grain boundary creep while the second term takes into account constant creep from intercrystalline gliding and movement of dislocations. Equation (1.4) represents a useful refinement of ice creep but is not easily incorporated into large-scale ice sheet and glacier flow behaviour.

1.3.5 Impurities and creep

Natural ice in glaciers and ice sheets is rarely as pure as that used in laboratory experiments but contains a variety of foreign matter. Chemical solutes and precipitates formed from undissolved solute atoms are often contained within the ice and usually located in liquid in veins and at grain boundaries. Gases including air may be present at high pressures in small bubbles, resulting from the closing-off of pores during snow metamorphism. Discrete solid particles may be distributed randomly throughout an ice mass or concentrated into distinctive horizons. The location and concentration of chemical and physical impurities will depend upon specific processes and on the history of the ice. For instance many substances are introduced (entrained) into the lowermost layers of a glacier as it moves over its bed by processes which include **regelation**. Freezing of sea water onto the base of floating ice shelves may produce a distinctive layer of salty ice. The upper surface of a glacier is a depository for impurities precipitated from the atmosphere (volcanic dust, acids, trace elements, etc.) and deflated from ice-free terrain adjacent to the ice. We shall deal in more detail with the characteristics of these impurities and entrainment processes in Chapter 7. The size of impurities varies from a few Ångström for dissolved atoms to kilometres in the case of large rafts of rock transported by ice

sheets such as the **'Schollen'** of Northern Germany. Typical concentrations of soluble and insoluble matter in glacier ice are reported in Chapters 5 and 7 respectively.

Few studies have been undertaken to determine the effect of impurities on creep rate in ice, although this is clearly important. For soluble impurities the addition of various chemicals to single ice crystals (**doping**) has yielded some information. In general the effect is to 'soften' the ice and enhance the creep rate, as well as to decrease the mechanical **relaxation time**. The effects of different materials are, however, quite variable as shown in Table 1.3.

For solid additives studies with admixtures of amorphous silica, quartz sand, fibreglass and kaolin to ice in concentrations about 0.5%–15% by volume demonstrate a significant increase in strength ('hardening'), possibly related to the decrease in grain size so that the secondary creep is many times smaller than that for clean ice. This result has been observed, for instance, in the laboratory deformation of dirty basal ice taken from the Camp Century ice core (1 368 m). The effects of debris, as with chemical impurities, are not uniform. A very fine dispersion of small-size particles may enhance the creep rate, as discovered in ice from the Byrd core and the Thule Ramp in Greenland.

As the volume of solid impurities increases the creep rate continues to decline and the peak

Table 1.3 Effect of chemical impurities on creep rate of single crystals of ice: temperature range −5°C to −7°C (from Jones and Glen, 1969; Nakamura and Jones, 1973).

Chemical	Creep rate effect
HF HCl H_2O_2	enhanced (ice 'softened')
NH_3 He	decreased (ice 'hardened')
HBr NH_4F NH_4OH NaF KF NaOH	No effect

strength of the ice–sediment mixture rises due to the increase in particle-particle contacts which raise internal friction (Figure 1.5). Nickling and Bennett (1984), from shear-box experiments, show that when the ice content is reduced to a quantity just filling the pores (~25%) peak shear strength is achieved. Thereafter strength drops quickly as the cohesive bonding effect of the included ice is lost (Figure 1.5 inset).

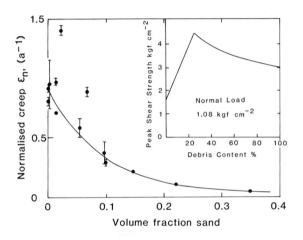

Figure 1.5 Dependence of creep rate on solid impurities (sand) in ice. $\sigma = 0.56$ MN m^{-2} at $-19°$C (after Hooke *et al.*, 1972). Inset shows variation of peak shear strength with ice content of sediments (after Nickling and Bennett, 1984).

1.4 Large-scale ice dynamics (motion in glaciers and ice sheets)

The large-scale flow of ice sheets and glaciers is governed by the non-linear flow law discussed in the previous section. The power-law for creep with $n = 3$, as originally determined by Glen, is the most useful and commonly used approximation. In addition, where conditions at the bed are favourable for basal melting, bulk sliding may also account for a significant proportion of the motion. It is thus possible to describe the motion of ice by the general equation:

$$U = U_c + U_s \qquad (1.5)$$

where U = total surface velocity
 U_c = velocity contribution from creep
 U_s = velocity contribution from basal sliding

These components are shown in Figure 2.1 for South Cascade Glacier.

Such modes of flow, expressed in terms of velocity and stress distribution within the ice, especially close to the substrate, are of central importance to understanding erosive and sedimentational processes. Yet despite this critical role exact solutions to many flow problems are not yet possible and often drastic simplifying assumptions have to be made to obtain even basic relationships. The adequate inclusion of such phenomenological factors as the effects of variable temperature, realistic ice flow law, the presence of water, characteristic bed roughness, etc. are continuing major problems in glaciology. Many of the questions inherent in the current understanding of glacier flow are discussed by Lliboutry (1964), Budd and Radok (1971), Paterson (1981) and Hutter (1983).

1.4.1 Creep velocity (U_c)

The flow of large ice masses is, at a very general level, the result of gravity forces which induce spreading of the ice by creep under its own weight. This action is resisted, on the small scale, by atomic (electrical) forces within the ice crystal and, on the large scale, by frictional forces at the glacier bed.

Figure 1.6 is an idealized section through a glacier or ice sheet. The upper surface is usually smooth or gently undulating relative to the bed which may exhibit considerable irregularity on scales of roughness ranging from a few tens of millimetres to some kilometres. An approximation often used in glacier flow problems is to assume the ice mass constitutes an inclined slab of uniform thickness. In this way the stresses related to flow are understood in terms of static forces in equilibrium. The driving or body force is the specific weight of the ice, $\rho_i g$ – where ρ_i is ice density and g is gravitational acceleration. For an ice mass

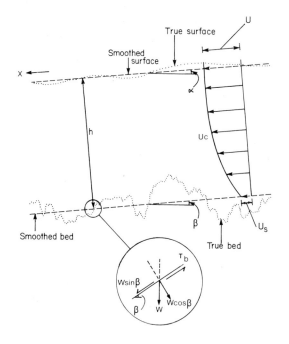

Figure 1.6   Idealized section through a glacier or ice sheet. The real ice and bed surfaces are shown dotted whilst the parallel slab approximation is shown dashed (with slopes α and β). A velocity profile is illustrated defining total surface velocity (U), creep (U_c) and basal sliding component (U_s). Forces at the bed are shown (W = weight of overlying ice [ρ_i g h], x = direction of ice flow). Ice thickness is given by h. τ_b is basal shear stress.

inclined at an angle, α_s, the downstream component of the glacier's weight is exactly balanced at the bed by a force, τ_b, resisting the motion (assuming that there is no **sliding**: ice is frozen to bedrock):

$$\tau_b = \rho_i\, g\, h\, \alpha_s \qquad (1.6)$$

where h = thickness of ice slab

The force, τ_b, is termed the **basal shear stress** and may be compared to a frictional force in soil or rock mechanics. Some workers also refer to τ_b, when calculated over a large area, as the **'driving stress'** since Equation (1.6) fundamentally describes the forces required to induce motion in

the ice. As the surface slope (α_s) of most glaciers is small (gradients of tan α = $1{:}10^2$–$1{:}10^3$) $\alpha_s \approx \sin\alpha_s$ and the more exact form of Equation (1.6) is unnecessary. If ρ_i g is constant, τ_b is determined by ice thickness and surface slope regardless of small-scale bed slopes.

If it is assumed that there is no basal sliding (say, in parts of large, cold ice sheets such as Antarctica) the vertical shear stress gradient yields variation of shear strain rate from below the surface to the bed of an ice mass. The velocity at the surface (U) and velocity at the bed, U_b, are related by the expression:

$$U - U_b = (\tau_b)^n\,[2B/(n+1)]\,h \qquad (1.7)$$

where B = ice hardness parameter in the flow law (Equation 1.1)

n = creep exponent

h = ice thickness

derived from an integration of a form of the Glen flow law ($\dot{\epsilon}_s = [\tau_b/B]^n$).
More generally the velocity (U(h)) at any depth h is:

$$U - U(h) = 2B\,(\rho_i\, g \sin \alpha_s)^n\, h^{n+1}/(n+1) \qquad (1.8)$$

The following approximations are predicted by Equation (1.8):

$$U \propto h^4$$
$$U \propto \alpha_s^3$$

Complicating factors
Several factors limit the use of Equations (1.6) and (1.7) for describing the flow of glaciers:

1   The temperature gradient through the ice mass will affect the temperature-dependent ice-hardness parameter, B. This can be accommodated in calculations if the thermal state of the glacier is known.

2   Surface slopes (α_s) have to be small as Equation (1.7) assumes that ice responds like a 'fluid', is at rest, and thus possesses a horizontal upper surface. Nye points out that the effect of deviation from these assumptions is only secondary.

3   Equations (1.6) and (1.7) apply more exactly to large ice sheets. In valley glaciers, however,

where channel width may be only of the order of a few ice depths the side walls tend to support the ice to some degree and introduce a drag term which reduces velocity and basal shear stress. The effects of channel sides on velocity is clearly observed from surface measurements such as those reported from the Blue Glacier (Meier *et al.*, 1974).

Nye has introduced a 'shape correction factor' (F′), typically between 0.5 and 0.75, by which the right-hand side of Equation (1.6) should be multiplied. F′ is related, in fluvial hydrology, to channel **hydraulic radius** (cross-sectional area divided by cross-sectional perimeter) and also to the basal boundary condition (sliding or frozen bed).

4   Variations in mass discharge of ice resulting from changes in **accumulation** and **ablation** have to be superimposed on any simply deduced flow regime. The effects of snowfall and melting perturbations can be understood through the **continuity equation**:

$$\frac{dQ_i}{dx} + \frac{dh}{dt} = A\,(x, t) \qquad (1.9)$$

where Q_i = mass flux or discharge in the x or down-glacier direction. It is equivalent to the **mass balance** (M′) integrated over the glacier area (a).

= $\iint_a M'\,da$
Where there is no ablation (as in Antarctica), $M' = A$.

Adding mass in the form of snow to the upper levels of a glacier, for instance, changes (dh/dt) and steepens the surface slope. This increases the driving stress (Equation 1.6) and thus accelerates the flow in order to discharge the excess mass. The increase in velocity downglacier produces longitudinal tensile strain. Conversely loss of mass downglacier gives rise to reduced flux and velocity and compressive strain.

5   Similar variations in **longitudinal strain** may be produced by deviations of the glacier bed from the simple slab approximation suggested earlier.

Table 1.4 shows how several of the factors mentioned above are reflected in changes to the strain rate (dU/dx).

Table 1.4 Factors influencing longitudinal strain rate on glaciers

Factor	Longitudinal strain rate (dU/dx)	
	Positive	Negative
Bed shape (radius of relative bed curvature)	Convexity	Concavity
Mass balance	Net Accumulation	Net ablation
Channel width (transverse strain rate)	Widening	Narrowing

6   Commensurate with longitudinal variations in strain rate are down-glacier stress gradients. From Equation (1.6) τ_b is a function of ice thickness and surface slope. Several workers have shown that dependence on surface slope is complicated and not constant. A correction factor has to be introduced which relates to gradients of the deviatoric stress components and may be applied at varying scales. On the short scale (equivalent to one or two ice depths) all corrections are necessary, while on the large scale (tens of ice depths) the effect of longitudinal stress gradients appear negligible. It would seem therefore, that when dealing with the large-scale patterns of driving stresses over the Antarctic and Greenland ice sheets Equation (1.6) may be found quite satisfactory (Cooper *et al.*, 1982).

7   It is likely that local stress concentrations on bed features related to glacier flow are important in erosional processes. Stress variations across bed hummocks will set up stress gradients in bedrock which will also affect rock strength and its erosional propensity. The ice itself at enhanced stresses encountered near the bed and in the vicinity of the bed irregularities, may behave with increasing departure from known

power-law creep. Several lines of evidence support this contention. First of all ice fabric studies close to the bed indicate bubble-free regelation ice with low-deformation crystal fabrics. Second, motions observed close to the bed suggest complex flow. Third, and at a large scale, **radio echo sounding** through large ice sheets indicates that basal ice flow is sufficiently disturbed to break up the continuity of **isochronous layers** (see Figure 8.2) (Robin *et al.*, 1977; Robin and Millar, 1982). Thermal and strain softening of ice in basal layers may thus result in the ability of a glacier, ostensibly frozen to bedrock, to accomplish modest erosion. We shall return to some of these aspects in later chapters.

1.4.2 Basal sliding velocity (U_s)

In the cases dealt with so far all motion is assumed to take place by power-law creep within the ice mass. This is equivalent to proposing that the ice is frozen to bedrock and that no additional movement can occur at this interface. Such conditions are indeed found over large areas of the Antarctic and Greenland ice sheets and in parts or all of a small number of valley glaciers where basal temperatures are below the pressure melting point. It is clear, however, that maximum surface velocities achieved by creep may be significantly smaller than those actually observed. Equation (1.7) shows that a glacier of h = 500 m and α_s = 2° will have a velocity of 50 m a^{-1} by creep (n = 3). Many glaciers exhibit average speeds of several hundred metres per year, while large outlet glaciers of ice caps and ice sheets achieve astonishing velocities of several km per year (Byrd Glacier, Antarctica – 0.8 km a^{-1}; Thwaites Glacier, Antarctica – 3 km a^{-1}; Jakobshavn Isbrae, West Greenland – 7 km a^{-1} (Weertman, 1983) – the fastest known ice mass. This latter value is approximately equivalent to 0.8 m hr^{-1}!). The difference between observed and calculated creep velocities is accounted for by basal sliding (U_s) and in many glaciers U_s may be the dominant term in ice flow accounting for up to 90% of the motion. In addition the contribution to U provided by U_s may vary from

place to place within a glacier or ice sheet, as well as exhibiting fluctuations over a wide range of time scales (from daily to annual) often in relation to the availability and penetration of water to the bed. It is probably true to say that sliding along the ice/substrate interface (either hard intact rock or deformable sediments) is the dominant active influence on erosion. Yet basal sliding is one of the least understood of fundamental glaciological processes. This lack of knowledge is not hard to comprehend and derives from an inability to penetrate the beds of ice masses and directly observe and record bed slip and associated phenomena. Only in favourable locations such as marginal tunnels (natural or man-made), in bore-holes or in thin, heavily crevassed ice falls have observations been made (Figure 1.7). Some of the more well-known glaciers to be investigated in this manner are the Blue Glacier in Washington State, Glacier d'Argentière in France and Bondhusbreen in Norway. It is often not possible to assess how typical measurements in these places are. A further problem arises from the complex mathematical treatment required to fully formulate sliding relationships.

Current working hypotheses on sliding derive from the early studies of Weertman (1957, 1964) and Lliboutry (1968, 1975) and more recent developments by Nye (1969b; 1970), Kamb (1970) and Morland (1976). The treatments given by these workers highlight three principal cases: sliding in contact with bedrock, sliding with a water film and sliding with the formation of cavities in the the lee of bed hummocks.

Weertman theory
Two processes, first recognized by Weertman (1957), are fundamental to basal sliding: regelation and enhanced creep. Irregularities in the bed give rise to local stress concentrations on the upstream side of hummocks and, following Equation (1.1), enhance the ice creep rate. For cubic-shaped bed hummocks of side l and spacing of λ_b the increase in stress against a bed protuberance is:

$$\Delta \tau = \tau_b \lambda_b^2 / 2l^2 \tag{1.10}$$

The velocity achieved by ice strain at these increased stresses is:

$$U_{sc} = l\, B\, (\tau_b\, \lambda_b^2/2l^2)^n \qquad (1.11)$$

where n = creep exponent
           B = ice hardness parameter.

It can be seen that U_{sc} is proportional to l. This means, for instance, that the smaller the size of the bed irregularity the less the creep enhancement: the ice velocity will consequently be lower. Hutter (1982a) considers U_{sc} is the dominant contribution to sliding.

A second mechanism derives from the fact that there is increased stress on the upstream side of the bed protuberance and a corresponding decrease on the downstream side, which gives rise to a difference in the pressure melting point at these two locations. The ice moves more rapidly around the obstacle as upstream melting and downstream refreezing take place, and velocity is proportional to the volume of ice melted per unit time divided by the cross-sectional area. As latent heat is transferred from the refreezing zone through the bed hummock and also through the ice to the melting plane (thus controlling the effectiveness of the regelation process), the ice velocity can be given as:

$$U_{sr} = (k\, K_r/L\rho_i l)\, \tau_b(\lambda_b/l)^2 \qquad (1.12)$$

where k   = constant
           K_r = thermal conductivity of bedrock
           L   = latent heat of fusion of ice
           τ_b = basal shear stress (against bed obstacle)

Weertman has suggested that when the size of bed hummocks reaches >1 m effective heat transfer is difficult to sustain and the regelation mechanism becomes inoperative.

The opposing nature of the two processes of enhanced creep and regelation with respect to size of bed protuberances gives rise to a unique, intermediate obstacle size ($\wedge$) which controls, to a large extent, the efficiency of the sliding process. Weertman shows that:

$$\wedge = (K_r\, k/L\rho_i\, Bl)^{1/2}\, \beta_t^{-n/2}\, (\gamma'\tau_b)^{(1-n)/2}\, (\lambda_b/l)^{1-n} \qquad (1.13)$$

Figure 1.7 Sliding of the base of the marginal zone of Mitdalsbreen, an outlet of Hardangerjökull, Norway. Basal ice (top left) is moving from right to left over water-saturated till. Note the grooving of the bed materials by irregularities and/or clasts held in the sole.

where β_t = a constant related to longitudinal tension on a bump

$(\gamma' \tau_b)$ = basal shear stress supported by obstacles other than the controlling size.

The basal sliding velocity for ∧ is

$$U_s = D\tau_b^{(n+1)/2} (\lambda_b/l)^{(n+1)} \qquad (1.14)$$

where D = constant.

Thus with n = 3

$$U_s = D\tau_b^2 (\lambda_b/l)^4 \qquad (1.15)$$

It can be seen that U_s is proportional to both the basal shear stress and bed roughness. For $\tau_b = 100$ kPa and $(\lambda_b/l) = 20$ (equivalent to a bed hummock size of 0.1 m at 2 m separation) the sliding velocity as given by the derivation is 100 m a^{-1}.

Nye–Kamb extension for watertight beds
A critical factor in Weertman sliding theory is an adequate description of bed roughness. The cubic protuberances of his original formulation are unrealistic and should be abandoned. More sophisticated beds have been considered by Kamb (1970) and Nye (1969b, 1970) who provide a fourier analysis of bed topography where slopes are small. Two types of roughness are identified: **'white roughness'** where the variance of bed elevations is the same over a range of wavelengths and 'truncated white roughness' in which the shorter wavelength component is absent. This latter may more nearly approximate real glacier beds where small-scale irregularities tend to be planed off by abrasion. By describing the bed in this manner Nye and Kamb have been able to exactly solve for τ_b as a function of sliding, and the relative contributions from regelation and creep. Critical to their solutions is a term equivalent to Weertman's **controlling obstacle size**, called the **transitional wavelength**, λ_t. With bed roughness on a scale $<\lambda_t$ sliding is principally by regelation, and by enhanced creep for $>\lambda_t$. Kamb considers λ_t of the order of 0.5 m. Based upon eight field examples from Blue and Athabasca Glaciers, where borehole

data have yielded basal sliding velocities and where τ_b can be calculated fairly accurately Kamb has shown, for white roughness, sliding velocities can be accounted for with bed roughness in the range 0.01 to 0.05 and a relationship of the form:

$$U_s \sim \tau_b^{(n+1)/2} \qquad (1.16)$$

and for truncated roughness

$$U_s \sim \tau_b^n \qquad (1.17)$$

as shown in Figure 1.8. Weertman's order-of-magnitude calculations compare surprisingly favourably with the rigorous, although simplified, analyses by Nye and Kamb.

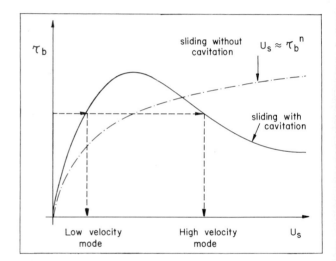

Figure 1.8 Relationship between sliding velocity and basal shear stress for sliding without cavitation (Equation 1.17). The possible double-valued characteristic of the sliding law giving a high and a low velocity mode for a single value of τ_b where there is cavitation is also illustrated (see text for further explanation) (modified from Hutter, 1982).

Sliding with cavitation
If the tensile stress due to ice flow exceeds the normal or cryostatic pressure (ρ_i g h) the ice cannot keep in contact with the bed as it passes downglacier of a hummock. This condition is given by:

$$\tau_b (\lambda_b/l)^2 s > \rho_i g h \qquad (1.18)$$

where s = parameter related to dimensions of the bed hummock

As a consequence of ice flow satisfying Equation (1.18) a gap or cavity will form in the lee of the obstacle at a rate:

$$C_{av} = \wedge l \, [\tau_b \, (\lambda_b/l)^2]^n \qquad (1.19)$$

where $\wedge$ = the controlling obstacle size.

The presence of **cavities** markedly influences the basal sliding velocity by reducing the area of bed contact and hence drag (or basal friction) while increasing τ_b at those points remaining in contact with and supporting the glacier. As τ_b rises there is enhanced creep and a higher U_s. Lliboutry (1979) suggests, however, that if the amount of cavitation is large τ_b is reduced as $U_s^{-1/2}$. The presence of cavities means that the relationship between U_s and τ_b may be multi-valued (Figure 1.8). In addition to simple physical separation the presence of water in cavities (resulting from pressure melting or in-flow from the glacier sub- and englacial water network) will have a significant effect on U_s by 'lubricating' the ice–bed interface, submerging bed irregularities, reducing roughness and, according to Lliboutry, if the water is under pressure further reducing τ_b by a decrease in cryostatic pressure. There is some uncertainty and disagreement between glaciologists as to the location and influence of subglacial water on sliding. The principal question is whether water is disposed in channels, cavities or sheets at the glacier bed and the consequent development of theory to calculate sliding behaviour.

Weertman's 'modified sheet'
According to Weertman (1964, 1972) basal water affecting U_s is confined in a thin film or sheet at the bed rather than being concentrated into cavities or channels. In his original formulation Weertman considered the water layer to be of uniform thickness but later modified the sheet theory to accommodate a film of variable depth and equivalent to interconnecting cavities. Hallet (1979a) has found evidence for the presence of a thin subglacial film of water from the investigation of recently deglaciated rock surfaces

exhibiting $CaCO_3$ precipitates and solution features. Weertman argues that the thin sheet of water submerges bed irregularities and thus reduces the bed roughness 'seen' by the glacier base. The result is an increase in sliding velocity. The pressure in the water layer is considered by Weertman to play *no* significant role in basal sliding.

Lliboutry cavitation
Lliboutry has made extensive investigation of problems of basal sliding with cavitation. He favours basal water concentrated into cavities in the lee of bed obstacles rather than in a simple or 'modified' sheet. He suggests that τ_b is almost independent of U_s when basal water pressures are taken into account:

$$\tau_b \approx 0.37 \, (\rho_i \, g \, h \, - \, \rho_w) - (0.27 \text{ Pa m}^{-0.1} \text{ s}^{0.1}) \, U_s^{0.1} \qquad (1.20)$$

where ρ_w = basal water pressure
and U_s = $0.017\tau_b^2 \text{m}^{-3}$
where m = mean quadratic slope in the direction of sliding

Figure 1.8 illustrates the relations between τ_b and U_s considered by Lliboutry. More recent improvements to the model have been made by Lliboutry.

1.4.3 Complications to sliding theory

Kamb *et al.* (1976) have shown that U_s may be governed, in part, by the debris content (C_o) in basal ice. The greater the sediment content the more substantial will be local bed–particle friction and frictional drag. The net result is for a decrease in basal sliding velocity.

Hallet (1981) has investigated this problem in detail and suggests a sliding relationship based upon Nye-Kamb theory but taking into account the concentration of debris in the ice at the interface:

$$U_s = [\tau_b/\eta_i \, (\xi \, + \, \Omega \, \mu_c \, C_o)] \qquad (1.21)$$

where η_i = ice viscosity
ξ = bed roughness factor (power

μ_c = friction between bed and clasts
C_o = debris concentration
Ω = viscous drag term

In addition the increased stiffness of the ice containing sediments will reduce the 'enhanced' creep effect around bed obstacles. Lower sliding velocities will inhibit cavitation. Such theoretical studies are supported by the observations of Engelhardt *et al.* (1978) from bore-hole investigations which showed significantly low velocities where there was appreciable basal debris. Sliding is more rapid over a smooth bed: there is increased viscous drag over irregularities of a rougher surface. Consequently debris plays a more important role in reducing U_s in the former case.

1.5 Thermodynamic processes

It will already be apparent that temperatures and temperature changes play important roles in the behaviour of ice masses. The influence of temperature on ice deformation, for instance, is shown by the Arrhenius equation (1.2), cold ice reacting more 'stiffly'. A 20°C temperature change, for instance, produces a 100-fold change in strain rate at a given stress. Changes through the melting temperature, induced by stress concentrations, are fundamental to basal sliding. Water can only be released from and flow freely in glaciers under temperatures close to the melting point. Such temperature-dependent processes are of major importance to erosion and sedimentation. Where ice is frozen to bedrock sliding and significant erosion will be inhibited. The subglacial release of sediments entrained in basal ice layers, as melt-out till, is controlled by heat flow at the ice–rock interface. Processes determining sedimentation beneath floating ice shelves, glacier tongues and icebergs are likewise governed by critical thermal conditions. In cold glaciers and ice sheets streamflow in subglacial and proglacial areas is negligible or non-existent: as a consequence glaci-fluvial and glaci-lacustrine sedimentation may be largely suppressed.

The changing pattern of the thermal regime of glaciers will control the mosaic of deposition and erosion in both space and time. Of interest to suceeding chapters, therefore, are temperatures towards and at the base of ice masses, as well as in subglacial sediments and rock.

1.5.1 Simple temperature distribution in a slab glacier with no motion

Figure 1.9 illustrates a stagnant ice mass with parallel upper and lower surfaces which possesses a temperature distribution in equilibrium governed solely by conduction. If the temperature at the ice surface is T_s and at the bed T_b the temperature gradient is:

$$(dT/dh) = (T_s - T_b)/h \qquad (1.22)$$

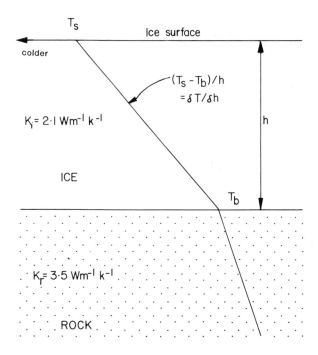

Figure 1.9   Simple temperature profile through a stagnant glacier of thickness h. Surface and bed temperature are T_s and T_b. Thermal conductivities (K_i, K_r) for ice and typical rocks are given. Note that the upper surface is also the abscissa for temperature scale with decreasing temperature from right to left.

At the glacier surface the temperature, T_s, will be determined by local meteorological and long-term climatic conditions. The heat flux through subglacial rock will govern the temperature at the base of the glacier. Thus the temperature profile in the simple glacier of Figure 1.9 will reflect the competition or balance between cold surface temperatures and warm basal conditions. The amount or flux of heat passing through bedrock to arrive at the glacier sole is:

$$\Lambda_g = - K_r \, (dT/dh)_r \qquad (1.23)$$

where K_r = **thermal conductivity** of subglacial rock

$(dT/dh)_r$ = temperature gradient in rock (i.e. geothermal gradient)

Typical values of K_r lie between 1.2 and 4.0 W m^{-1} K^{-1}, that of ice is 2.1 W m^{-1} K^{-1} (a list of thermal properties of ice is given in Table 1.2).

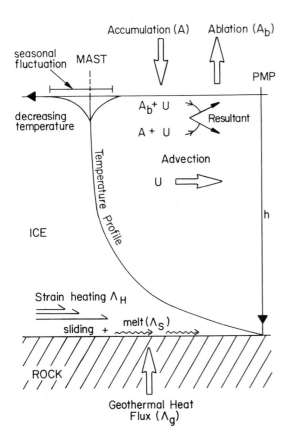

1.5.2 Temperature in a moving ice sheet or glacier

The distribution of temperature within an ice mass, and more specifically basal thermal conditions which are critical for erosional and depositional processes, can rarely be determined by simple calculations as given in Subsection 1.5.1. Several other factors need to be taken into consideration and are summarized in Figure 1.10.

Mechanical deformation of ice during flow is a source of heat, strain heat, and may be considered as a flux proportional to the ice velocity. The heat due to internal shear is:

$$\Lambda_H = \epsilon_{xy} \, \tau_{xy} \, / \, J \, \rho_i \, c_i \qquad (1.24)$$

where ϵ_{xy}, τ_{xy} = shear strain and shear stress in ice

J = mechanical equivalent of heat

c_i = **specific heat capacity**

When sliding takes place frictional work gives rise to an additional heat term:

$$\Lambda_s = \tau_b \, U_s \, / \, J \, \rho_i \, c_i \qquad (1.25)$$

where U_s = basal sliding velocity

1.5.3 Simple temperature distribution in a moving ice mass with accumulation and ablation

Variable snow accumulation or ice ablation provides a further complicating factor in the distribution of temperatures within ice masses (Figure 1.10). Annual snowfall carries cold temperatures down into the body of a glacier or ice sheet so that, for instance, where accumulation is high lower temperatures are found at a greater depth than where accumulation is small (Figure 1.11). The combination of high ice velocity and high

Figure 1.10 Factors affecting the vertical temperature profile in an ice mass. Seasonal variations and mean annual surface temperatures (MAST) are shown. PMP = pressure melting point. The effects of surface snow accumulation (A) and ablation (A_b), and ice motion (U) are also indicated. The resultant direction of ice movement and thus heat transport is shown.

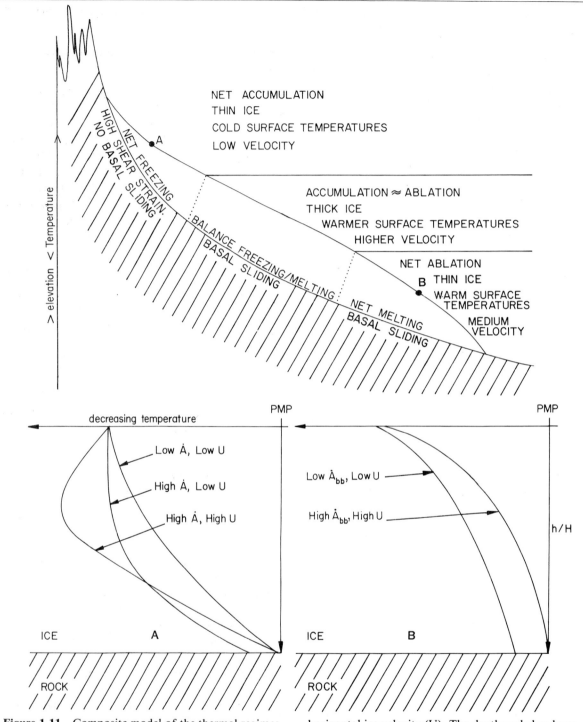

Figure 1.11 Composite model of the thermal regimes of a valley glacier showing the variety of conditions that are possible and controlling parameters. Typical vertical profiles for points A and B are shown beneath. **(A)** Temperature profiles in the accumulation zone of an ice mass for specified rates of accumulation ($\dot{A}$) and horizontal ice velocity (U). The depth scale has been normalized.
(B) Temperature profiles in the ablation zone of an ice mass for specified rates of ablation ($\dot{A}_{bb}$) and horizontal ice velocity (U).

snowfall is to rapidly replace ice warmed by **geothermal heat** conduction with colder ice from up-flowline (Figure 1.11). The reverse effect is experienced in the ablation zone of **temperate** and **sub-polar glaciers**. Loss of ice at the surface aligns particle paths upwards, and warmer ice is brought towards the ice surface (Figure 1.11).

A one-dimensional heat transfer equation, combining conduction, advection and strain heating is (after Cary *et al.*, 1979):

$$\frac{d^2T}{dh^2} - \frac{A}{k_i}(1 - h/H)\frac{dT}{dh} + \frac{2B}{K_iJ} \cdot$$
$$\exp(-Q/R_gT)\tau_{xy}^{(n+1)} = \frac{1}{k_i}\frac{dT}{dt} \quad (1.26)$$

where H = maximum ice thickness
R_g = gas constant

Changes in temperature with time (dT/dt) are also important and may be affected by surface mean annual temperatures which can vary due to long-term climatic oscillations or short-term cycles of a diurnal or seasonal nature.

1.5.4 Temperature control of erosion and sedimentation

The thermal boundary conditions at the ice-bedrock interface are some of the most important in the whole of glacial geology. Depending upon whether basal ice is cold (ice below the pressure melting point) or temperate (at the pressure melting point) erosion and sedimentation processes may be severely affected. Bottom temperature characteristics and changes through time thus provide rate-control of erosional and depositional processes.

Three fundamental boundary conditions have been recognized (Weertman, 1961; Boulton, 1972):

$$\Lambda_s + \Lambda_g > K_i(dT/dh)_i^* \quad (1.27)$$
$$\Lambda_s + \Lambda_g \approx K_i(dT/dh)_i^* \quad (1.28)$$
$$\Lambda_s + \Lambda_g < K_i(dT/dh)_i^* \quad (1.29)$$

where $(dT/dh)_i^*$ = the temperature gradient in the lowermost layers of the ice mass

In Case 3 above (Equation 1.29) the temperature gradient is more than sufficient to conduct all the heat from the bed (i.e. that produced by sliding (or straining) motion and geothermal heat). The ice remains firmly frozen to the bed and any water migrating towards or along the ice–rock interface will become frozen. Case 2 (Equation 1.28) describes the situation where the temperature gradient is just sufficient to conduct heat from the bed: there is an approximate balance between melting and freezing. Under such conditions regelation takes place readily and the glacier is able to slide.

The temperature gradient in the first case (Equation 1.27) is insufficient to drain away heat supplied to the basal zone and melting at the ice–rock interface will occur. Basal sliding will be a dominant process. In topographic lows in the bed, where rock permeability is also low, water released by melting may become ponded and form subglacial 'lakes' (see Section 2.5). Elsewhere a thin film or discrete channel may result. It is quite possible for one or for all three of the conditions given in Equations (1.27)–(1.29) to exist within a single ice mass (Figure 1.11) or at the margins of ice caps and parts of large ice sheets.

1.6 Transient flow phenomena: surging behaviour

Most treatments of glacier and ice sheet dynamics begin with the simplest case and assume steady-state – based upon the premise that nature normally favours a stable configuration to systems. Small deviations around an equilibrium are considered normal. Thus Hutter (1982a) argues that if an ice mass is in a condition of non-steady flow, which is usually (but not necessarily) unstable, any perturbation may cause it to revert to a stable situation.

At certain times, however, large perturbations may occur to otherwise steady ice flow which cause marked or catastrophic changes in the behaviour and configuration of an ice mass. Usually termed **'surges'**, these dramatic alterations are typically manifested in a rapid and substantial increase in ice velocity by sliding (from, say, a few m a⁻¹ to ten, hundreds or even thousands of m a⁻¹). A pronounced advance of the glacier terminus with associated crevassing is typical and a change in the longitudinal surface profile as mass is transformed from upper glacial reaches (experiencing surface lowering) to lower sections (with accompanying surface increases and ice thickening). Such activity is normally periodic with quiescent or even stagnant phases of several decades separating short surge-periods (a few months to a few years). Figure 1.12 shows some of these surge characteristics. Many valley glaciers are known to surge as described by Meier and Post (1969). Several ice caps and their outlet glaciers exhibit surging behaviour (e.g. certain drainage basins of Vatnajökull, Iceland; Barnes Ice cap, Baffin Island, NWT and Austfonna, Nordaustlandet, Svalbard where the largest recorded surge took place between 1936 and 1938. The southwestern margin of this ice cap moved forward 20 km along a 30 km-wide front to create a new ice lobe – Bråsvellbreen (Figure 1.12 and 1.13)). No large ice sheet has yet been shown to surge, although several theoretical and modelling studies suggest the possibility (Wilson, 1964; Budd and McInnes, 1974).

Numerous mechanisms have been proposed to explain surging activity but these will not be elaborated in detail – interested readers are referred to Chapter 13 of Paterson (1981) and the discussion by Clarke *et al.* (1985) and Kamb *et al.* (1985). Most reasonable theories entertain creep instability, thermal instability, changes in basal shear stress and in basal water regime and mobility in water-saturated basal sediments. Detailed observations were made during the 1982–83 surge of Variegated Glacier in Alaska which demonstrate the critical role of the building of high basal water pressure (Kamb *et*

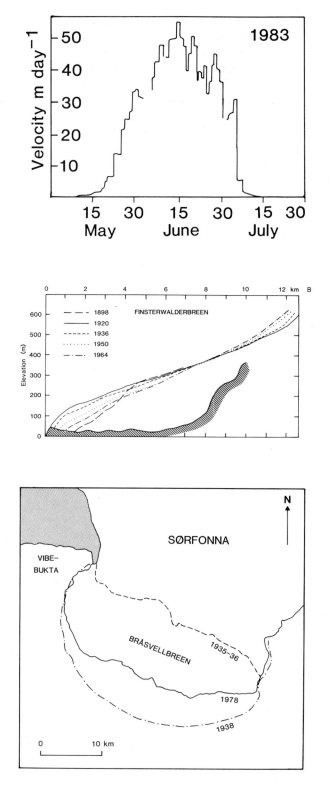

Figure 1.12   Some characteristics of surge activity. **A**: Short period velocity increase for lower Variegated Glacier, Alaska in 1983 (from Kamb *et al.*, 1985), **B**: changes in the long-profile for Finsterwalderbreen (from Liestøl, 1969), **C**: changes in position of the glacier front for Bråsvellbreen, Svalbard between 1936 and 1978 (from Drewry and Liestøl, 1985).

al., 1985). The important point for glacial geologic processes is that during a surge rapid basal sliding occurs, associated with basal water lubrication. In some glaciers and ice caps there is a change from frozen bed conditions to pressure melting and full sliding. In others the sliding component of flow rises by an order of magnitude. Such switchover can be related to the double or multivalued relationship between U_s and τ_b shown in Figure 1.8. Hutter (1982, 1983) provides an additional, phenomenological, set of relationships which may be used to explain surge-type behaviour. Unique associations arise at the extremes of low and high basal shear stress but in between up to three values of U_s may be possible. These are probably unstable so that U_s increases continuously as τ_b rises to a critical separation or threshold whence the sliding velocity 'jumps' to a new, stable but much higher value. The same phenomena, in reverse, are experienced for diminishing τ_b.

By analogy with hydrological processes, where most erosive activity is associated with major flood events, it is possible to suggest that during surging with very high basal sliding velocities, considerable bedrock erosion will take place and substantial quantities of sediment be transported to the low parts of the valleys.

Figure 1.13   Front of Bråsvellbreen, Svalbard after the 1936–38 surge (courtesy Norsk Polarinstitutt, Olso).

2 Water in glaciers

The role of water in glaciers and ice sheets is important and at times critical to processes which control erosion and sedimentation. Water can usually only exist in the glacier environment if temperatures are close to the melting point. Although water may be supercooled, especially when charged with **solutes**, it is then thermodynamically unstable. In cold glaciers and ice sheets surface ablation is negligible yet basal melting may occur and the small quantities of water thus produced exert a major influence on ice dynamics. In this chapter the presence and movement of water in glaciers are examined specifically; erosive and depositional aspects are discussed in Chapters 5 and 10.

2.1 The role of meltwater

Water is of primary importance to glacier flow, both by creep and basal sliding. Duval (1979) has demonstrated a clear relationship between water content and creep rate. For a change in the amount of water held in the ice from 0.1% to 1% he observed an increase in strain rate by an order of magnitude at a given stress.

At the glacier bed a water layer which reaches sufficient thickness to submerge low-amplitude roughness elements will give rise to a significant decrease in τ_b and an increase in sliding velocity, U_s (as predicted by Weertman theory). This is shown in Figure 2.1 by the correlation of sliding speed with meltwater discharge for the South Cascade Glacier, Washington. In as much as basal water may regulate sliding there is a close relationship with erosive processes, especially abrasion (see Chapter 4). Where semi-consolidated sediments are present at the bed high **interstitial water** pressure will decrease sediment, **shear strength** and hence have a direct effect on erosion and the deformation of basal materials.

The flow of meltwater at the bed, at times with high velocity and under great pressure, causes significant mechanical erosion with the production of channels and potholes (Chapter 5).

Meltwater also plays an important role in flushing chemical impurities through the glacier, while at the sole, flow in films and channels in contact with the bed and crushed particles allows chemical exchange and **ionic** enrichment. In addition basal water flow provides a means of removing and transporting debris produced by abrasion, crushing and comminution processes. Water discharge from the glacier terminus is an important source of stream flow, and is affected by the input and storage of water within the ice. Glacial waters can be seen, therefore, not only to play a part in geological activity (proglacial sedimentation/erosion) but may be an important commercial source of water, especially for power generation.

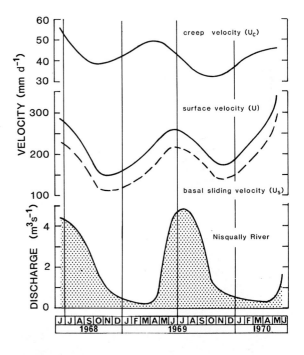

Figure 2.1 Relations between subglacial water and ice flow (total, creep and sliding velocities) for the South Cascade Glacier, Washington between July 1968 and June 1970. Velocity values are averages of 19 stakes. Water discharge is for the Nisqually River (after Hodge, 1976).

2.2 Sources of water in glaciers and ice sheets

2.2.1 Surface melting

Surface melting varies both spatially and temporally due to variations in incoming solar radiation and convective and turbulent heat tranfers. There are distinctive diurnal cycles. Up to 70% of surface melt is usually accounted for by solar radiation. The remaining 30% is termed non-radiational melt and arises from near-surface turbulent heat transfer, latent heat of condensation and net long-wave radiation. The quantity of melt or net ablation is in general more vigorous at lower elevations and low latitudes and with favourable aspect.

Platt (1966) found that 90% of the observed melt on central portions of Lewis Glacier on Mount Kenya (latitude 0.2°S) was due to radiation.

2.2.2 Dissipation of mechanical heat

Water is produced by melting due to the heating effect of rapidly straining ice, as found deep within ice sheets and glaciers, or from basal sliding. The amount of melt can be estimated from Equation (1.25):

$$M_{sm} \approx \Lambda_s / \rho_i L \qquad (2.1)$$

where L = latent heat of fusion
  ρ_i = ice density

Typically M_{sm} is of the order of 0.01–0.02 m a^{-1} but will rise rapidly under fast outlet glaciers and ice streams. For a sliding velocity of 2 km a^{-1}, M_{sm} = 0.4 m a^{-1}.

2.2.3 Geothermal heat

Heat entering the base of a glacier from the earth's crust will raise its temperature and melt ice under favourable conditions (see Subsection 1.5.1). The amount of melt is given from Equation (1.23):

$$M_{gm} \approx \Lambda_g / \rho_i L \qquad (2.2)$$

where Λ_g = geothermal heat flux

M_{gm} is usually of the order of 0.005–0.01 m a^{-1} but may rise significantly in areas of elevated heat flow (e.g. glaciers in active volcanic regions – such as Iceland, Kamchatka, Aleutians and parts of Antarctica).

2.2.4 Groundwater flow and surface runoff

There is an input of water to a glacier system from runoff from surrounding valley sides (due to spring snowmelt or rainstorms). In addition contributions may be made from groundwater issuing beneath the glacier. In the latter case, however, very high overburden pressures and associated pressure gradients created by a glacier or ice sheet may well force water at the sole to flow *into* **permeable** bedrock.

2.2.5 Liquid precipitation

A final source of water is that added directly to the glacier surface. Haakensen and Wold (1981) report mass balance measurement from Bondhusbreen, Norway which show that over half (34 M m³) of the total water **discharged** from the glacier in 1979 (i.e. 60 M m³) resulted from liquid precipitation within the glacier catchment (~15 km²). In polar ice sheets, ice caps and glaciers, however, surface melting, surface run-off and contributions from precipitation will be minimal.

2.3 Location and flow of water at and from the glacier surface

At the glacier surface water is usually confined into a network of streams and in places ponded in surface lakes (Figure 2.2). Flow also takes place in near-surface layers of permeable snow and firn. The downward percolation of meltwater in the upper porous zone is usually slow and complicated due to the presence of many inhomogeneities such as ice lenses, layers and fractures. These result from refreezing of meltwater early in the ablation season or by slow surface freezing of rain which is subsequently buried. One of the consequences of ice layers is the diversion of infiltrating meltwater demonstrated in Figure 2.3 for the

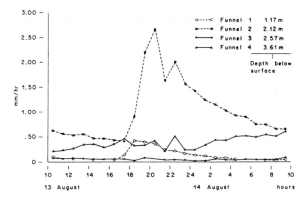

Figure 2.3   Variations in the infiltration of meltwater into surface snow on Ewigschneefeld due to the presence of ice lenses and layers. Percolating water, during a diurnal melt cycle, was intercepted by funnels (numbers 1–4) located at varying depths. Funnels 1 and 2 showed a marked increase in flow whilst at 3 and 4 no significant change was observed due to inhibition of percolation by ice layers (from Lang *et al.*, 1977).

Figure 2.2   Surface meltwater streams on glaciers in East Greenland. **A**: stream descending a moulin on Bersaerkerbrae.
B: small stream on Roslin Gletscher.

Ewigschneefeld, Switzerland. Four 4-mm diameter funnels, located at depths between ~ 1.2 and 3.6 m below the surface collected varying quantities of meltwater during a summer period. Two funnels registered an insignificant amount of water due to the presence of ice layers which cut off the water supply. The results of Austrian investigators suggest that during the summer period the firn body probably retains and stores the largest amount of water within the whole glacier system.

Meltwater percolates slowly downward through snow and firn at a rate controlled by gravity and the **hydraulic gradient** and can be modelled by some form of **Darcy's law** (Colbeck, 1973; Male, 1980):

$$Q_{wf} = -[N_p / \eta_w] \, dP_{cg}/dz - \rho_i \, g \, \alpha_s \qquad (2.3)$$

where Q_{wf} = vertical water flux
α_s = slope of the snow surface
dP_{cg}/dz = capillary pressure gradient
N_p = permeability
η_w = water viscosity

At a depth of a few metres the porous-medium type of flow changes as small channels develop coalescing with increasing depth to form a

vertical drainage network. At the firn/ice transition ice becomes impermeable as pores in the firn are sealed off. Saturation occurs at this level and a continuous 'water table' or **potentiometric surface** is usually present. On the Ewigschneefeld of the Grosser Aletsch Gletscher, Switzerland and the South Cascade Glacier, Washington (Figure 2.4) measurements in boreholes show a 'water table' exists continuously at depths of 14–32 m and ~25–75 m respectively below the surface and in the former case, is closest to the surface in areas with surface depressions. Downward movement of water into the deeper body of the glacier continues either through intergranular veins and capillaries, or in cracks, fissures and moulins. Nye and Frank (1973) show that flow through the three-dimensional vein system (diameters in the order of 0.1 mm) is proportional to the fractional volume occupied by the water, and could reach a value of up to 0.1 $m^3 a^{-1}$. Interconnecting veins will be cut by deforming ice and ice recrystallization will lead to a coalescence of the liquid inclusions. Complementary flow

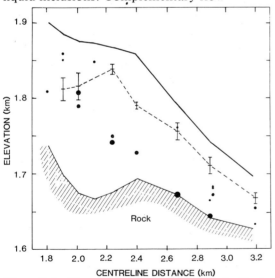

Figure 2.4 Water within the South Cascade Glacier, Washington. Water is located in cavities encountered during drilling operations (solid circles and proportional to cavity size. Mean summer (July and August) borehole water levels are shown by vertical bars and joined by dashed line–the potentiometric surface. (after Hodge, 1976).

experiments on Medenhall Glacier, Alaska by Wakahama et al. (1973) indicate water drainage through the intergranular vein-tube network of between 10^{-7} and 10^{-5} m s^{-1}. Raymond and Harrison (1975) and Berner et al. (1978) found percolation rates of 7.6×10^{-8} m s^{-1} for water penetrating to a glacier sole.

Cracks and fissures allow descending water to develop a vertical network of conduits, pipes and cavities (similar to those in the firn zone) which by coalescing decrease in number, but increase in size and the associated quantity of throughflow towards the bed.

It seems clear from **dye-tracer** studies, theoretical considerations and radar observations that an extensive network of sub-horizontal channels cannot be supported within the body of a glacier. Water descends rapidly to the bed taking the shortest possible route without being stored in any considerable volume nor for any significant length of time. Collins (1977), from work on Görnergletscher, Swizerland considered that two, largely independent systems may be present. He argues that water, principally from summer surface ablation, is discharged through high-level (englacial) conduits which do not join a distinctive subglacial system until near the glacier terminus. In the former network water has only a low ionic content whilst the latter it is enriched in solutes from flow through and storage in subglacial sediments (see also Chapter 5).

2.4 Location and flow of water at and from the glacier bed

Water reaches the bed in a number of relatively large channels and thereafter, if the bed is impermeable, flows out towards the glacier snout under a generalized pressure gradient which, if considered at a point (see Figure 2.5) is:

$$
\begin{aligned}
P_g &= -dP_w/dx^* \\
&= -\rho_i\, g\, d_w\, h/dx^* \\
&= -\rho_i\, g\, (d_w\, (H^* - h_z)/dx)(dx/dx^*) \\
&= \rho_i\, g\, (\tan\alpha_s - \tan\beta)\cos\beta \qquad (2.4)
\end{aligned}
$$

For small surface and bed slopes (Figure 2.5):

$$P_g = \rho_i \, g \, \alpha_s + (\rho_w - \rho_i) \, g \, \beta \qquad (2.5)$$

where d_w = thickness of a water layer or conduit diameter at the bed

h_z = height (of channel) above datum

α_s = ice surface slope

β = bed slope

x = distance measured horizontally

x^* = distance measured along bed

H^* = $h + h_z$ (h being the ice thickness)

The pressure gradient is thus some function of the ice overburden pressure. It is also possible to consider flow in terms of a **fluid potential** (Φ_f) and Equation (2.5) may be rewritten in the form:

$$\Phi_f = \rho_w \, g \, h \, Z + \rho_w \qquad (2.6)$$

where Z = depth coordinate (normal to x–y plane).

The location and flow of basal water is of considerable interest and importance for glacier sliding and subglacial erosion. Several modes of flow have been proposed: in thin sheets or films, in various types of channel or conduit, through interlinked cavities and ponded water bodies.

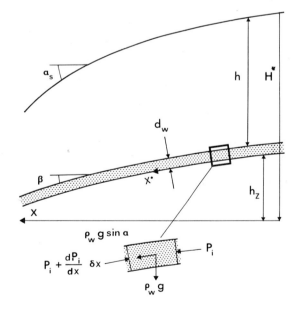

Figure 2.5 Schematic representation of the factors involved in producing a generalized pressure gradient in one–dimensional sheet flow. See text for further explanation (after Weertman, 1966 and 1972).

2.4.1 Water layers and films

Weertman (1966, 1972) has argued that where the bed of a glacier is impermeable basal water will discharge in a thin film, sometimes only a few μm thick. Walder (1982) has examined the stability of such sheets and finds that they are probably quasi-stable only at thicknesses <4 mm. Meltwater will flow in such a layer if the basal water pressure (P_w) is equal to the cryostatic pressure (P_i). The effect of the generalized **pressure gradient** (Equation 2.5) is to direct the flow of water perpendicular to the surface slope. A fundamental result of Weertman's analysis is that the glacier surface slope is 10 times more effective in directing water flow than the local bed slope and will favour flow as a thin sheet. It is quite possible, however, for bed slopes to significantly exceed those at the surface and to control the water flow direction favouring flow in channels. The sheet flow of water, therefore, will not necessarily parallel the long axis of a valley glacier nor the ice flow line. In two dimensions, for instance, where $\beta \neq 0$, $\alpha_y = 0$ and $\beta_y \neq 0$ (the subscript, y, refers to slopes transverse to the x direction) water flow must be weakly directed towards the glacier centreline. For other configurations of convex and concave upper slopes, basal water will be forced to flow to the sides or centre of a valley respectively. The thickness of the basal water film (Π) can be calculated as a function of the pressure gradient and distance from the glacier terminus, given a value for the volume of basal water being transported per unit time per unit width (V_m) (Weertman, 1966):

$$\Pi = (12\eta_w \, V_m \, x/P_g)^{1/3} \qquad (2.7)$$

where η_w = viscosity of water

Although Weertman's analysis is plausible the likelihood of sheet flow over substantial areas of real, rough glacier beds is open to question (Nye, 1976; Walder, 1982). Nevertheless ice melting at the **sole** due to sliding, deformation or geothermal heat input has the highest probability of flowing in a thin '**Weertman film**' as evidenced by several studies.

Dye-tracer experiments of Stenborg (1969, 1970) showed that water entering some **moulins** on Störglacieren, Sweden travelled extremely slowly and was almost undetectable at the glacier outlet. One interpretation is that the water could have been forced into a thin sheet rather than a channel. Work by Hallet (1979a) has confirmed the presence of thin water films beneath some glaciers. Walder and Hallet (1979) examined deglaciated, calcareous bedrock at Blackfoot Glacier, Montana, which revealed a cover of precipitates (spicules and furrows parallel to the former ice flow direction). They interpret these features as being formed with ice and rock in almost intimate contact, but separated by a very thin film of water (of order $10\mu m$ thick). Up to 80% of the sole of Blackfoot Glacier displayed evidence of such a thin water film.

Irregularities in the glacier bed will give rise to variations in water layer thickness, at times reducing it close to zero and creating high pressure zones. Once a significant pressure difference is created in a layer and water pressure is no longer cryostatic there will be a tendency for basal water to accumulate. When variations in water film thickness occur, water velocities will vary, and where they increase due to constrictions the amount of turbulent heat will rise and has the effect of locally melting more ice, which again concentrates the water. Nye (1976) has demonstrated that where water is twice as thick as the average the rate of concentration of water into a basal film (equivalent to the growth of a distinctive channel) goes up by a factor of eight.

Water descending from the surface to the bed will arrive at very specific locations and in large quantities. It is unlikely that this water will readily dissipate into a thin film but will continue to flow in hydraulically conducting channels at the bed. Tracer studies show this latter situation to be quite common – both from the velocity of the water and from dye concentration – time curves which exhibit marked similarities with open channel flow (Behrens *et al.*, 1975).

2.4.2 Cavities

In order to accomodate several of the critical points raised in connection with the stability of

'Weertman films' and the conclusions reached in Chapter 1 that, according to sliding theory, water will accumulate on the downglacier flank of bed obstacles (Equations 1.18 and 1.19), water may be considered to be located in and flow through cavities connected by thin channels. Under these circumstances pressures are controlled by the sliding parameters, P_n, τ_b and bed roughness. Walder and Hallet (1979) Hallet and Anderson (1980) have described cavities from the recently deglaciated bed of Blackfoot Glacier, Montana and Castelguard Glacier, Alberta which have a vertical relief of between 0.1 and 1.0 m. These authors also demonstrate that such cavities may constitute up to 20–24% of the ice–rock interface of these glaciers.

Weertman (1972) has attempted to take account of water in cavities by modifying his sheet flow theory. Films of water are allowed to thicken thus forming minor pools – a geometry identical to that envisaged by Lliboutry (1965, 1969). Such 'Lliboutry pools' are present only when $P_w < P_n$. The depth of water in the pools (i.e. 50–100 mm) is 1–2 orders of magnitude greater than the thickness of 'Weertman films' and, according to Weertman, depends upon the pressure gradient driving the water flow. Lliboutry maintains that the pressure in hydraulically conducting cavities and tubes is an independent variable, whereas Weertman (1972) considers that pressure is governed by P_n, τ_b and bed roughness, in keeping with Nye–Kamb sliding theory. Although Weertman has attempted to justify his modified sheet flow theory it has not gained widespread acceptance and is considered as one particular case of basal water flow.

2.4.3 Channels and conduits

If the pressure in subglacial water is sufficiently low (i.e. less than cryostatic) it is to be expected that the water will flow into channels – a suggestion confirmed by dye-tracer experiments. Detailed treatments of water flow in subglacial channels have been given by Weertman (1972), Röthlisberger (1972) and Nye (1976). Two fundamental types of channel are recognized: **Röthlisberger (or 'R') Channels** in the ice and **Nye channels** in sub-ice rock.

'Röthlisberger channels'

'R' channels comprise a series of conduits or pipes at the bed of a glacier incised upwards into the ice (Figure 2.6). Röthlisberger (1972) has shown that closure of such an ice tunnel by creep will be proportional to the cryostatic force (P_n). Tunnel melting is due to heat generated by turbulent water flow (i.e. frictional heating). For a channel in steady-state these processes just balance to keep the conduits open. Detailed treatments of these tunnel processes have been given by Nye (1976), Clarke (1982) and Spring and Hutter (1982). A typical 'R' channel is shown in Figure 2.6. Closure of the tunnel progresses due to power-law creep (Equation 1.1) and the rate of tunnel closure is:

$$T_c = R_t B (3^{1/3}/2)(3^{1/3} \Delta P'/n)^n$$
$$= R_t B (\Delta P')^n \qquad (2.8)$$

where $\Delta P'$ = difference in pressure between water at the tunnel wall and in the ice

R_t = tunnel radius

B = ice hardness parameter in ice flow law

n = viscoplastic parameter in ice flow law

Nye (1976) has given an alternative version of Equation 2.8 expressed in terms of changing tunnel cross-sectional area.

The rate of ice melting is given by:

$$M_t = Q_w P_g / 2 \pi R_t L \qquad (2.9)$$

where Q_w = water discharge

L = latent heat of fusion

In most circumstances water pressure in the 'R' channel (P_w) will be less than P_n and it might be expected that water melted from the base of the glacier will tend to migrate towards the channel. Weertman, however, found that this effect is limited to a certain distance out from the channel and beyond that the pressure forces water away from the conduit. The critical distance is in the

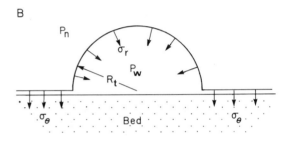

Figure 2.6 Röthlisberger or 'R' channels. **A**: a typical channel in the slow moving marginal zone of the Schuchert Glacier, East Greenland. Note some of the scalloped features on the outer part of the ice roof. K.J. Miller as scale. **B**: idealized transverse section of a 'R' channel. Water fills the tunnel of radius R_t and with water at pressure P_w. The tunnel attempts to close by radial creep under cyrostatic pressure P_n the radial stress (σ_r is equal to $\Delta P(R_t/r^{**})^{2/n}$ where r^{**} is the distance from the tunnel centre. σ_θ is an azimuthal stress which is given by $[(n-2)/n]/\sigma_r$. (after Weertman, 1972).

order of 100 R_t given by the term $R_t (\Delta P/\tau_b)^{n/2}$. 'R' channels are thus maintained by water descending to the bed via moulins and crevasses. It would appear that 'R' channels occur primarily along the central parts of the bed of valley glaciers, and are certainly observed issuing from their snouts in regions of more stagnant ice (Figure 2.6). Such conduits are less likely to develop beneath ice sheets or ice caps where the focusing effect, created by surface gradients, is less pronounced.

Water pressure in an 'R' channel tends to rise at times of high meltwater production and discharge. This may cause $P_w > P_n$ (Figure 2.6) which has a twofold effect. One is to enlarge the existing tunnel by creep (due to the increased pressure), but is a relatively slow process. The second effect is more immediate – water is forced out of the 'R' channel beneath the surrounding ice and possibly into any adjacent water film. Although relations between meltwater input and water pressure are complicated and may, at times, be the reverse of the above situation, it is clear that increases in water pressure and flow at the bed will have a tendency to cause increased separation of ice from bed and must, therefore, strongly influence sliding velocity – a fact demonstrated by several studies of meltwater discharge and surface velocities of glaciers (Iken, 1981) (see also Figure 2.1).

'Nye channels'

A second suite of water-discharging basal conduits are those incised into bedrock and termed 'Nye channels' after their description by Nye (1973). Nye channels may be coupled, temporarily, with a superimposed 'R channel', and may form a mildly arborescent network. It is difficult to be certain of the durability of Nye channels – whether they may be destroyed by glacial erosion or whether basal water can flow for a sufficiently long time in the same location to cut a bedrock channel. On theoretical grounds it is possible to consider that such channels would parallel the ice flow direction and must be of moderate to large size with depths greater than the amplitude of undulations of obstacles controlling sliding (>0.005 m). This is necessary in order to prevent ice being injected into them under high pressure, thus expelling the water.

Detailed investigation of recently deglaciated bedrock surfaces show quite clearly the presence of significant incised channels. In the front of retreating Nigardsbreen, an outlet of Jostedalsbreen in southern Norway homogeneous gneiss, recent ice retreat has uncovered a sequence of meltwater drainage channels and potholes (Figure 2.7).

In the forefield of Blackfoot Glacier,

Figure 2.7 Nye channels incised into bedrock. **A:** channel cut by meltwater beneath Nigardsbreen, Norway and now exposed by glacier retreat during the period 1970–1980.

B: A subglacial meltwater channel exposed in the forefield of Castleguard Glacier, Canada. Note the scallops produced by solution of the partly dolemitized limestone. Scale is given by knife (right) (photo courtesy B. Hallet).

Montana, Walder and Hallet (1979) describe narrow, elongate depressions which they believe were formed by subglacial water. Typically they are 25–50 mm deep, 100–200 mm wide, 2–5 m length, 5 m apart and parallel to the former ice flow direction or local bed slope. The presence of precipitate coatings suggests they may have been intermittently occupied by basal ice. Significantly the channels do not form an arborescent network but comprise short, often unlinked segments which may have been eroded out along the lines of abrasional features such as striations or gouges.

Walder and Hallet (1979) suggest that the bulk of the Nye channels beneath Blackfoot Glacier would have been inactive at times of no surface melting and that during periods of significant surface melt with water penetrating to the bed the entire Nye channel system could discharge all the meltwater produced. This conclusion is based upon the calculation that a channel of average size could discharge $\sim 10^4 - 10^5$ m³ a⁻¹ which is equivalent to the total discharge produced by basal melting alone. The total discharge for Blackfoot Glacier with summer melt included would be $> 10^7$ m³ a⁻¹.

2.5 Lakes

In favourable locations meltwater may be ponded to form large bodies of water, or lakes. Such reservoirs occur at the surface of a glacier (Figure 2.8) or where a glacier dams a suitable topographic depression (such as Graenalon and Vatnsdalen in Iceland, Märjelensee and Görnersee in Switzerland or Summit Lake, British Columbia, Canada). Water may also accumulate in hollows or under favourable pressure gradients at the glacier bed as occurs at Grimsvotn on Vatnajökull or as shown by radio echo sounding beneath the Antarctic Ice Sheet.

Björnsson (1964, 1975, 1977) and Nye (1976) have investigated the development of **subglacial lakes**. Lakes form where pressure gradients drive basal and interstitial water into a central catchment and where the escape of water along the ice–bedrock interface is prevented by a **potential** barrier. Consider the subglacial lake shown in Figure 2.8A. The level of the lake is given by L_o. By visualizing the whole of the ice as floating in a sea of infinite depth with a surface at L_o yet maintaining the present surface elevation it is possible to construct a curve depicting the bottom of the ice mass. This is accomplished by reflecting the ice surface in L_o and multiplying its vertical scale by $\rho_i/(\rho_w - \rho_i)$,

≈ 10 to obtain hydrostatic equilibrium: shown by the curve ABC. Those sections of the glacier lying below 'sea level', L_o, would require a compensating 'negative' thickness. The part of the curve between X and Y represents a seal or barrier preventing hydrostatic communication between the lake and the glacier snout. ABC represents an **equipotential** curve for the water pressure within the glacier:

$$\Phi_f = \rho_w \, g \, (hZ - L_o) + P_w \qquad (2.10)$$

where Z = height above an arbitrary datum
P_w = water pressure

If $P_w = P_n = \rho_i \, g \, h$

$$\Phi_f = (\rho_w - \rho_i) \, g \, h - \rho_w \, g \, L_o + \rho_i \, g \, Z_s \quad (2.11)$$

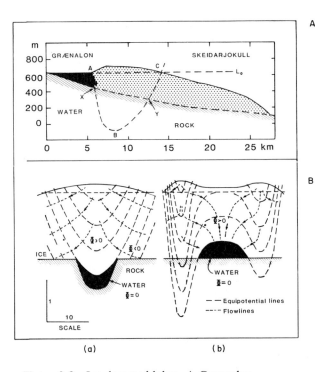

Figure 2.8 Ice-dammed lakes. **A** Graenalon, Vatnajökull illustrating aspects of the potential barrier (A-B-C) as envisaged by Björnsson at the beginning of the 1935 jökulhlaup (from Björnsson, 1977). **B**: Water flowlines and equipotentials for two subglacial lakes: (a) in a bedrock hollow and, (b) in an ice cavity above bedrock. Note how the equipotential lines and hence water flow are governed by surface slopes (from Björnsson, 1977).

where Z_s = height of the glacier surface

Thus the curve ABC is equivalent to a potential $\Phi_f = 0$ as discussed above which is given by:

$$h_2 = L_o - [\rho_i/(\rho_w - \rho_i)](Z_s - L_o)$$
$$= L_o - 10(Z_s - L_o) \qquad (2.12)$$

and was used in constructing Figure 2.8. Water in the glacier (i.e. in intergranular veins and capillaries) will flow from high to low potential, orthogonal to the equipotentials as shown in Figure 2.8 while in the water reservoir the pressure is hydrostatic ($\Phi_f = 0$).

Some of the more remarkable sub-ice water bodies or 'lakes' occur beneath interior portions of the Antarctic ice sheet. A large number of these lakes have been identified by radio echo sounding from anomalously strong bottom echoes with significant lateral continuity (Oswald and Robin, 1973). They occur in areas of low surface slope, low ice velocity, low surface accumulation and where bedrock hollows exist. An example is shown in Figure 2.9. While most sub-ice lakes are of small horizontal dimensions (same order as the ice thickness, ~ 5 km) several very large lakes have been observed (up to 8 000 km²) with important consequences for ice sheet internal deformation and surface slopes (Robin *et al.*, 1977). Because $\tau_b \to 0$ over the large lakes, surface slopes flatten significantly

and velocities increase. Internal deformation is also affected as the ice moves from non-sliding to fast-sliding conditions and back again.

Although it is not possible to determine the depths of water in these lakes from radar studies the minimum thickness of a freshwater layer can be estimated by considering the **'skin depth'** necessary for radio reflection. Taking a value for the electrical conductivity (Ω_w) of fresh water (10^{-4} mhos m^{-1}) and the radio frequency (f) the skin depth is given by (Drewry, 1981):

$$\xi_s = 0.5\pi\sqrt{(0.1f\,\Omega_w)} \qquad (2.13)$$

and is 6.5 m at 60 MHz.

2.6 Discharge of meltwater from glaciers and ice sheets

Water which has travelled through the glacier is discharged in one or several large melt streams (Figure 2.6). Such discharges are usually restricted to a short summer season with the peak following the maximum snow melt. Østrem

Figure 2.9 Subglacial lake beneath the East Antarctic ice sheet as recorded by SPRI 60MHz airborne radar system. Note the typical rock reflections (weaker and more irregular) to either side of the lake, and the internal ice sheet layering (from Drewry, 1981).

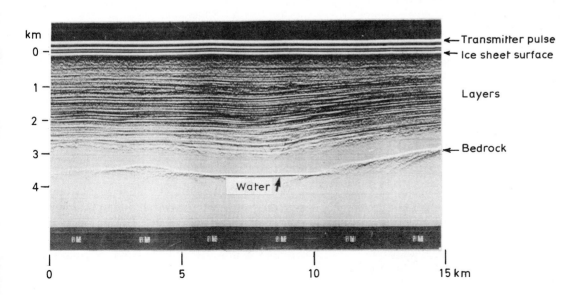

(1974) indicates that, in general, Scandinavian glaciers yield 85% of their total annual discharge within the months of July, August and September. Water issuing from a glacier **portal** effectively integrates the water balance over the whole glacierized basin (area, a):

$$Q_w = \iint_a (M \pm C_e + P_a)\, da \pm S_w \pm g_w \pm P_z \cdot L_t$$
(2.14)

where Q_w = water discharge (at snout)
  M = surface melt (geothermal and sliding deformation)
  C_e = condensation/evaporation
  P_a = precipitation over the glacier in liquid form
  S_w = internal glacier water storage
  g_w = contribution from groundwater
  P_z = precipitation (liquid) on watershed
  L_t = lag time = distance

Studies have shown that some of these components can be effectively separated by careful hydrochemical analyses, especially of environmental isotopes. Using fluctuations in Deuterium (^2H), Tritium (^3H) and ^{18}O in water samples Behrens *et al.* (1971) and Moser and Ambach (1977) found meltwater from glacier ice formed prior to thermonuclear weapons testing (i.e. 1952) is practically free of Tritium while snow accumulating since 1952 has a Tritium content an order of magnitude greater and which reached a maximum in 1963. Subglacial water possesses much lower Deuterium (^2H) levels than water melted at the glacier surface as it is made up of meltwater percolation from the winter snow cover. Groundwater on the other hand can often be characterized by high electrical conductivity values.

There are pronounced fluctuations in meltwater discharge from glaciers. These occur principally on a diurnal and seasonal periodic basis although superimposed are the effects of semi-random catastrophic events such as sudden release of water stored in glacial lakes or by a particular combination of meteorological conditions.

A typical daily discharge curve is shown in Figure 5.12 for the Gornergletscher, Switzerland

(Collins, 1979b) with components of englacial and subglacial flow calculated from conductivity measurements. Figure 2.10 shows the discharge record for Nigardsvatn, the outlet lake of Nigardsbreen in central Norway. The mean curve shows the seasonal rhythm – the maximum and minimum extension show the high degree of variability both within a season and from year to year.

Elliston (1973) has demonstrated that as the melt season progresses a correspondingly smaller fraction of the total flow is responsible for daily peaks – these disappear if there is any substantial snowfall on the glacier by reducing ablation.

At the beginning of a melt season discharge builds up with strong diurnal fluctuations of increasing amplitudes superimposed upon a more slowly increasing base flow (Elliston, 1973). This results from the opening-up, under increasing water pressure, of old narrow conduits in the ice. The volume of the previous winter's snowfall will also determine the magnitude of the summer melt. A small winter snowfall, for instance, is rapidly melted off exposing old firn and ice of much lower albedo early in the season with subsequent increased production of meltwater. Table 2.1 shows these relationships for the South Cascade Glacier, Washington, USA.

Table 2.1 Seasonal meltwater production and winter snowpack depth, South Cascade Glacier (from Meier, 1974)

Season	Snowpack depth (m)	(% annual mean)	Summer runoff (m)	(% annual mean)
1957–58	2.31	77	5.76	158
1963–64	3.53	117	2.31	68

Both Elliston (1973) and Collins (1979b) show that summer snowfall significantly reduces the discharge of meltwater but that flow continues for several days after snowfall due to draining of water reservoirs in the glacier. During such **'recession' flow** the subglacial component is increased to upwards of 90% from its usual value of 20–25%.

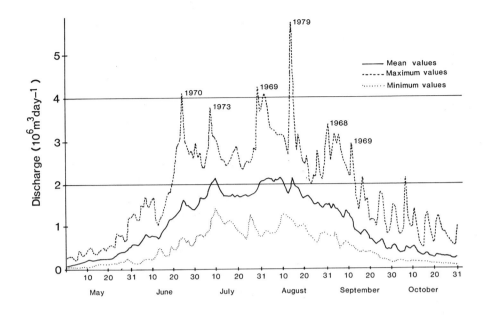

Figure 2.10 Water discharge from Nigardsvatn, Norway (1968–79) illustrating maximum and minimum flow conditions (with dates). The heavy line is the mean value (from Kjelsden and Østrem, 1980).

Not all water melted or added to the glacier is discharged immediately. Some may be stored within the ice, but this may vary considerably from glacier to glacier. On South Cascade Glacier Hodge (1976) reports that water levels in a drill hole showed an overall decrease to the end of the ablation season. Thereafter in October and November the waterlevel rose suggesting seasonal storage since the water does not drain away during the winter. Hodge found that by mid-January (1974) the waterlevel in another hole was at the same height below the glacier surface (~22 m) as in the previous summer. Similar storage has been postulated by Mathews (1973) and Vivian and Zumstein (1973). As suggested by Collins the bulk of the water stored on an annual basis is likely to be subglacial, in cavities or in permeable **till**, while in the summer considerable quantities of water are held within the firn. Elliston (1973) concludes that up to 50% of meltwater in Gornergletscher is stored for at least 24 hours, while in Burroughs Glacier, Alaska Larson (1978) found ~20% of melt was added to storage and the average lag time for discharge was 3 hours.

2.7 Sudden, high-magnitude discharge events (Jökulhlaups)

Numerous streams and rivers discharging from glaciers show sudden flood events. Many of the smaller floods result from heavy rainstorms or periods of strong ablation on the glacier. The larger floods, however, are usually caused by the rapid emptying of glacial water bodies. In Section 2.5 it was shown that water may be stored in subglacial reservoirs, in lakes dammed at the margins of glaciers, or in depressions on the ice surface.

Glacier outburst floods are known as jökulhlaups in Iceland, where some of the most spectacular and disastrous water releases occur. Jökulhlaups from Grimsvotn, a subglacial lake in the central part of Vatnajökull, Iceland have occured regularly, about twice each decade. Björnsson (1974) indicates that the average water volume discharge from Grimsvotn is 3–3.5 km³. The **hydrograph** of the 1954 jökulhlaup, which

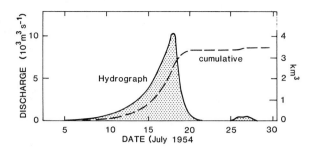

Figure 2.11   Hydrograph for the 1954 jökulhlaup, Grimsvotn, Vatnajökull, Iceland (from Rist, 1954).

reached a maximum discharge of $\sim 10^4$ m^3 s^{-1}, is shown in Figure 2.11. The 1922 jökulhlaup was one of the largest ever recorded, discharging ~ 7.1 km^3 water with a maximum flow of 5.7×10^4 m^3 s^{-1}. Nye (1976) has remarked that this is more than the flow of the River Congo and about one quarter of that of the Amazon! Figure 2.11 indicates that there is a steady, exponential rise in discharge, followed by an abrupt cut-off as all the water in storage is drained.

The mechanisms by which Jökulhlaups occur are not well understood but have formed the focus of several studies by Björnsson (1974, 1975), Nye (1976), Clarke and Mathews (1981), Spring and Hutter (1981), and Clarke (1982). Meier (1972) summarized several of the hypotheses for the triggering of jökulhlaups: these include the vertical release of an ice dam by flotation consequent upon raised water levels; overflowing of an ice dam (especially in ice marginal locations); destruction or fissuring of an ice dam by earthquakes; drainage at the ice–bedrock interface through pre-existing tunnels by either creep-enlargement with increased water pressures or melt-enlargement from turbulent heat transfer; and subglacial melting by heat from volcanic activity.

One of the most widely accepted trigger mechanisms can be understood in relation to Figure 2.8. Water within certain regions of the glacier, and at parts of the bed, is driven along potentials and forms a reservoir. If subglacial leakage into bedrock is less than this water influx the lake level will rise. This has the effect of

reducing the potential barrier by floating the ice dam, and when point B reaches a critical distance from the bedrock surface a hydrostatic connection is made and water will drain out of the reservoir. Nye (1976), Spring and Hutter (1981) and Clarke (1982) have considered in some detail the mechanism by which, when the ice dam is lifted, water will escape by enlarging subglacial tunnels. The release of thermal energy from the relatively warm water held in the reservoir is considered to be the principal factor in the development of drainage tunnels. The melting of conduits has been discussed in Subsection 2.4.3 and the evolution of the jökulhlaup is again governed by the balance between creep-closure and melt-enlargement of the tunnel. The change in cross-section of the tunnel (S_*) with time, given that creep-closure is insignificant, is:

$$\frac{dS_*}{dt} = \frac{M_t}{p_i} - k\,S_*(p_i - p_w)^n \qquad (2.15)$$

where M_t = ice mass melted from tunnel wall per unit time
p_w = water pressure
p_i = ice overburden pressure
k = constant

The discharge of water increases with such enlargement until, as the reservoir drains, the water pressure falls, increasing the pressure difference between ice and water. This eventually results in the creep-closure rate exceeding the melt-enlargement rate and drainage is cut off – and not necessarily when all the water has been discharged.

Mathews (1973) estimated the increasing diameter of a subglacial tunnel beneath Salmon Glacier, Canada during a jökulhlaup from ice-dammed Summit Lake. The conduit appeared to enlarge from about 1 m to 12–13 m over a period of as little as 7 days as discharge increased from a few metres per second to almost 3 000 m s^{-1}.

It is clear that high-discharge events as exemplified by jökulhlaups will be enormously important for the erosion of bedrock and transport of sediments in and beyond the glacier. These aspects are further considered in Chapters 5 and 10.

3 Glacial erosion: subglacial bedrock failure by crushing and fracture

3.1 Introduction

There is as yet no quantitative theory of glacial erosion, and development of the necessary and fundamental unifying concepts has been slow. Progress so far has been in two disparate directions. One has involved the description and classification of the morphology of erosive features, principally by geographers and geologists. In retrospect the product of this approach has been the formulation of gross, genetic models which, at best, provide a simple qualitative understanding of form.

During the last 20 years a second approach has gained prominence which attempts to understand the physical relationships involved in geological activity thus linking erosion closely with glaciological processes. An early study of this genre was Crary's (1966) analysis of **fjord** formation based upon detailed geophysical measurements from Skelton Glacier, in Antarctica. More recently the work of Boulton (1974) and Hallet (1979, 1981) in developing hypotheses of glacial abrasion has been of major importance.

Several factors are responsible for the deficiencies in our knowledge of erosive activity. *First*, and most important, is the highly complex nature of the erosive process which requires understanding of ice and rock material properties, ice dynamics, thermodynamics, friction and lubrication, physico-chemical effects and subglacial hydrology. In each of these fields knowledge is incomplete so that physical relationships may be only partially formulated.

The relative youthfulness of glaciology results in serious deficiencies in existing physical theories such as those of ice flow, heat transfer, and the role of water. In addition the glacial environment is particularly complex when compared with the study of sea and air which possess simpler structures and boundary conditions. In glaciology, therefore, only some of the simplest cases are reasonably well understood (e.g. ice deformation) and in the region of the glacier bed

where all the erosive and indeed much of the sedimentational activity is focused physical conditions may be sufficiently different and extreme for generalized solutions to the inapplicable. This is certainly true in the case of sliding, which is of fundamental importance to erosion.

A *second* problem involves the collection of measurements close to the bed of a glacier or ice sheet. These are essential to adequately model processes by provision of input data and boundary conditions. Two factors are here important – the difficulty of deploying instruments and sensors at the bed without major disturbance to the ice conditions to be measured, and the typically slow rates at which processes operate hindering the speedy collection of an adequate data set.

A *final* consideration relates to the scale of analysis. The real world presents a continuum of form and process from microscopic (10^{-6} m) to global (10^6 m). It is impossible to provide adequate coverage throughout these 12 decades so that examples and discussion in this and the following two chapters will attempt to be representative if not comprehensive.

3.1.1 Mechanisms of glacial erosion

A conventional treatment of glacial erosion would be to deal with three or possibly four primary mechanisms: Abrasion; Plucking (Quarrying); Subglacial water action; Subglacial dissolution.

Of these, abrasion is the easiest to define on *a priori* grounds. Sugden and John (1976), for instance, consider abrasion to be 'the process whereby bedrock is scored by debris carried in the basal layers of the glacier'. Such erosion by glacier ice, charged with basal debris and sliding over bedrock, may also be viewed as an engineering problem of 'wear', 'Abrasion' is then only one of several well-known wear mechanisms.

Plucking has been described as 'joint-block

removal' but is a complicated process. Röthlisberger and Iken (1981) and Sugden and John (1976) suggest that plucking may be made up of several mechanisms – loosening of rock fragments, evacuation of those fragments from the bed and finally their entrainment in the ice.

The erosive role of subglacial water, often under high pressure has only recently received widespread recognition as a fundamental agent in glacial erosion. The mechanical action of high velocity streams with a substantial sediment load (in suspension and traction) differs little from that in open-channel flow. Chemical processes associated with subglacial water are also identifiable: they include the dissolution of bedrock material of suitable composition and removal of solutes.

In this book we deal with the mechanics of glacial erosion under the following general headings:

- *Subglacial Bedrock Failure* (Chapter 3)
  - Crushing and fracture
  - Cyclic loading and fatigue
  - Evacuation
- *Subglacial Wear* (Chapter 4)
  - Abrasion
- *Subglacial Water Activity* (Chapter 5)
  - Mechanical erosion (abrasion and cavitation)
  - Chemical erosion (dissolution, carbonation and cation exchange)

In addition to discussing the primary physical controls of each of these modes of erosion it will be important to assess the relative, though highly variable, importance of each one. This speculative task forms the topic of Chapter 6.

3.2 Subglacial bedrock crushing and fracture

Glaciers and ice sheets reach thicknesses of ≤ 5 km. The cryostatic or normal stress at the glacier bed may thus reach a value of ~ 45 MPa. Table 3.1 gives a selection of values for the **unconfined compressive strength** of rocks. Typically they lie in a range well above cryostatic

Table 3.1 Unconfined compressive strength of typical rock samples (σ_p) (from Goodman, 1980)

Rock type	σ_p (MPa)
John Day basalt	355.0
Baraboo quartzite	320.0
Solenhofen limestone	245.0
Palisades diabase	241.0
Pikes Peak granite	226.0
Navajo sandstone	214.0
Hackensack siltstone	122.7
Nevada Test Site granite	141.1
Nevada Test Site basalt	148.0
Cedar City tonalite	101.5
Tavernalle limestone	97.9
Lockport dolomite	90.3
Oneota dolomite	86.9
Monticello Dam s.s. (greywacke)	79.3
Micaceous shale	75.2
Berea sandstone	73.8
Tensleep sandstone	72.4
Cherokee marble	66.9
Taconic marble	62.0
Quartz mica schist $\perp$ schistocity	55.2
Bedford limestone	51.0
Flaming Gorge shale	35.2
Nevada Test Site tuff	11.3

stresses. Nevertheless a superimposed ice load may be sufficient to occasionally cause failure in some rocks, and frequently to weaken subglacial strata which may subsequently be removed by ice or water flow. Where rock fragments lie at the sole the pressure they transmit to bedrock may be significantly increased by point contacts as irregular clast edges and surfaces impinge on bedrock. That very high stresses are experienced by subglacial rocks is clearly demonstrated by fracture marks on exposed glacier beds (Figure 3.1). A whole range of such features are known (cracks, **chattermarks, crescentic gouges (sichelbrüche)**) and have been described by Chamberlin (1888) and Wintges and Heuberger (1980). Furthermore the repeated passage of basal clasts over a portion of bedrock will give rise to loading cycles which significantly reduce rock strength.

Loosening of bedrock by the development of high-density fractures is also common and can frequently be seen on the lee sides of bedrock hummocks. While it is true that pre-existing weaknesses such as joints and faults will be exploited, clasts entrained in ice will also initiate cracks and fractures in an homogeneous bed.

To understand the role of ice and basal rock debris in the crushing of bedrock several factors require consideration: the nature of the applied stresses, the physical characteristics of rocks such as the presence of pre-existing weakness (joints, cracks and **foliations**) and their influence on **hardness** and strength, the effects of water and finally the manner in which loosened fragments are evacuated.

Figure 3.1 Effects of glacier crushing: fracture marks on gneiss bedrock in the forefield of Nigardsbreen, Norway. Palaeo-ice flow was approximately in the direction of the ice axe shaft and towards the lake.

3.2.1 Rock strength and applied loads

Consider a cube of bedrock material subject to stress. Each stress will result in corresponding strains. The stress can be resolved into nine components – three perpendicular to each of the three axes of the cube. Three forces are normal to the faces of the cube (σ_{xx}, σ_{yy}, σ_{zz}) and six are tangential or shear forces (τ_{xy}, τ_{xz}; τ_{yx}, τ_{yz}; τ_{zx}, τ_{zy}). It is usual to consider the normal stress as positive in tension and negative in compression. There are a set of strain components for each of these stresses. Thus for normal stress:

$$\sigma_{xx} \rightarrow \epsilon_{xx} , \gamma_{xy}$$
$$\sigma_{yy} \rightarrow \epsilon_{yy} , \gamma_{yz}$$
$$\sigma_{zz} \rightarrow \epsilon_{zz} , \gamma_{xz}$$

where ϵ_{nn} = **normal strain**
γ_{nn} = **shear strain**

For anisotropic materials such as rock the relationships are more complex as a given stress does not necessarily produce corresponding pure strain. Twenty-one independent elastic constants or moduli are needed to completely define an anisotropic material body which has no symmetry (derived from the generalized form of **Hooke's Law**). With increasing symmetry the number of elastic moduli is reduced. For tetragonal symmetry six are required, for hexagonal five, cubic three and only two for fully isotropic bodies. Hooke's law can be further manipulated to derive three moduli of elasticity related to the main stress fields of compression, tension and shear. They are:

Young's modulus, E, defined as:

$$\sigma = E \epsilon$$

Shear modulus, G:

$$\tau = G\gamma$$

Bulk modulus, K:

$$P_n = - K \frac{\Delta V}{V_o}$$

where P_n = hydrostatic pressure
V/V_o = volumetric strain (ΔV = change in volume of V_o, the original volume)

Some of these elastic moduli are interdependent such that

$$G = E/2 (1 + v)$$

where v = **Poisson's ratio** (ratio of lateral strain to longitudinal strain).

As all the strains are dimensionless, the moduli have dimensions identical to the stress and are given in N m^{-2}.

The inset in Figure 3.2 represents a sample of bedrock subject to three principal normal stresses: σ_1 is the primary or dominant stress, σ_2 is an intermediate stress and σ_3 is minor, often called the **confining** or hydrostatic **stress**. Rocks under load, beneath a glacier bed for instance, are subject to all three stresses but it is usual to ignore σ_2 and consider only the effects of the principal and the confining stresses (σ_1 and σ_3). Figure 3.2 illustrates modes of strain behaviour in rocks under a stress σ_1 but at different confining pressures. The concepts already introduced in exploring ice creep (Section 1.2) should be recalled. Where stresses are atmospheric ($\sigma_3 = 0$) the rock exhibits a predominantly elastic response, with a short plastic creep phase beyond the **yield point** (σ_y). At higher strains the rock fractures or fails, that is, the **cohesion** between mineral constituents is overcome. This response, shown in curve (a) of Figure 3.2, is termed **brittle behaviour**. If the confining pressure is raised it is possible to inhibit fracture and after the failing point there is continued deformation without a drop in stress. Such behaviour, shown in curve (c), is termed **ductile**. Curve (b) demonstrates a transitional (semi-brittle) type of stress–strain behaviour, neither completely brittle nor ductile. The transition between brittle and ductile behaviour (BDT) is affected by, in addition to confining pressure, temperature and strain rate. In general where temperatures are low, confining pressures small and strain rates high, brittle-type activity is favoured. Such conditions are those most likely to be encountered in subglacial environments. Although the BDT is ill-defined it usually occurs at stresses well beyond those normally encountered at or near the surface of the earth. It is possible, however, for ductile

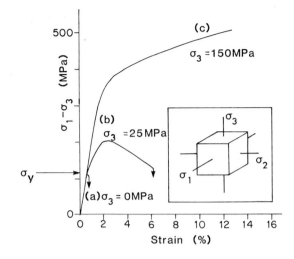

Figure 3.2   Strain behaviour of rocks under stress at several confining pressures (σ_3). (a) brittle behaviour: elastic response followed by failure at the yield point (σ_y); (b) semi-brittle behaviour with a short period of plastic creep (ductile deformation) beyond the yield point and followed by failure; (c) at higher confining pressures there is a marked increase in ductile deformation and specimen strength (after Jaeger and Cook, 1979). Inset shows stresses acting on a cubic element of bedrock: σ_1 = principal stress; σ_2 = intermediate stress; σ_3 = minor or confining stress.

behaviour to take place in certain 'soft rocks' such as clays, shales and evaporites. Ductile behaviour may also arise on a very limited scale at the tips of rocks indenting bedrock – where pressures may reach very high magnitudes.

3.2.2. Rock strength and resistance to failure

The resistance of a rock to fracture and crushing can be given by the peak stress in Figure 3.2 in either confined or unconfined mode. Values for this unconfined compressive strength (σ_p) lie in the range 30–100 MPa for most sedimentary rocks, and up to 400 MPa for some very strong metamorphic and igneous rocks as indicated in Table 3.1. Several factors influence the strength of rock:

1 **Stereology,** geometry, orientation and continuity of pre-existing discontinuities

such as joints, faults, foliations, bedding planes, etc. and friction along them

2 Degree of weathering
3 Presence and properties of material infilling discontinuities
4 Presence and flow of water
5 Residual stress (if any)
6 Loading cycle

The role of discontinuities in rock mechanics is complex and a detailed discussion beyond the scope of this book. Goodman (1980) suggests that, in general, they make the rock mass weaker, more susceptible to deformation and very anisotopic since shear strength is reduced and the permeability of the rock is increased parallel to the discontinuities. Perpendicular to them compressibility is increased and tensile strength reduced. Of fundamental importance are the size and orientation of fractures and joints. A simple but usable relationship is to correlate increased density of discontinuities with reduction in rock strength as suggested by Müller (1963) and developed by Selby (1980).

A significant weakening of rock may result from the addition of water. Under conditions of compressional loading where pore water is not free to drain away the pore-water pressure (p_w) rises and reduces the peak stress σ_p. This effect is shown in Figure 3.3 by curve b. Following the work of Terzaghi we can define an *effective stress* which takes account of this water-pressure:

$$\sigma' = (\sigma_3 - p_w) \qquad (3.2)$$

The water pressure within fissures and pores may be sufficient to initiate failure from an initial state of stress given by σ_1 and σ_3. The value is given by:

$$p_w \fallingdotseq \sigma_3 - \left[\frac{(\sigma_1 - \sigma_3) - \sigma_p}{\tan^2 (45 + \phi/2) - 1} \right] \qquad (3.3)$$

where σ_p = unconfined compressive strength
(= peak stress)
ϕ = angle of internal friction

It is quite clear that the geometry of discontinuities and their role as a site for groundwater in bedrock will be fundamental in controlling the failure of substrata subject to stresses imposed by a glacier or ice sheet. This was originally realized by Matthes (1930) and has been reiterated by Boulton (1974).

Beneath a glacier the loading of bedrock is liable to considerable change – as the ice thickens and thins in response to climatic or other factors and with the passage of cavities and clasts of variable size, shape and composition. Combined with the presence of discontinuities, loading cycles can be expected to give rise to non-elastic behaviour. This means that as bedrock experiences periodic loading a certain amount of the resulting strain is not recovered. The process

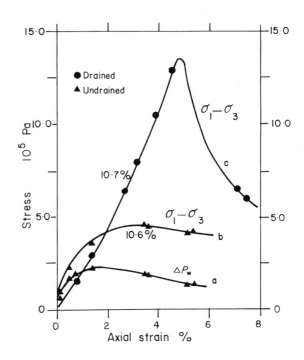

Figure 3.3 Effects of pore water on rock strength. The results of drained and undrained tri-axial compression tests on two samples of Carboniferous shale (with initial water content of 10.7 and 10.6%) are illustrated. Under drained conditions excess pore-water could escape but with undrained conditions this was not possible and the corresponding pore-water pressure (p_w) is also shown. A significant loss in strength is indicated (after Mesri and Gibala, 1972).

may, over many cycles, severely weaken the rock and give rise to failure at much lower loads than during initial loading. Cyclic loading and resulting fatigue failure are discussed further in Subsection 3.6.2.

3.3 Pattern of loading in bedrock

A more precise view of the effects of loading on bedrock is achieved by plotting in two dimensions, the stresses created as lines of constant maximum shear stress (τ_{max}) (these are sometimes referred to in civil engineering as 'bulbs of pressure'). The pattern of τ_{max} can be helpful in identifying zones where the propensity for rock failure and fracture will be greatest. Such analyses can be carried out for both a basal clast in contact with subglacial bedrock or for a section of the glacier bed subject to large-scale stress variations.

3.3.1 Elastic contact between clasts and bedrock

Regardless of whether significant plastic deformation occurs at the tips of an **asperity** in contact with bedrock an elastic response will normally occur. Considerable theory has been developed from the early treatment of Heinrich Herz for such elastic behaviour – termed **Herzian contact** and readers are referred to Bowden and Tabor (1964), Halling and Nuri (1975), and Sarkar (1980).

It is possible to simplify the nature of the contact of asperities by considering them as spheres of radius (r') as shown in Figure 3.4. The

contact area is thus a circle of diameter 2a such that:

$$2a = 1.75 \left[P_n r' \left(\frac{1}{E_1} + \frac{1}{E_2} \right) \right]^{1/3} \qquad (3.4)$$

where: E_1 and E_2 = Young's moduli for the two materials (clast and bedrock)

P_n = the normal load

Poisson's ratio is taken as 0.3.

The mean pressure over the surface is:

$$\overline{P} = 0.42 \, P_n{}^{1/3} \left[r' \left(\frac{1}{E_1} + \frac{1}{E_2} \right) \right]^{-2/3} \qquad (3.5)$$

This pressure creates a shear stress (τ) in the xz plane such that:

$$\tau_{max} = - \left(\frac{P_n}{\pi \, r'} \right) \cos \theta \qquad (3.6)$$

and is shown in Figure 3.4. The distribution of this shear stress may be plotted as a series of lines of constant τ_{max} and is shown schematically.

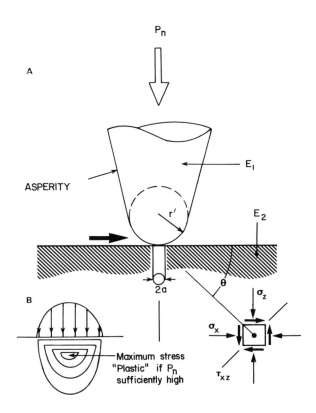

Figure 3.4 Contact of a spherical asperity with bedrock under a load (P_n). **A** The asperity tip has a radius r'. Asperity and bedrock possess Young's moduli E_1 and E_2. The pattern of loading through a cross-section of the contact area (2a) is shown. Shear (τ) and normal (σ) stresses in bedrock resulting from surface loading are also indicated. The load is separated into normal and horizontal components. **B** shows the contours (isolines) of constant maximum shear stress beneath the loaded point (for a normal load only).

Figure 3.5 shows the results of photoelastic stress experiments which enable the stresses set up in a material to be studied by double refraction effects under polarized light. The fringe patterns, or **isochromatics**, produced in Figure 3.5 demonstrate that the greatest value of τ_{max} occurs below the surface which theory suggests is at a depth of 0.66 a.

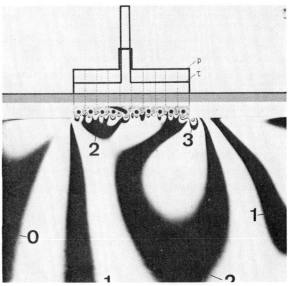

Figure 3.6   Isochromatic fringe patterns resulting from the loading of an epoxy plate by normal and horizontal stresses. The normal and horizontal loads at the plate surface are given by plots P and T. The stress contours are numbered. Note the complex pattern immediately beneath the plate (from Ficker *et al.*, 1980).

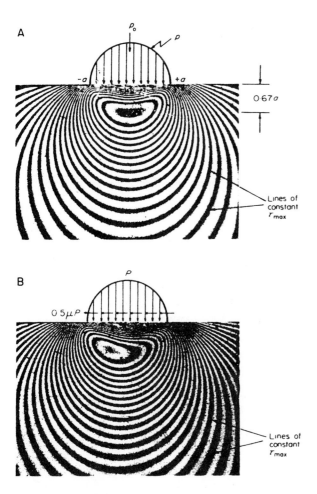

Figure 3.5   Photo-elastic representation of stress pattern induced in a plane surface in response to loading effects (from Halling and Nuri, 1975).
A fringes (isochromatics) produced in and below a surface in response to a normal load (P_n). Note the maximum stress at 0.67a depth.
B isochromatics produced in response to normal and tangential loads with a cylindrical contact.

Ficker *et al.* (1980) and Ficker and Weber (1981) have undertaken a series of interesting experiments and numerical calculations to understand the fracturing of subglacial bedrock from superimposed loads. Figure 3.6 is a photo-elastic illustration of the stress patterns generated during one of the Ficker experiments. A 10 mm-thick sheet of epoxy resin was loaded by a stress (resolved into normal and horizontal components) as shown, to simulate the effects of a single clast impinging on the glacier bed. The resulting complex pattern of loading should be compared with the results of simple studies shown in Figures 3.4B and 3.5. Ficker and his colleagues also undertook numerical calculations to derive the stress patterns in 'bedrock' for a variety of models: (i) a single irregular clast embedded in ice and touching the bed, (ii) cavities (empty or filled with a variety of materials such as water, sand and till), and (iii) a clast–cavity combination. This latter case is illustrated in Figure 3.7 with the corresponding numerical calculations based on a form of Equation (3.6) for a matrix of points within the bedrock. For each point the two principal stresses, σ_1 and σ_2 are depicted and the direction of potential failure and fracture is portrayed by the line representing σ_2, since the largest strains are typically orthogonal to σ_1. It is likely that similar, and yet more sophisticated, finite element models of bedrock loading will provide a valuable means of studying the manner in which glaciers and entrained debris crush and weaken their beds.

3.3.2 Plastic contact between clasts and bedrock

If the load P_n shown in Figure 3.4 is increased, elastic behaviour changes to plastic deformation at a critical shear stress. It should be remembered, however, that most rocks are brittle materials and only under limited conditions will there be plastic deformation. The value of the critical shear stress (τ_p) is usually at about 0.5 the yield strength in tension. Plastic yield will first take place at some depth below the surface: the result shown in Figures 3.4 and 3.5A. It can be seen that the 'plastic' centre is enclosed by an

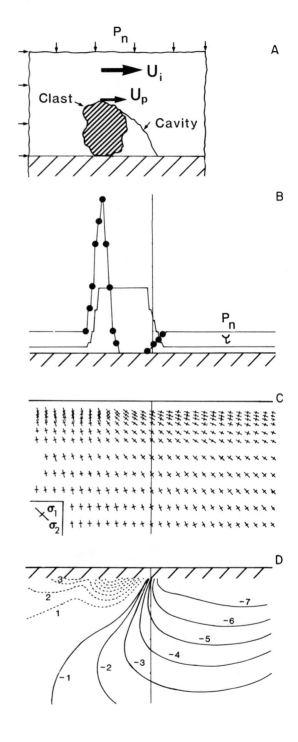

Figure 3.7   Calculated pattern of stress in a loaded rock layer (after Ficker *et al.*, 1980) for variations in the normal and horizontal stress components induced in bedrock by a clast held in the ice and an associated cavity.

A: The cavity is formed on the downflow side of the clast in response to a significant difference in velocity between ice (U_i) and clast (U_p). The latter is retarded by clast-bed friction (see Subsection 4.4.3 and Figure 4.5).

B load distribution (normal (P_n) and horizontal (τ)) for the clast–cavity model.

C Directions of principal stresses. Principal and secondary stresses (σ_1 and σ_2) shown in small inset bottom left and indicating the direction of potential failure (the long line, σ_2).

D Contoured values of τ_{max}, the maximum stress in the rock layer.

'elastic' zone which inhibits penetration of the asperity even at contact pressures in excess of the yield value of the materials. Plastic deformation is found to occur only when:

$$P_n / \pi a^2 = 3\sigma_y \qquad (3.7)$$

The quantity $3\sigma_y$ may also be defined as the *hardness* (H_d) of the material. If tangential stress is applied to the asperity contact the result is shown in Figure 3.5B. The zone of τ_{max} now approaches the surface so that plastic deformation may take place at lower loads. This is usually the case of basal clasts transiting a bedrock surface.

3.3.3 Large-scale bedrock stress patterns

Subsections 3.3.1 and 3.3.2 treated the contact between basal clasts and bed and examined the stress distribution thereby created. At a greater scale an ice mass will stress the whole bedrock surface and important stress differences will be set up due to variations in pressure at the ice–rock interface. In a first analysis of this problem Morland and Boulton (1975) modelled a

simple homogeneous bed undulation assuming elastic behaviour of the rock. The cryostatic (normal) pressure change (ΔP_n) over the bed hummock was calculated using the analysis of Nye (1970) which treats bed roughness as a Fourier series.

$$\Delta P_n \approx 10 \, \eta_i \, \overline{U} / \lambda_u \qquad (3.8)$$

where η_i = ice viscosity
$\overline{U}$ = mean ice velocity
λ_u = wavelength of bed undulation ($\gg 1$ m)

Figure 3.8 shows the results of the Morland and

Figure 3.8   Stress patterns induced in a subglacial bedrock undulation for various values of normal pressure (P_n). Inflow is from left to right (after Morland and Boulton, 1975)
A distribution of normal pressure at the interface;
B maximum shear stress in surface rock layers of the undulations;
C maximum shear stress contours (τ_{max}) in the undulation (in MPa given Equation 3.8)
D orientation of principal stress axes.

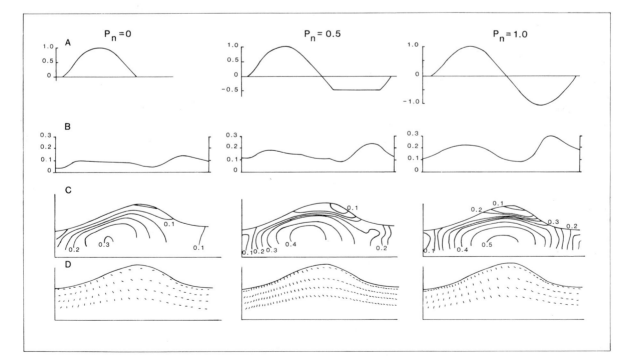

Boulton (1975) study for various overburden pressures. τ_{max} in the near-surface rock layer attains its greatest values where cryostatic pressure is equivalent to the pressure fluctuation over the hummock (ΔP_n). τ_{max} lies towards the centre of the downglacier flank. In Figure 3.8 the only pressure is that of the ice moving against the face of the bed undulation and τ_{max} peaks in the centre of the downglacier flank. Figure 3.8 plots τ_{max} for a cryostatic pressure equal to $\Delta P_n/2$. Under these circumstances **cavitation** occurs on the downglacier side of the hummock and τ_{max} lies close to the zone of cavity close-off.

3.4 Modes of failure

When bedrock experiences its peak stress due, for instance, to indentation by the passage of clasts held in basal ice, failure will usually take place. This will depend, however, upon the confining pressure (or the minimum principal stress (σ_3)) a relationship which can be initially understood with recourse to the simple and well-known **Mohr–Coulomb** failure criterion.

3.4.1 Mohr–Coulomb failure

The shear resistance (shear strength represented by peak shear stress τ_p) may be considered to be made up of a cohesive or residual shear strength component (τ_o) and frictional resistance to failure along potential fractures (the product of effective normal stress, σ', across the plane of fracture and the coefficient of internal friction or angle of internal or residual friction). Thus:

$$\tau_p = \tau_o + \sigma' \tan \phi \qquad (3.9)$$

The slope of τ_p. σ' describes a line tangent to all the 'Mohr circles' for combinations of principal stresses (σ_1). It can be shown (see Roberts, 1977; Jaegar and Cook, 1979 for a full explanation of the Mohr construction) that the Mohr circle has a diameter of $(\sigma_1 - \sigma_3)/2$ and a centre at $(\sigma_1 + \sigma_3)/2$ which gives the value for τ_p on fracture planes inclined at an angle 2θ to the direction of σ_1. In general a Mohr envelope will not be a straight line but convex upward indicating that $\tan \phi$ is

Table 3.2 Residual Shear strength (τ_o) and angle of internal friction (ϕ) for typical rock samples (from Goodman, 1980)

Rock type	Porosity (%)	τ_p (MPa)	ϕ	Range of confining pressure (MPa)
Berea sandstone	18.2	27.2	27.8	0–200
Bartlesville sandstone		8.0	37.2	0–203
Pottsville sandstone	14.0	14.9	45.2	0–68.9
Repetto siltstone	5.6	34.7	32.1	0–200
Muddy shale	4.7	38.4	14.4	0–200
Stockton shale		0.34	22.0	0.8–4.1
Edmonton bentonitic shale (water content 30%)	44.0	0.3	7.5	0.1–3.1
Sioux quartzite		70.6	48.0	0–203
Texas slate: 90 degrees to cleavage		70.3	26.9	34.5–276
Georgia marble	0.3	21.2	25.3	5.6–68.9
Wolf Camp limestone		23.6	34.8	0–203
Indiana limestone	19.4	6.72	42.0	0–9.6
Hasmark dolomite	3.5	22.8	35.5	0.8–5.9
Chalk	40.0	0	31.5	10–90
Blaine anhydrite		43.4	29.4	0–203
Inada biotite granite	0.4	55.2	47.7	0.1–98
Stone Mountain granite	0.2	55.1	51.0	0–68.9
Nevada Test Site basalt	4.6	66.2	31.0	3.4–34.5
Schistose gneiss: 90 degrees to schistocity	0.5	46.9	28.0	0–69

not constant. By undertaking triaxial compression tests in rock it is possible therefore to derive σ_1, σ_3 and θ and thence to calculate cohesive strength.

Typical values of τ_o range from a 10–20 MPa for sedimentary rocks up to 50–100 MPa for crystalline specimens (see Table 3.2). ϕ ranges between about 10° for moderately ductile rocks to 60° for brittle samples (Handin, 1969). This suggests that the angle of most faults will be between 15 and 40° with an average near to 30°.

Boulton (1974) and Morland and Boulton (1975) use the Mohr–Coulomb criterion to estimate the pattern of failure in large-scale bed hummocks based on the pattern of τ_{max} in near-surface layers from the ice overburden pressure (see Figure 3.8). Failure will occur if:

$$\tau_o / \tau_{max} \leq 1.$$

3.4.2 Cycle loading and fatigue failure

When bedrock is loaded and unloaded it may eventually exhibit fatigue failure. The intuitive relationship that the lower the applied cyclic stress the longer the life of the material prior to failure has been confirmed in numerous engineering tests on metals; the period to failure is typically inversely proportional to some power of the applied load (Teer and Arnell, 1975b).

At the base of a glacier loading cycles will be somewhat irregular as clasts of various sizes pass over an element of the bed. Nevertheless over periods of years (ranging from a few to hundreds) loadings may produce failure. The type of motion exhibited by clasts will also be variable. Some will slide whilst others may roll. The latter, giving rise to rolling contact, are significantly more important in fatigue failure. Englehardt et al. (1978) observed, by down-borehole video photography, rolling motion of clasts in the basal sediment layer beneath Blue Glacier, Washington. Although it was not possible to objectively separate the effects induced in the sediment horizon by the presence of the borehole it was considered by these workers that rolling motion represented a real basal phenomenon, and driven by sliding. An analogy with balls in a ball bearing is appropriate. Such rolling of clasts is favoured under conditions of reduced confining pressures induced by the ice overburden. These may be brought about by the presence of gaps at the sole due to flow over irregular bedrock, and by water pressurization of basal sediment layers.

During rolling bed material is compressed at the front of the zone of contact followed by release as the clast rolls forward. It was shown in Subsection 3.4.2 that the maximum stress occurs

a small distance below the surface. Fatigue failure, therefore, frequently occurs initially below the surface. If the clast exhibits some sliding as well as rolling motion the position of failure moves upwards towards the surface. On rock, however, the exact locus of failure is difficult to determine, being governed by size and shape of individual mineral grains, voids and existing microcracks. Fractures are produced which lead to the removal of relatively large fragments of rock – as individual crystals, or **lithic clasts**. This process gives rise to 'pitting' of the rock surface.

3.4.3 Macroscopic consideration of fracture and failure

Bedrock may fracture at considerably lower values than its yield stress. This is achieved principally by the development and propagation of microfractures in the host rock and the crushing, granulation and shearing of individual mineral grains **(cataclasis)**. Goodman (1980) suggests that upon initial loading there is a slight compression of the grains and closure of pre-existing fissures in the rock. At higher stresses deformation of pore spaces occurs and continued grain compression. After this stage, however, there is development of new cracks and exploitation of crystallographic and **cleavage planes** in the rock fabric due to the increasing importance in lateral strains (i.e. assuming strain is not simple and axial). Such cracks grow uniformly to a point where they begin to intersect and coalesce. This may then lead to a semi-continuous 'rupture' or shear surface developing and thus leading to ultimate failure of the rock.

Under conditions of subglacial stress where there is little confining pressure, such as at the top of a bed hummock, or where there is good jointing, the exposed bedrock may split longitudinally – possibly the result of tension in the direction normal to the principal stress (σ_2 and σ_3) (Figure 3.9).

Where the confining pressures are more significant, semi-random longitudinal cracking is suppressed in favour of a single fracture plane at

45° inclination to the principal stress (σ_1), and along which some shearing may take place.

A whole complex of interlocking shear fractures may result under very high confining pressures and the rock may behave in a 'ductile' manner. Crack formation may be progressive during a *single* loading phase when, for instance, a large rock held within ice is pressed against bedrock, or more typically during several loading cycles over a long period with the passage of numerous clasts. Microcracks tend eventually to coalesce and define some large-scale zone of weakness through the rock along which catastrophic failure may occur (Figure 3.9).

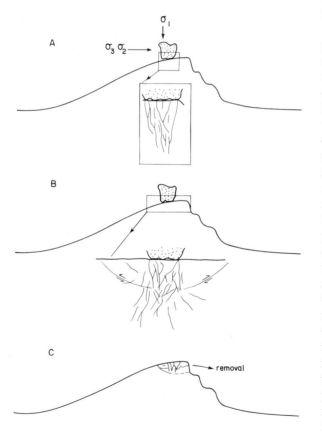

Figure 3.9   Failure response of a subglacial bed undulation to the crushing effects of a clast held in basal ice. **A** the initial development of longitudinal fractures; **B** coalescence of fractures and development of a large-scale shear plane; **C** removal of weakened bedrock along shear planes by ice flow.

3.4.4   Crack propagation and fast fracture

The theoretical or ideal strength of rock, considered as a brittle material, may be expressed as the force required to overcome interatomic bonds:

$$\sigma_c \approx E/j \qquad (3.10)$$

where E   = Young's modulus
      j   = constant related to interatomic potential ($15 > j > 8$)

This is a treatment similar to that for ice considered in Section 1.3. Rocks possess actual yield strengths far below the theoretical level.

The discrepancy may be explained by the presence in a rock of microscopic defects (surface cracks, scratches, notches, angles or internal flaws) which give rise to very high, local stress concentrations when the material is subject to loading, such as that imposed by clasts held within the base of an ice mass. The stress concentrations at crack tips may be in excess of the ideal strength of the rock and, therefore, sufficiently great to break interatomic bonds. The microcracks tend to propagate through the material under certain conditions, leading to full fracture and failure. The speed of fracture propagation may be very rapid – up to the speed of sound. A theory of brittle fracture based upon these ideas proposed by Griffith in 1924 has, with subsequent modification, found widespread theoretical and practical support in the study of rock mechanics. Elaboration of the complex Griffith theory of crack propagation lies outside the realm of this chapter, nevertheless it is important to understand the general nature of **fast fracture** in rocks and a brief outline of the process is given below. Interested readers are referred to Jaeger and Cook (1979, Chapter 10) for a more detailed discussion.

Consider an elliptical crack tip in rock. The crack will grow (propagate) if the strain energy stored in the rock, by virtue of an externally applied stress, exceeds the energy required to create two new surfaces within the material governed by its material properties – in other words we can understand fast fracture through an energy budget equation. For the extension of a crack by an amount δl_c (Ashby and Jones, 1980):

$$\delta W^* \geq \delta E + G_c H_r \delta l_c \qquad (3.11)$$

where $\delta W^* =$ change in load
H_r = thickness of rock
δE = change in elastic energy
G_c = elastic energy absorbed per unit area of crack generated

G_c is a fundamental material property often called **'toughness'**. The higher the value of G_c the more difficult it is to propagate a crack and cause fracture. Typical values of G_c for rocks fall in the range 10–1 000 J m^{-2} (granites, for instance, are about 100 J m^{-2}).

Fast fracture commences when a rock containing microcracks of a given size is subject to a critical applied stress (σ_n). This can be expressed in the form:

$$\sigma_n \, (\pi l_c)^{1/2} = (EG_c)^{1/2} \qquad (3.12)$$

where E = Young's modulus.

The term $\sigma_n \, (\pi l_c)^{1/2}$ is called the 'stress intensity factor', K_f, and (EG_c) the 'critical stress intensity factor' or the **'fracture toughness'** of the material (K_c). Thus fast fracture takes place when: $K_f = K_c$. For rocks, K_c lies between 0.5 and 5 MN m$^{-3/2}$.

Griffith provided an extension of his theory to deal with biaxial stress conditions and considered a plane which contains a large number of randomly oriented elliptical flaws. Even under compressive stress regimes local tensile stresses of high magnitude may be generated at the tips of some of the more suitably oriented of these microflaw ellipses. Fast fracture will take place when this local stress reaches σ_2. The failure criterion is thus:

$$(\sigma_1 - \sigma_3)^2 + 8\sigma_t (\sigma_1 + \sigma_3) = 0 \qquad (3.13)$$

where σ_t = tensile strength

If σ_3 is taken as zero (i.e. confining pressure is atmospheric) then $\sigma_1 = 8\sigma_t$, that is, the uniaxial compressive strength is equivalent to a tensile strength.

Murrell (1958, 1963) has shown that the fracture criteria can be expressed in terms of a Mohr envelope of the form:

$$\tau^2 + 4 \, \sigma_t\sigma_n - 4 \, \sigma_t^2 = 0 \qquad (3.14)$$

McClintock and Walsh (1962) have further extended Griffith's theory to take account of the collapse or partial close of internal microcracks under a dominantly compressive stress regime. By introducing a shear stress along the closed part of the microflaw dependent upon a kinetic friction (μ_k) term these workers were able to derive a new criterion for fracture:

$$4\sigma_t = \mu_k \, (\sigma_1 + \sigma_3) + (\sigma_1 - \sigma_3)(1 + \mu_s)^{1/2} \qquad (3.15)$$

where μ_s = coefficient of static friction

Equation 3.15 yields a straight-line relationship between the principal stresses in a Mohr diagram *and* retains the Murrell extension in the tensile stress range (since cracks will not be closed under such conditions)

Price (1966) has drawn attention to the fact that most rocks subject to stresses are anisotropic to some degree and microflaws are rarely uniform in shape or size. Recognition of these realistic heterogeneous conditions leads to a concept of fracture with a pronounced hierarchy of stress concentrations and crack propagations. Only at the very highest applied stresses will all flaws propagate and run together to generate macroscopic shear failure.

3.5 Evacuation of crushed and fractured rock

Once bedrock has been weakened by fracturing and crushing activity of the glacier it is necessary to remove the loose material in order to achieve effective erosion. A detailed and interesting study on the problem of evacuation has been presented by Röthlisberger and Iken (1981). They propose that loosened bedrock can be evacuated by a combination of rapid opening of subglacial cavities by high water pressures and **heat-pump** effects, which are described in the work of Lliboutry (1964) and Robin (1976) respectively.

The general incorporation of debris into ice, whether produced by crushing and fracturing or by abrasion, is usually termed entrainment; various mechanisms will be discussed in detail in Chapter 7. **Entrainment** thus plays a central role in the continuum between erosion and transportation. There are some factors, however, which are more directly related to aspects of rock mechanics and are considered under the heading of **evacuation**.

3.5.1 Heat pump (Robin effect)

The removal of rock at the edge of steps above a cavity may be reinforced by a heat pump mechanism proposed by Robin (1976). As ice moves over an irregular bed the stress distribution in the ice and at the glacier bed will change (due perhaps to variation in basal water supply). If the pressure increases the ice may melt with expulsion of any free water in veins and at crystal boundaries, with corresponding lowering of the pressure melting point. When the pressure is released the absence of the expelled water requires that the heat lost from the ice is restored by conduction through the rock leading to production of cold patches. The change in temperature (T) with pressure (P_i) can be calculated using the **Clausius–Clapeyron equation** which gives:

$$T = P_i \Delta v_1 / \Delta v_2 \qquad (3.16)$$

where Δv_1, Δv_2 = changes in volume and entropy due to melt.

Thus a change in pressure of 13 MPa will change the melting point by 1°C. The size of the cold patches, according to Robin, will be of the order of a few hundreds of millimetres to a few metres across.

The effect of a developing cold patch is both adhesion of ice to rock and an increase in ice hardness. Experiments by Jellinek (1959) and Röthlisberger and Iken (1981) suggest the tensile strength of a junction due to ice adhesion may be up to 5×10^5 Pa. Such adhesion will enable loosened rock fragments to be extracted from bedrock by ice flow. Ficker *et al.* (1980) describe some simple experiments which simulate the erosive role of ice when frozen to a uniform,

isotropic bed by the removal of fragments of rocks due to ice adhesion. The process may explain the production of gouge-marks. The experiments were carried out using glass and epoxy resin, the latter having an elastic modulus (E) some 20 times smaller. A 8 mm-thick glass plate was cleaned and coated with 10 mm epoxy resin and polymerized at ~ 100°C for 24 hours. The glass–epoxy specimen was then allowed to cool slowly. Since the coefficients of linear heat expansion of the epoxy and glass differ by almost 10:1 a significant shear stress is created at the cooling interface which results in separation of the two materials at the edges of the specimen. A little way in from the edge separation appears to cease and adhesion between epoxy and glass is very strong. Cracks appear in the glass at this point, moving outward from the junction between adhesion and separation. Ficker *et al.* (1980) indicate that any direct analogy with the stress caused by ice adhering to rock applies only to this junction. The result of the experiment was for flakes of glass to be removed by the epoxy and for crescentic, gouge-like cracks to form with steep break-off edges: simulated sichelbrüche.

Ice hardness rises by a factor of 2–3 for a temperature change of only a few tenths of a degree through the pressure melting point. Such stiffening of the ice will further assist in the fracturing of rock steps above cavities.

3.5.2 Hydraulic jack effect

Röthlisberger and Iken (1981) discuss the mechanism whereby sudden rapid changes in basal water pressure act like a hydraulic jack acting horizontally and move the glacier forward by opening large cavities at the bed, a process first proposed by Lliboutry (1964). The edges of fractured steps in the bed (Figure 3.10), upstream of cavities, may be broken off by ice flow over them and rock fragments deposited in the cavities (Figure 3.11). This process has been confirmed by Anderson *et al.* (1982) who observed the freshly fractured surface of several ledges beneath Grinnell Glacier, Montana. The blocks and flakes resulting from this activity were 0.01–1.0 m in size. The critical water pressure (P_{wc}) at which such cavities open by the hydraulic mechanism is:

$$P_{wc} = \overline{P}_n \left(\frac{\sin \beta_w}{\cos \overline{\beta} \sin (-\beta_w)} \right)$$
$$= \overline{P}_n - \tau_b / \tan (-\overline{\beta}_w) \qquad (3.17)$$

where P_n = mean overburden pressure at bed
(equivalent to normal stress)

β_w = angle at which bed steps are tilted
up-glacier

$\overline{\beta}$ = mean slope of bed

τ_b = basal shear stress.

3.5.3 Evacuation by water

A certain proportion of crushed and fractured bedrock may never communicate with the ice but will be evacuated by meltwater flowing at the bed. As crushing results in a predominantly coarse grade of material only the larger and discrete subglacial streams will be capable of its removal.

Figure 3.10 Typical rock steps in veined gneiss exposed by recent glacier retreat, Nigardsbreen, Norway.

Aspects of the transport of sediments in glacial meltwater are further discussed in Chapter 10.

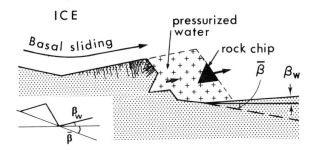

Figure 3.11 Hydraulic jack effect with removal of fractured bedrock by the opening of cavity resulting from high water pressures. Rock fragments are detached by freezing to the ice by the Robin heat pump mechanism. At intermediate water pressures freezing and hydraulic jacking combine to remove rock fragments. Step geometry shown by $\overline{\beta}$, mean bed slope and β_w angle at which steps are tilted up glacier (based upon Röthlisberger and Iken, 1981).

4 Glacial erosion: abrasional wear

4.1 Introduction

An analogy often employed in discussing glacial erosion is to liken an ice mass containing basal sediments to a piece of sandpaper rubbing over bedrock. This resemblance, although facile, is useful in a number of respects. Rather than consider glacial erosion as a complex and unique natural phenomenon the analogy immediately raises the possibility of understanding erosion as a problem in **'tribolgy'** – the science of interacting surfaces in relative motion (*tribos*, Gr. = rubbing). Tribology is a branch of engineering and surface physics which has developed a substantial body of theory supported by experimental results that can be applied with advantage to the study of glacial erosion. Some attempts in this direction have already been made (e.g. Boulton, 1974, 1979; Riley 1982). In general terms glaciers and ice sheets slide over a bed of rigid or deformable substrata. Both rocks held in basal ice and bedrock posses certain measurable material properties such as hardness and surface roughness. Being in relative motion they are subject to widely applicable laws of friction, with temperature, load, phase changes as additional important factors. The result of rubbing contact is damage to or erosion of the surfaces (both bed and tractive particles) called *wear* with production of wear debris (rock fragments and **rock flour**). It is possible to extend these ideas from the experience gathered in excavation technology especially in the ripping and cutting of rock by machines. In the sections that follow we attempt to build on tribological and engineering concepts to obtain an integrated view of glacial erosion.

4.2 Abrasion by 'pure' ice

In most conventional treatments of the erosive, especially abrasional role of glaciers and ice sheets, little effectiveness is attributed to ice devoid of rock clasts or inclusions. Embleton and King (1975) quote Chamberlin (1888): 'the ice of a glacier has of itself but little abrading and practically no striating power. It can neither groove nor scratch. . . .' (p. 208). Laboratory studies, however, in which pure ice slides over a rock surface, indicate that significant wear may occur in the absence of rock fragments. Budd *et al.* (1979) describe experiments of sliding ice over a rough granite slab. Wear was found to increase as velocity and normal (P_n) and shear (τ) stresses increased.

Table 4.1 Wear of rough granite surface by sliding ice at the melting point (from Budd *et al.*, 1979)

P_n [MPa]	τ [kPa]	U_s m a^{-1}	Wear rate mm a^{-1}	Debris flux kg m^{-1} a^{-1}
0.5	180	320	2.5	2.1×10^3
2.0	100	12	0.81	2.6×10
2.0	75	4500	33.0	4.0×10^5
4.0	13	13	5.0	1.8×10^2
4.0	25	63	12.0	2.1×10^3
4.0	50	400	55.0	60×10^4

The data in Table 4.1 from Budd's experiments can be fitted to a relation:

$$W \propto \tau\, P_n\, U_s^{1/3} \qquad (4.1)$$

where W = wear

τ = shear stress

P_n = normal stress

U_s = sliding velocity

From sliding experiments it was found that $U_s^{1/3} \propto \tau\, P_n^{1/3}$ so that:

$$W \propto \tau^2\, P_n^{2/3} \qquad (4.2)$$

At sliding speeds in the range 50–300 m a^{-1} wear is of the order of several mm a^{-1} and quite in keeping with measured values (see Chapter 6). These results suggest that ice alone may be an effective agent of erosion. Riley (1979), however, has commented that there can be substantial quantities of pre-existing and microscopic loose material on 'clean rock' prior to commencement of an experiment of the type described by Budd *et al.* (1979). This will be removed as ice commences to slide giving a spurious wear rate.

The problems in glaciology for laboratory studies are, however, severe and it is rare for the results of such experiments to be directly comparable with nature. In experimental work in fluids, for instance, the conditions under which tests are performed are frequently applied to other sets of conditions in the real world. For instance density or other material properties may be changed, such as the use of lignite for rock in flume experiments. This approach is justified and made possible by the laws of similarity in which scale, **kinematics** (i.e. motion without reference to mass or force) and forces possess defined and corresponding quantities. In glaciology, however, it is impossible to find a material to represent ice in a scaled experiment (with increase in volume on freezing) so that processes which involve ice deformation and regelation (such as ice abrasion) may be very difficult to model effectively. In addition problems may arise from the poorly determined or yet unknown magnitudes of the real forces involved preventing characteristic ratios from being applied. It is likely, however, that as theory advances, the techniques of *field*

measurement improve and logistic provisions enable certain difficult determinations to be made, many physical quantities will be better defined. Laboratory experiments will then become an important and necessary area of glaciological research.

4.3 Abrasion by rock particles in sliding basal ice: a first simple model

Basal ice in glaciers and ice sheets usually contains debris in varying concentrations and sizes. Consider a single clast held in the glacier sole. The clast has a number of rough, angular edges (asperities) which come into contact with the bed. If the clast is composed of material harder than bedrock it will indent the substratum and, as the clast moves with sliding basal ice, the asperities will plough out striations or grooves (Figure 4.1). This is abrasive wear.

Figure 4.1 **A** Glacial grooves in gneiss in the forefield of Franz Josef Glacier, New Zealand.
B: Glacial grooves cut into Tertiary Granite, Arran, Scotland.

4.3.1 The single asperity

Consider a single asperity of a basal clast as a conical cutting tool (Figure 4.2A). If the asperity is pressed downwards it will indent the bedrock. The size of the indentation will depend upon the load and relative hardness of the clast (H_1) and the substream (H_2). If $H_1 > H_2$ the area of the indentation can be shown to be:

$$\pi\, d^2 = W^*/\sigma_y \qquad (4.3)$$

where d   = radius of indentation
W^* = load
σ_y = yield strength of bedrock

Thus:
$$d = \sqrt{W^*/\sigma_y\pi} \qquad (4.4)$$

The depth of indentation h_i is:

$$h_i = d \cot \theta_a$$
$$= \sqrt{W^*/\sigma_y\pi} \cdot \cot \theta_a \qquad (4.5)$$

where θ_a = half-angle of asperity tip.

If the clast with the above asperity is now moved horizontally by an amount l_g the rock tip will plough out a volume of the softer substratum (Figure 4.2C). The volume of the groove is:

$$V_g = (d^2 \cot \theta_a)\, l_g \qquad (4.6)$$

As, however, only half of the conical asperity tip makes contact during sliding (i.e. leading half-cone) as opposed to simple indentation we may now redefine the radius of indentation as:

$$d = (2W^*/\sigma_y\,\pi)^{1/2} \qquad (4.7)$$

If the clast is moving at a velocity U_p (Figure 4.2C) the abrasive wear rate is:

$$A_b = (2 \cot \theta_a/\pi)(W^*/\sigma_y)U_p \qquad (4.8)$$

Equation (4.8) is a first model of abrasive wear and can be used to estimate subglacial erosion. Figure 4.3 plots the cross-sectional area of a groove cut by clast asperities against bedrock yield strength (or hardness) according to Equation (4.8) for various loads and half-angles. It assumes that the indenting asperity is significantly harder than bedrock ($H_1 \gg H_2$) and remains undamaged during sliding motion. It can be seen that abrasion rate (simply obtained by multiplying the ordinate by the clast velocity)

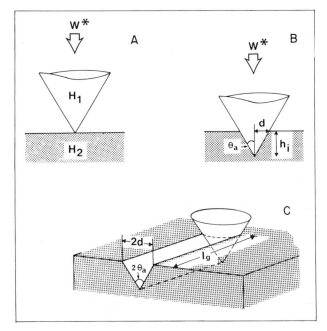

Figure 4.2 Abrasion by a single conical asperity. **A** and **B** show a single asperity (of half-angle, θ_a) of a clast indenting bedrock under load (W^*). Clast is of hardness (H_1), bedrock (H_2). Depth and radius of indentation are h_i and d respectively. **C** Ploughing action of a single conical asperity sliding horizontally over bedrock, l_g = distance travelled by the asperity.

declines exponentially with bedrock hardness. For a given yield strength (say 100 MPa) abrasion rate falls with increasing angle of the asperity tip. This could assist in understanding what happens as an asperity is blunted during erosion. For θ_a = 25° and a load of 1 tonne the abrasion rate for a medium hardness rock (σ_y = 50 MPa) is of the order of 2×10^{-3} m³ a⁻¹ for a sliding velocity of 100 m a⁻¹. The depth of the groove is 6 mm.

According to engineering usage Equation (4.8) leads to three principal axioms for abrasive wear:

1 The volume of wear debris is proportional to the distance travelled.
2 The volume of wear debris is proportional to the load.
3 The volume of wear debris is inversely proportional to the yield stress or hardness of the softer material.

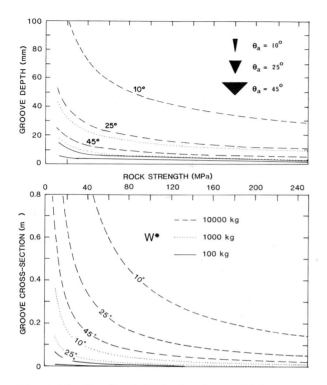

Figure 4.3 Results of simple abrasion model (groove depth and cross-sectional area) as a function of rock strength for a sliding velocity of 100 m a^{-1} with $H_1 \gg H_2$, according to Equation (4·8). Each plot shows load as parameter with three angles of the asperity tip. Wear rate (volume per unit time) is obtained by multiplying cross-sectional area by the sliding velocity.

Experiments with metals have shown that if the ratio of hardness of substratum to hardness of abrading clast (H_2/H_1) > 0.8 the resistance to abrasion increases rapidly, whilst below this value abrasion takes place much more easily.

It is important to realize that the process is not continuous but comprises jerky steps. As grooving commences there is a build-up of elastic strain at the asperity tip. This is released giving rise to the impact of the asperity against the rock surface with subsequent production of rock chips. Stress again builds up with continued crushing and displacement of rock debris until the asperity effectively bears on an unbroken step in bedrock. This will subsequently fail, usually creating a large fragment. The sequence then recommences. This type of jerky abrading

motion can be demonstrated from the force-displacement curves generated during **drag-bit cutting** of rocks. In Figure 4.4 the thrust force on a cutting tip oscillates as minor rock chips are created, but then builds up to a higher peak just prior to the formation of a major fragment. Immediately after the stress falls, almost to zero. Several complications may be considered to the simple model presented above.

Wear debris may build up in front of the cutting asperity which will cause a change in the effective value of d. Neither all the asperities nor all clasts will be cutting bedrock at any one time because they are composed of softer material or because they have already been worn. In addition it is quite obvious that not all debris produced by abrasion will be removed. Various coefficients can be introduced to account for these effects.

4.4 Components of the simple abrasion model

From the foregoing discussion a simple wear model has been given which upon substitution of realistic values for rock hardness and load does appear to give abrasion rates and striation dimensions which accord with those observed in

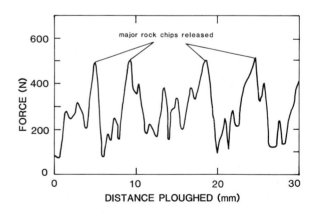

Figure 4.4 Force on the cutting tip of an industrial drag-bit during rock cutting (Darley Dale Sandstone). Note various oscillations of the thrust force as the drag-bit is displaced and major excursions as rock chips are broken off. Speed of cutting ~ 4 mm s^{-1} (from Fairhurst, 1964).

nature (see Chapter 6). It is now time to examine in turn each of the major factors illuminated by the general model:

$$A_b = f [\Delta H_d, F, U_p, C_o, c', S_r] \qquad (4.9)$$

where

ΔH_d = relative rock hardness of clast and substratum (alternatively $1/H_2$ where H_2 = hardness of substratum)

F = force pressing clast against bed

U_p = velocity at which clast is dragged over bed

C_o = concentration of debris in ice at the abrading surface

c' = parameter related to number of clasts cutting bedrock and ratio of wear particles removed

S_r = area of bed in contact with clast of given shape and size (ratio of bed roughness to clast diameter).

4.4.1 Hardness and clast–bed friction

Hardness is defined as resistance to local deformation and can be explained by reference to the surface area of a permanent indentation made experimentally by an indenter (a_i). Thus $H_d = (W^*/a_i)$. The term (W^*/a_i) can be taken as $3\sigma_y$, where σ_y is the yield strength of the material (i.e. the point at which elastic deformation goes into plastic behaviour). In relation to **Mohs' scratch hardness** (Mh) familiar to geologists, Bowden and Tabor (1964) have shown that:

$$H_d = k (1.2)^{Mh} \qquad (4.10)$$

where k is a constant

Frictional drag

Friction between clasts and the glacier bed will retard particles in traction. It is thus important to both clast velocity U_p and the force pressing clasts to the bed F. Friction is made up of two components – adhesion and asperity interlocking. The former is related to the adhesive strength of junctions created by plastic deformation at the tips of the asperities of rough clast and bedrock surfaces in contact. In order to initiate sliding

these junctions have to be sheared which requires a force:

$$F_s = \mu_s P_n \qquad (4.11)$$

where μ_s = coefficient of static friction
P_n = normal stress (equivalent to W^*).

Thus

$$\mu_s = \frac{\text{Shear strength}}{\text{yield stress}}$$

$$\mu_s = \frac{F_s}{P_n} = \frac{A_r \tau_p}{A_r P_n} = \frac{\tau_p}{\sigma_y} \qquad (4.12)$$

where A_r = real contact area
σ_y = normal yield stress of bedrock in compression
τ_p = yield strength in shear.

We have already seen when considering rock hardness that $H_d \approx 3\sigma_y$, and $\tau_p \approx \sigma_y/2$ so that $\mu_s \approx 0.5\sigma_y/3\sigma_y \approx 1/6$. This solution is valid when the two rocks are of the same hardness and the surface between them extremely smooth. Under most rock–rock contact conditions μ_s is likely to be in the order of 0.5 to 0.7.

Once sliding commences and any junctions have been sheared the force (F_s) required to maintain motion drops due to a reduction in time available to allow renewed **junction growth** at the asperity tips:

$$F_s = \mu_k P_n \qquad (4.13)$$

where μ_k = coefficient of kinetic or dynamic friction ($\mu_k < \mu_s$)

Friction created by the interlocking of asperities is more complicated. The problem is shown in Figure 4.5 where it is necessary to resolve both the normal force P_n and the tangential force T_f in moving the surfaces relative to one another (up-and-across motion).

Figure 4.6 shows in general how both the normal (P_n) and frictional force (F) vary as two rough interlocking surfaces slide over each other when the mean junction angle, $\theta_j = 10°$. P_n and F initially increase due to junction growth and plastic deformation.

In the cases of both asperity interlocking and

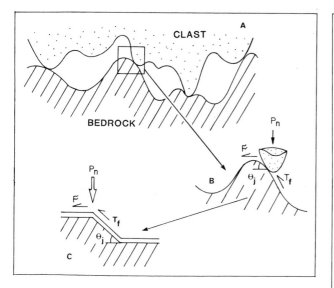

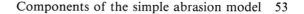

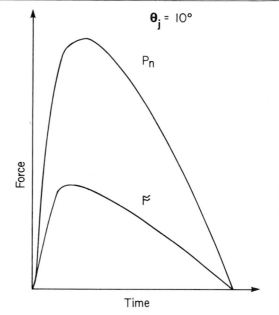

Figure 4.5 Asperity interlocking which creates friction between particles, or particles and bedrock. **A** General view of interlocking; **B** more detailed view of the interactions between two asperities; **C** schematic representation of the interaction. P_n = normal load; F = frictional force; T_f = tangential force, θ_j = junction angle.

Figure 4.6 Variation in P_n and F for two rough surfaces sliding over one another with a mean junction angle of 10° (after Teer and Arnell, 1975a).

adhesional friction considered above, it is assumed that the clast and substratum possess similar hardness. This may occur from time to time especially where blocks, removed from fractured bedrock, are incorporated into basal ice and slide across rock of the same composition. In other cases a hard exotic clast may be brought into contact with and slide over a softer substratum causing significant indentation and **'ploughing'**. This latter process will add to the force resisting motion and hence contribute to the total frictional resistance.

There is, therefore, a mild paradox, in that where lubrication is absent, a glacier charged with basal debris sliding over a hard bed may move more rapidly than a glacier sliding over softer materials, due to the increased friction from 'ploughing'.

4.4.2 Debris concentration (C_o)

The notion, used for modelling purposes, of a single striating rock inclusion held in ice close to the bed is clearly unreasonable as most glaciers possess dirty basal layers (up to several metres in thickness) in which the concentration of debris may rise to 50–60% by volume (see Chapter 7). Both the presence of substantial quantities of debris in the ice and their **inertial interactions** introduce considerable complications to a simple erosional hypothesis. It has already been shown that particle content stiffens ice and reduces the creep rate (Subsection 1.4.3). Beyond a critical concentration this will reduce basal sliding velocities due to enhanced bed friction. Particles also exhibit a great variation in geometry, density and hardness.

Hallet (1981) has made an initial theoretical attempt to investigate the problems of debris concentration on abrasion. The conclusion of his hypothesis suggest that for a given bed roughness the abrasion rate, although variable, may be highest at low values of C_o. Figure 4.7 indicates that an abrasion maximum results at a debris concentration of about 10–30%. Hallet (1981) gives a derivation for this optimum debris content in terms of the proportion of the bed covered by fragments:

$$I_d \approx 2.7[R' \, k_t^{*2} \, (r^{*2} + r_2^2)/r_c \, k^*] \qquad (4.14)$$

where k_t^* = transitional wave number in Nye sliding theory (see Equation 1.21)

k^* = wave number of bed undulations

r^* = transitional particle radius

r_c = particle radius

R' = amplitude to wavelength ratio of bed (a^*/λ_u).

Hallet has indicated that for a given particle size and concentration the abrasion rate falls with increasing bed wavelength.

4.4.3 Clast velocity (U_p)

A rock fragment pressed against the bed and whose hardness is the same or greater than materials composing bedrock will accomplish wear only if it is moved along the ice/rock interface. Calculation of particle velocity, and hence rate of abrasion, is complex but motion is essentially dependent upon components of the sliding mechanism which urge the particle forward against the resistance from clast–bed friction. Thus:

$$U_p = U_b - U_r \qquad (4.15)$$

where U_p = clast velocity at the bed

U_b = ice velocity at the bed (usually the sliding velocity, U_s)

U_r = relative decrease in particle velocity due to frictional drag ($\propto F$)

Hallet's U_p

Hallet (1979b, 1981) has proposed a model of glacial abrasive wear of similar form to Equation (4.9). The contact force terms and particle sliding velocity terms are, however, closely linked: Hallet considers the force pushing a particle against the

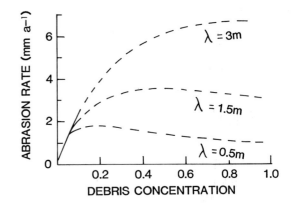

Figure 4.7 Abrasion rate as a function of the concentration of basal debris (i.e. proportion of fragments covering the bed) and for three different wavelengths of bed undulation (λ). Fragment size = 0.2 m; bed roughness = 0.05 (from Hallet, 1981).

bed is directly related to glacier sliding velocity, U_s, as calculated from Nye-Kamb theory.

For a glacier bed of low roughness Hallet finds that since the frictional retardation of clasts moving over the bed is roughly equal to the melting rate in the regelation process the velocity of particles is approximately equal to the component of the sliding velocity parallel to the bed: $U_p \approx U_s$, and substantially independent of clast geometry. His general derivation for U_s has already been given (Equation 1.21):

$$U_s = [\tau_b /\eta_i \, (\xi + \Omega \, \mu_c \, C_o)] \qquad (4.16)$$

where ξ = bed roughness term

Ω = viscous drag term

μ_c = friction coefficient between bed and clast

C_o = debris concentration

η_i = ice viscosity

τ_b = basal shear stress.

We shall more closely examine aspects of the basal ice velocity especially that normal to the bed, and the role of regelation in a later section.

Boulton's U_p

Boulton (1974) has used Weertman's (1957) sliding theory to derive motion, initially for a single clast, which he views as equivalent to a quasi-stationary bed protuberance. In contrast to Hallet's result velocity (U_p) is, according to Boulton, strongly controlled by clast geometry. There is a critical particle size for which the mechanisms of regulation and enhanced creep are maximized (the 'critical particle size' cf. critical obstacle size in Weertman sliding) and at which rock fragments are moved at their greatest speed. Below this critical value regulation is highly effective in increasing the relative velocity between ice and clast. Above the critical value enhanced creep likewise reduces particle velocity. The 'critical particle size' in which U_p is at a maximum is found by Boulton to lie between -7ϕ and -3ϕ. The velocity is given as:

$$U_p = U_s - \left[B2r_c \, (P_n \, \mu_c \, A_a / A_2)^3 + \left(\frac{\psi \, K_i \, P_n \, \mu_c \, A_a}{L \, \rho_i \, 2r_c \, A_2} \right) \right] \qquad (4.17)$$

where B = thermally activated ice hardness parameter in the flow law
ψ = coefficient related to pressure melting point
K_i = thermal conductivity
L = latent heat of fusion
A_a = apparent area of clast contact
A_2 = area of clast in transverse plane.

A complicating factor arises from the possibility that different velocities may be exhibited by clasts of differing size: interference of basal fragments in motion may slow down the faster moving particles (see also Subsection 9.2.3). Again velocity is dependent upon aspects of regulation theory which are discussed in the next section.

4.4.4 Effective contact force (F)

The rate of abrasion is closely related to the force pressing a clast against the substratum. When one surface, such as a rough rock fragment held in the ice, is pushed against another (bedrock) contact will occur only at certain points (asperities) due to the roughness of those surfaces. Some of these concepts have already been outlined in Subsection 4.3.1 and it is

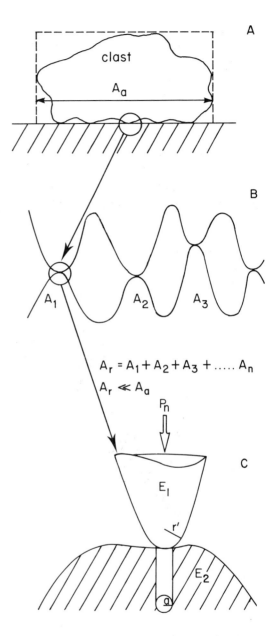

Figure 4.8   Loading of bedrock by a superimposed clast. **A** Clast resting on bedrock with an apparent area of contact (A_a) but actually touching at only a limited number of asperities. **B** A series of asperity contacts. Note that the real contact area (A_r) is very much less than A_a. **C** Schematic of the disc of contact (radius, a,) between a conical asperity (radius of tip = r') and bedrock under a normal load (P_n). Clast asperity and bedrock possess Young's moduli E_1 and E_2, respectively (cf. Figure 3.4).

relatively easy to see that the real contact area (A_r) will be a small fraction of the 'apparent' interface area A_a. Any load applied across the surface is transmitted only at those contacts, where high stress concentration will occur (Figure 4.8).

At low loadings the asperities yield elastically. Experiments with metals show that under higher loads plastic deformation occurs which results in an increase in the contact area (which then just supports the normal load) – a phenomenon called junction growth by **cold welding**. Rocks and minerals, however, are brittle materials in which such plastic deformation is limited, and surfaces would be expected to crack and fragment at the contacts. Nevertheless Bowden and Tabor (1964) suggest that the very high compressive stresses generated in the contact region inhibit brittle fracture and the materials may behave plastically: the contact area is proportional to the load. Strong adhesion may also occur. Because ductility is limited, however, large-scale junction growth is not possible and the value for the coefficient friction μ is never high (i.e. never >1). It is quite clear that shattering and crushing of asperities does occur but Bowden and Tabor suggest this eventuates only *after* sliding has taken place. In addition it is reasonable to suppose that the higher asperities will deform with limited plasticity while lesser contacts will behave elastically – in other words the real contacts are a mixture of elastic–plastic reactions.

Boulton's F

Boulton (1974, 1979) envisages the load (F) pressing the clast to the bed as the sum of the buoyant weight of the clast and the weight of the overlying ice column less the effects from basal water:

$$F \approx (P_n - P_w)\, 2r_c^2 + (\rho_r - \rho_i)\, g\, 2r_c^3 \qquad (4.18)$$

where P_n and P_w = cryostatic and water pressures respectively

ρ_r, ρ_i = densities of rock and ice respectively.

It is not certain whether in this formulation the water pressure arises within a thin film or a cavity. If the calculation is dependent upon cavitation then its applicability to deriving a force pressing a clast to bedrock is doubtful. In addition, Equation (4.18) does not incorporate any forces acting on the clast from the deforming ice. Hallet (pers. comm.) argues that since Equation (4.18) implies that pressures act only on the uppermost part of a clast it is difficult to imagine how ice is prevented from deforming around the rock particles.

Wear is fundamentally controlled by the value of F in Boulton's abrasion theory so that for a given ice sliding velocity wear increases with respect to F at an exponentially decreasing rate to a critical quantity and thereafter falls as F continues to increase but clast-bed friction rises, according to the full abrasion model:

$$\overline{A}_b \propto \frac{C_o F}{H_d}\left[U_s - B2r_c \left(\frac{\mu_c F A_a}{A_2} \right)^3 + \left(\frac{\psi K_i F A_a}{L\, \rho_i\, 2r_c\, A_2} \right) \right] \qquad (4.19)$$

Measured abrasion rates from three locations at Breidamerkurjökull, Iceland are shown in Figure 4.9. The fit of the data to the model is, at best, uncertain as Boulton (1979) admits. It is unfortunate that no error bars are shown against measured abrasion rates as these must, by virtue of the exigencies of subglacial experimentation, be considerable. Whether such uncertainty fundamentally questions the relationship shown in Figure 4.13 is difficult to evaluate. One reasonably unequivocal result, however, is the presence of a distinctive hierarchy of abrasion rates related to bed hardness with the most resilient materials exhibiting least abrasion.

Hallet's F

In contrast to Boulton, Hallet (1979b, 1981) finds that in common with the early work of Gilbert (1906) and more recently McCall (1960), Röthlisberger (1968), and Weertman (1979) the contact force (F) is independent of cryostatic pressure and thus ice thickness.

Ice is considered to behave like a viscous fluid which imparts a degree of buoyancy to rock

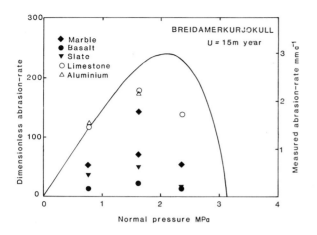

Figure 4.9 Measured abrasion rates at three experimental locations beneath Breidamerkurjökull, Iceland. Platens of different material characteristics were used (see key) and of the following hardnesses: marble = 450–510 kg mm^{-2}; basalt = 865–905 kg mm^{-2}; slate = 605–660 kg mm^{-2}; limestone = 180–215 kg mm^{-2}; aluminium = 50–60 kg mm^{-2}. Abrasion rate predicted by Boulton theory is shown as the solid line (after Boulton, 1979).

fragments (which 'float' within it) proportional to the weight of the 'fluid' replacing the clast and thus the ratio of the densities of ice and rock.

Hallet views the downward force pressing a clast to the bed as resulting specifically from the **buoyant weight** of the clast and viscous ice drag by flow towards the bed (U_n). Unless the particle reaches a critical dimension (say > 300 mm diameter) the contribution to the force from its buoyant weight is small and F is proportional to U_n. U_n is made up of three components:

1 Geothermal heat and sliding friction give rise to relatively uniform basal melting which Röthlisberger (1968) showed could contribute vertical velocities of 5–10 mm a^{-1} and ≪ 10–100 mm^{-1} respectively.
2 Regelation processes in sliding are calculated by Hallet from Nye (1969) sliding theory as the sum of velocities both parallel and normal to the bed.
3 Vertical straining, as a result of large-scale longitudinal extension, which Röthlisberger

(1968) considered to depend upon the distance between the centre of the clast and bedrock and to vary between + 10r_c and − 10r_c a^{-1} where r_c is the radius in mm.

Hallet (1981) suggests that as ice flows against bed irregularities regelation and straining will form the principal components of U_n so that $U_n \approx bU_s$, the parameter b varying with bed geometry. The contact force is thus proportional to the sliding velocity.

Irregularity in the shape of clasts, the interference that occurs between fragments and the steep gradients in ice velocity close to the bed present major problems in the derivation of the viscous drag force using conventional regelation theory (Morris, 1979). Nevertheless Hallet (1979b, 1981) has examined approximate solutions for use in his abrasion model. Viscous drag on a spherical particle due especially to proximity of the bed was found to be ~ 2.5 greater than when more than one particle radius from the bed.

Taking the various elements already discussed into account the full Hallet abrasion model is:

$$A_b = C_o U_p U_n H_d' \qquad (4.20)$$

where H_d' = coefficient related to rock hardness
C_o = debris concentration
$U_p \approx U_s$
$U_n = bU_s$ (see Equations 4.16 and 1.21).

Figure 4.10 shows the abrasion rate calculated by Equation 4.20 above in relation to clast size and wavelength of bed irregularities.

Several general conclusions can be made from the Hallet model. Abrasion rates are independent of ice pressure and depend upon U_n, the ice flow towards the bed. A_b does not increase monotonically with clast size but increases with the contact force, F. Hallet points out that where glaciers are actively sliding U_p is not sensitive to F. Particle velocities (U_p) will approach U_s, the sliding velocity. This is because U_n is only a small fraction of U_s with the factor b usually less than 0.1.

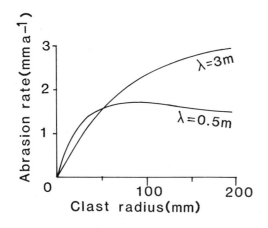

Figure 4.10 Abrasion rate from Hallet theory in relation to clast size for two bed undulation wavelengths. Bed roughness (0.05) and a debris concentration of 10% (from Hallet, 1981).

4.4.5 Removal of fines

A hard clast cutting into soft bedrock will create wear particles which, if not removed, may tend to clog the interface between asperities. A build-up of finely abraded material will thus inhibit continued abrasion by producing a contaminating layer. This layer may act as a 'lubrication interface' reducing clast–bed friction:

$$\mu_c = \tau_{ci}/\sigma_y \qquad (4.21)$$

$$= \frac{\text{critical shear stress of the interface}}{\text{yield stress of bedrock}}$$

As the shear strength of the interface is much lower in the presence of weak unconsolidated wear debris the clasts slide more easily over the bed and achieve less erosion.

Fines are usually removed by water at the bed – either a thin film which removes the μm-size material, or by discrete channels of varying sizes which have the capacity to remove large debris from areas accessed by the flow.

4.4.6 Discussion

We have examined, albeit in a condensed manner, the principal factors in a simple model of glacial abrasive wear. The opportunity has also been taken to contrast some of the ideas presented in the only two complete hypotheses of glacial abrasion to date – those of Boulton and Hallet. In is clear that these two glaciologists have contributed measurably to our understanding of abrasion but many questions remain unanswered.

There are major problems in the present glacier sliding theories. These are inherited in models which require a sliding parameter in the formulation of abrasion theories and must raise questions of model validity.

A critical aspect of deriving clast velocity and the force pressing the clast to the bed, in both Hallet and Boulton derivations, is the need to incorporate regelation processes. Morris (1976, 1979) has drawn attention to the inadequacy (by up to an order of magnitude) of classical regelation and plastic flow theories for calculating the ice forces on a clast. She points out that even with the most simple assumptions, of a flat bed and a circular clast, it is not possible to attain the correct stress and temperature fields which allow both their surfaces (bed and clast) to be at the pressure melting point. This is essential for effective regelation. Conditions at the glacier bed are usually complex with the presence of water, variable concentrations of sediments of different sizes and thermal properties, and geothermal heating.

4.5 Applications of Boulton–Hallet theories

Boulton (1974) has applied his model to understanding the abrasive erosion of a variety of small-scale bedforms as well as to larger-scale glacial features such as **cirques** and U-shaped valleys. Figure 4.11 illustrates a bedrock hummock of sinusoidal form over which there is a change in normal stress, due to ice load, (ΔP_n) of magnitude (Nye, 1970):

$$\Delta P_n = [2a'/\pi]^{1/2} \eta_i U_b k_* \qquad (4.22)$$

where a' = roughness constant ($4\pi r^2/0.148$)
$\quad\quad\ \ k_*$ = material constant
$\quad\quad\quad\quad$ typically $k_* = 0.01$ m^{-1}.

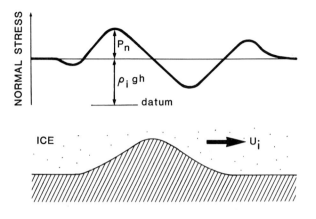

Figure 4.11 Distribution of normal stress over a bed undulation under sliding conditions (after Boulton, 1974).

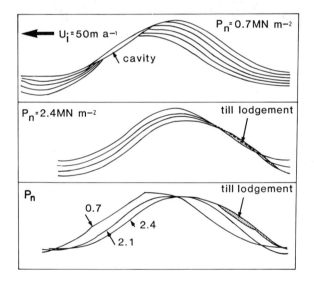

Figure 4.12 Abrasion of a sinusoidal undulation as a function of effective normal stress according to Boulton theory. Low normal stresses generate a downglacier asymmetric profile with the production of stepped forms and with an associated basin on the downglacier flank. High normal stresses induce lodgement of till (see Chapter 9) on the upglacier side and an upglacier asymmetry is produced (after Boulton, 1974).

Boulton used this stress change arithmetically added to the cryostatic pressure (i.e. $\Delta P_n + \rho_i g h$) in his abrasion model (Equation 4.19) to show erosion of the bed hummock for varying values of his most critical parameter – the normal stress (P_n).

At low stresses the upglacier surface (Figure 4.12) exhibits substantial abrasion (there is little modification on the crest and none on the leeside where cavities may arise). Further abrasion occurs at the base of the lee slope. According to Boulton theory relative abrasion gives rise to the evolution of an asymmetric step-like bedform.

At higher cryostatic pressures, the stresses on the **stoss face** may be of sufficient magnitude (Equation 4.19, Figure 4.12) for abrasion to be inhibited and debris in motion in the glacier bed to be lodged. A small amount of erosion takes place on the hummock crest whilst abrasion peaks on the lee face. There is only minor wear at the foot of the lee surface. Through time the hummock evolves an asymmetry which is the mirror image of that created at low values of P_n and the whole bedform migrates upglacier.

Boulton has extended his ideas to deal with larger bedforms, such as cirques and valleys (Figure 4.13). In the former case, maximum velocities and normal effective stresses P_n occur subjacent to the glacier equilibrium line where the greatest abrasion rates are inferred, thus giving rise to the overdeepened cirque basin form.

For overdeepened valleys occupied by active glaciers with thicknesses of several hundred metres and sliding velocities of several tens of metres per year Boulton suggests that both subglacial abrasion and quarrying (crushing and fracture) will be important and highly effective. The evolutionary sequence described by Boulton (1974) is again critically dependent upon P_n, in turn controlled by the depth of the valley glacier. Although Boulton's illustrations appear qualitatively persuasive they raise numerous problems. Many shallow glacial valleys are very wide (depth to top-width ratio is small) – whereas Boulton theory argues that most erosion should be concentrated in the valley bottoms. Perhaps steady-state has not been reached in the real-world examples? It is also difficult to explain

Figure 4.13 A Overdeepened glacially eroded valley, Isterdalen, Norway. **B** Cirque basin eroded into horizontal Tertiary strata, Svalbard.

very deep fjords by the Boulton mechanism: erosion should be inhibited at very great ice thicknesses.

Hallet (1979b) has discussed how bedforms may change with abrasion by considering the relative erosion at points over a bed hummock. The rate of abrasion in the bed troughs is A_T and that on the crest of hummocks A_c. A damping factor (D_f) is derived which describes the relative erosion of a hummock and bedform evolution:

$$D_f = (A_c - A_T)/ a* \qquad (4.23)$$

where a* = trough–crest amplitude.

When $D_f = 0$ abrasion is equal at trough and crest and reduces the feature uniformly. When $D_f > 0$ abrasion will tend to eliminate the hummock. As D_f is always positive the bed will always be smoothed by abrasion. Hallet finds that D_f is affected by particle size and bed wavelength; it decreases as these two factors decrease and increase respectively.

The size of rock fragments is also important. An increase in size of clast will achieve considerable wear of the hummock crests and leave the troughs almost untouched. For larger glacial features such as valleys Hallet suggests that their production is dependent upon high sliding velocities and a moderately high debris concentration (of the order of 10–30%).

4.6 Abrasion by cold ice

It is usually accepted that only where basal sliding can take place will abrasion be an effective erosive mechanism. This flow condition requires basal ice at the pressure melting point. Does this indicate that ice, frozen to bedrock under cold conditions, will have absolutely no abrasional potential?

Consider the clast shown in Figure 4.14, resting upon bedrock but embedded in ice which is

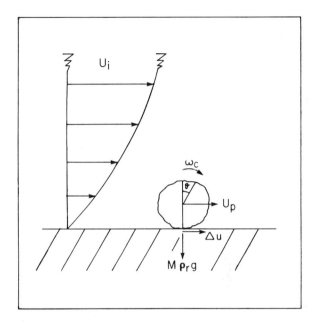

Figure 4.14 Rotation and net forward motion of a particle embedded in the basal layer of a cold glacier frozen to bedrock. The vertical velocity profile is shown (left) with zero forward movement at the bed but with finite horizontal flow, due to creep, increasing with distance above the bed (hence a shear zone). The clast experiences angular velocity (ω_c) and forward velocity (U_p).

frozen to the substratum. Although there will be no sliding motion at the interface, sensu stricto, at some small distance above the bed there will be a some horizontal movement. This is due to the strong shear created in the lowermost ice layers to balance the velocity (from creep) within the body of the glacier with zero motion at the bed. A clast of sufficient size will penetrate into this shear zone and will, as a consequence, be subject to a small but demonstable forward movement of the ice. The result is for the clast to experience a **torque** giving rise to an angular velocity (ω_c). The net forward motion of the particle (U_p) by rolling is:

$$U_p = \omega_c r_c \qquad (4.24)$$

All the motion may not be by rolling due to constraints of the immersing ice so that some dragging along the bed may occur and hence the ability of the clast to score the bedrock. Holdsworth (1974) noted the rotation of a peg in the lowest basal horizons of Meserve Glacier, Antarctica (ice temperature is at about $-18°C$). Over a period of one month the peg rotated approximately $57°$. The net forward motion given by Equation (4.24) was 0.17 mm day^{-1} (60 mm a^{-1}).

More recently Shreve (1984) has suggested that even where the sole of a glacier may be freezing, very small basal sliding speeds may be achieved. This contention is based upon ice regelation experiments well below the pressure melting point. Although the velocities are probably below present instrumental detection the distances traversed during such sliding over periods of several thousand years may amount to a few or tens of metres thus enabling a very limited form of abrasion by rock particles to occur.

5 Mechanical and chemical erosion by glacial meltwater

5.1 Introduction

The role of meltwater in glacial erosion is of great significance but only recently has it been recognized as a distinctive and integral glacial process. The mechanical erosion of bedrock proceeds in a manner similar to the denudation by sub-aerial streams exhibiting open-channel flow. There are, naturally, complicating factors in the glacial situation such as the effects of water **viscosity**, and the presence of **frazil ice**, and these are outlined in the proceeding sections. In some cases consideration of processes in the glacial environment elaborates activities which are otherwise overlooked in more conventional discussions of hydraulic erosion. The treatment in this chapter, for instance, of the erosion of hard bedrock might be found a useful ancillary to standard fluvial texts. The efficiency of chemical erosion under glacial conditions has been often a contentious issue. Some early considerations suggested that low temperatures and often low precipitation in polar regions inhibit chemical exchange and reduce the importance of glaci-chemical denudation (Strakhov, 1967). Recent work has indicated the importance of chemical activity in glaciated regions (e.g. Eyles *et al.*, 1982). It is hoped that the sections in this chapter will attempt to clarify some of these issues.

5.2 Mechanical erosion by glacial meltwater

Meltwater moving at high velocity and heavily freighted with abrasive debris can achieve considerable wear or erosion of bedrock both beneath glaciers and, after discharge from the portal, in the proglacial zone. The presence of Nye channels, potholes and large areas of smooth, water-worn bedrock attest to such activity (Figure 5.1). In general, there is little to distinguish the mechanics of erosion in hydraulically well-conducting subglacial channels from those in 'normal' open-channel flow.

Figure 5.1 Potholes eroded in gneiss by subglacial meltwater activity now exposed in the forefield of Nigardsbreen, Norway. Johan Ludvig Sollid as scale.

Erosion will thus be influenced by geotechnical properties of bedrock such as hardness (expressed in terms of **lithology**, fractures and foliations, general topography), shear strength, plasticity index and moisture content. Hydrological factors are also important and include water velocity and **turbulence**, as well as the quantity of **suspended sediments** and bedload in traction. Of specific importance in the subglacial and proglacial environment are the high velocities and **discharges** encountered in meltwater streams, their pronounced temporal variations, and low temperatures which increase water viscosity.

5.2.1 Hydrological variability

Fluctuations in the discharge of glacial streams were outlined in Chapter 2. During summer months water flow is usually high while during the winter flow may be extremely low or even absent (Figure 2.1). Haakensen and Wold (1981) report up to two orders of magnitude increase in subglacial flow between winter and summer as measured in a bedrock water intake tunnel beneath Bondhusbreen – an outlet glacier of Folgefonni in central Norway. Winter discharge amounted to about 0.1 m³ s⁻¹. Discharge commenced rising in mid May reaching about 1 m³ s⁻¹. By the end of June maximum daily discharge was experienced at ~11 m³ s⁻¹. Boulton and Vivian (1973) describe similar flow variations beneath the Glacier d'Argentière (winter flow: 0.1–1.5 m³ s⁻¹, summer flow: 10–11 m³ s⁻¹).

Superimposed on the high summer discharge are flood events. These abrupt increases in discharge, sometimes of only a few hours or days duration, are of considerable importance for erosion and sediment discharge and are probably some of the most distinctive characteristics of the glacial environment. The jökulhlaup of the Görnersee in 1967, for instance, produced 100 mm of measured erosion.

5.2.2 Velocity characteristics

Glacial meltwater streams usually flow at high velocities and are violently turbulent. This typically arises from strong pressure gradients in englacial and subglacial conduits, steepness of the channel in mountainous terrain, and high channel boundary roughness. Velocities in subglacial channels may be in the order of 0.2–2 m s⁻¹ while those on and in front of the glacier may be up to an order of magnitude greater (5–10 m s⁻¹). Such velocities are frequently exceeded during flood conditions (i.e. discharge of storm water or during jökulhlaups).

5.2.3 Viscosity–temperature effects

The viscosity of water increases with a decrease in temperature, 0.8 mN s m⁻² at +30°C to 1.8 mN s m⁻² at 0°C. Most subglacial meltwaters exist at temperatures of < +2°C. At such temperatures viscosity is, in fact, higher than that of mercury! It should also be borne in mind that water may be supercooled to temperatures well below 0°C (due for instance to the presence of impurities and solutes).

The principal effects of increased viscosity are to reduce particle **fall velocities** – important for sedimentation (see Subsection 11.4.2) and to

increase suspended sediment concentration. The quantity of sediment typically in suspension in glacial streams my reach several g l⁻¹ (see Subsection 10.3.1). In addition greater viscosity has the effect of allowing the transport of coarser bedloads by rolling, sliding or saltating along the channel bottom.

5.2.4 Mechanisms of erosion

As mechanical erosion is, in part, dependent upon both water velocity and sediment concentration, cold waters encountered in glacial streams will have a greater propensity for erosion.

Two primary mechanisms of bedrock erosion are recognized. Abrasion takes place by the striating or grooving of the channel surface by rock particles carried in the water. Small flakes of rock or individual mineral grains may also be gouged out by the impact of large particles in the water. At velocities of ≥ 12 m s⁻¹ an additional mechanism becomes operative. Pitting of the rock surface and loosening of mineral grains results from cavities forming and collapsing in the water, a process known as cavitation.

5.3 Abrasive water erosion of bedrock channels

Sediments in suspension will cause damage to the channel margins only if the two are brought into contact. In streams of low turbulence the probability of bed contact with suspended particles will be low but increases with turbulence and particle size. Channel shape is also of critical importance as changes in orientation (i.e. bends) and cross-sectional area will affect the angle of attack of the fluid flow and its velocity.

5.3.1 Abrasion by impinging suspended particles

As far as is known no comprehensive study of the erosion of meltwater channels in hard bedrock has been undertaken although rates of abrasion by glacial waters have been recorded (see Vivian,

1975). To develop a better understanding of abrasive processes and their physical controls it is necessary to apply results from related fields where analogous erosive wear has been investigated. The principal areas are in the pipeline transport of **slurries** and the erosion of concrete spillways and flood drainage channels.

Erosion by striation is similar to the abrasion of bedrock discussed in Chapter 4. A tangential or shear stress is imposed on the channel surface by particles as they are transported. Gouging out of small flakes from the surface is analogous to fracturing outlined in Chapter 3 – where a normal stress is imposed on the surface by the impact of the larger clasts. For brittle materials such as most igneous, metamorphic and many sedimentary rocks (but excluding clays and shales) impacting particles produce a series of concentric cracks at the surface of the channel. Only later, with increasing number of impacts, do the fractures intersect and enable material to be removed. Erosion may thus be related to (Figure 5.2):

- Suspended sediment properties
  – hardness and strength, particle size and shape, concentration
- Fluid properties
  – velocity and flow regime, angle of attack
- Properties of the channel surface
  – mineralogy, hardness and strength, surface roughness and orientation of facets.

Laboratory tests have shown particular relationships between these variables and wear rate and are discussed below.

Velocity (U_w)
Most studies indicate an increase in erosion with mean water velocity. The form of the relationship is exponential:

$$E_r = k_m - \overline{U}_w^{m*} \tag{5.1}$$

where k_m = constant
U_w – water velocity

The exponent, $m*$, has been shown to lie between about 2.0 and 3.3, in steel pipes. For more brittle materials such as rock the exponent could well be much higher (Mills and Mason, 1975). Lower

Figure 5.2 Factors controlling channel erosion in hard bedrock by meltwater.

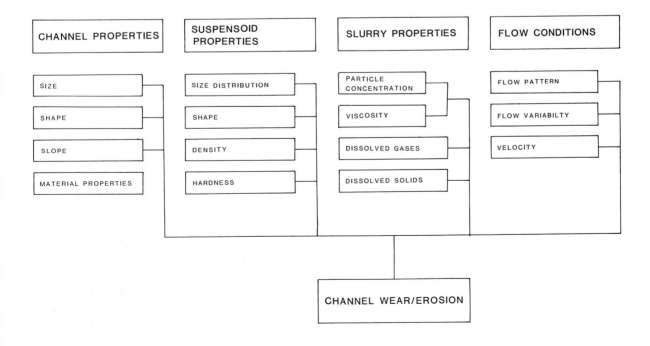

exponent values are usually characteristic of the sides of a channel while a larger value relates to the bottom of the channel where abrasion appears to be greatest.

In general, it should be expected that sediments in suspension should become more uniformly distributed in the channel cross-section at higher velocities (Raudkivi, 1976). This might suggest a decrease in erosion rate to a steady or constant value for the channel sides at least. Karabelas (1978) offers as explanation the notion that while the distribution of particles may be uniform gravitational effects may affect clasts in such a way as to cause them to strike the pipe bottom at an angle which inflicts maximum surface damage – and with impact velocities greater than any other part of the channel. We deal with critical angles of attack below.

Angle of attack (α_i)

It is well recognized that surface erosion by impinging clasts will be greatest at certain angles of incidence. Although no controlled tests have been undertaken on natural rock-lined channels results of engineering studies are qualitatively useful. Tilly's (1969) experiments shown in Figure 5.3 for brittle and semi-brittle materials indicate maximum erosion at normal incidence. Figure 5.4 indicates how angle of attack influences the pattern of wear at a bend in a channel.

Particle size may have an important secondary effect in such erosion. While larger suspensoids give rise to typical brittle erosion behaviour, smaller particles may produce a more ductile response due to the higher stresses experienced over the smaller loaded area.

Particle size (r_c)

The effects of particle size on erosion are not well known. General exponential expressions are often used of the form:

$$E_r \propto r_c^k \qquad (5.2)$$

where the exponent k has been found experimentally in pipes to lie between 0.75 and 2.15. It is unlikely, however, that such relationships are applicable across a wide range

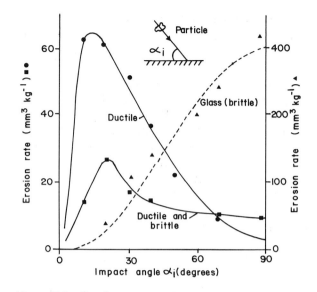

Figure 5.3 Erosion rate as a function of angle of attack for ductile, semi-brittle and brittle materials (glass). The brittle case is more indicative of hard bedrock substances whilst the ductile case would correspond to cohesive sediments. Note the much higher erosion rates for the brittle case at high angles of incidence (after Tilly, 1969).

of sediment sizes and especialy in the coarser fractions (Karabelas, 1978). A more realistic consideration of the effects of particle size is to assume that for a given set of circumstances (surface and particle hardness, angle of attack, velocity, etc.) abrasion rises with particle size to a critical size when further size increases have little effect on wear (Tilly and Sage, 1970). The initial size will increase linearly with velocity. However, data compiled by Kawashima et al. (1978) for wear in dredging pipes indicates pronounced increase in wear by coarse particles.

Particle concentration (C_s)

Abrasive surface wear is found to increase with the concentration of solids in suspension in the water. The rate of erosion, however, appears to fall at concentrations over 20% by weight (Faddick, 1975) due to the increasing importance of inter-particle collisions over channel surface impacts. The typical concentration of suspended

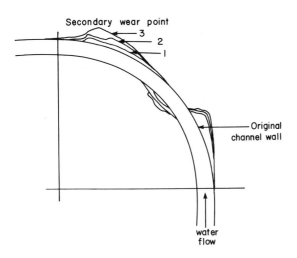

Figure 5.4 Experimental development of wear patterns in a pipe bend. The pipe was of perspex and the impacting solids of aluminium. Erosion commences at angle of 21° which becomes the primary wear point. After a certain depth of wear particles were deflected sufficiently to give rise to wear on the inside of the bend and a secondary wear point at a bend angle of 76°. A minor wear point was also established at an angle of 87°. A weak meandering pattern is emergent (after Mills and Mason, 1975).

sediments in meltwater streams rarely reaches 1% even under exceptional circumstances so that the exponential relationship at sub-critical phase densities suggested by pipe experiments is probably more appropriate, and of the form:

$$E_r \propto C_s^k \quad (5.3)$$

where the exponent k lies between 0.85 and 2.0.

Particle hardness/surface hardness
Important for abrasion of channels is the ratio of particle to rock surface hardness. Chapter 4 suggested that little wear will take place if clast hardness is less than that of the rock materials. Unfortunately, however, there has been minimal investigation of the influence of particle hardness on erosion. It would appear, on experimental grounds, that for a given particle channel erosion resistance will rise with an increase in surface material hardness (Goodwin *et al.*, 1969–70).

Total abrasion models
Several attempts have been made to combine the variables described above into a single model of channel wear by suspended particles. Karabelas (1978) found a least-squares fit to data, from pipe experiments using silica sand slurries eroding brass pads, which gave:

$$E_r' = k_d \, r_c^{2.15} \, U_w^{3.27} \quad (5.4)$$

(where the dimensional constant k_d accounts for the other variables such as channel size, particle hardness, etc.). E_r' is in mm a^{-1}. Rates of between 0.06 and 4.02 mm a^{-1} erosion predicted by Equation 5.4 compare very favourably with measured erosion rates from a variety of independent experiments.

Kawashima *et al.* (1978) have used an index proposed by Zandi (1971) which accommodates velocity (U_w), particle concentration (C_s), particle density (ρ_s), channel size (W_x) and a drag term (C_d):

$$E_r' = (U_w^2 \, (C_d)^{1/2})/(C_s \, W_x(\rho_s - 1)) \quad (5.5)$$

5.3.2 Abrasion by bedload transport

The very coarse sediment fraction found in glacial meltwater streams comprises cobbles, gravel and coarse sand. These may never become suspended but roll, slide and occasionally **saltate** along the channel bottom. Because of their greater weight abrasive damage can be quite severe. Appropriate analogies for understanding the mechanics of this process are found in the wear of hydraulic structures such as spillway aprons and **stilling basins** in dams. Here coarse rocks are trapped incurring considerable abrasive grinding to the dam concrete with their continued circulation. Depending upon flow conditions erosion has been reported to range from a few centimetres to metres. Concrete provides a good artificial test material for comparison with rocks as its short-term compressive strength lies within, although at the lower end of, the range exhibited by typical rocks (Table 3.1) and the aggregate is frequently of natural rock materials. The abrasive action of bedload on concrete has been studied in the laboratory by Liu (1981). Several rock types were used as aggregate in the preparation of concrete. The mean water velocity in an experimental tank was 1.8 m s^{-1} – comparable with many meltwater streams. Abrasion was calculated as per cent mass lost:

$$W_r = (M_i - M_f) \, 100/M_i \qquad (5.6)$$

where M_i = mass of specimen before testing
 M_f = mass of specimen after testing.

Equation 5.6 provides a *relative* gauge of abrasional resistance since the scaling-up from such experimental studies to the real world is often difficult, and absolute rates are probably also markedly different. The results of Liu's experiments show, as expected, that abrasion is directly proportional to the strength and hardness of bed materials (Figure 5.5). The weaker aggregate concrete displays an initial rapid erosion which then declines. Harder materials display a more constant abrasion rate.

5.4 Cavitation erosion

During the turbulent flow of meltwater at high velocity over rough bedrock local areas of low pressure may be created in the water. If the pressure falls as low as the **vapour pressure** of the water at bulk temperature macroscopic bubbles of vapour (cavities) will form. The cavitation bubbles grow and are moved along in the fluid flow until they reach a region of slightly higher local pressure where they will suddenly collapse. If cavity collapse is adjacent to the channel wall localized but very high impact forces are produced against the rock. This action may give rise to mechanical failure of the channel by **fatigue** resulting in severe damage to the rock surface in the form of pitting and etching. This

Figure 5.5   Relative abrasion rate of various concretes composed of different aggregate materials. Concrete strength as parameter (from Liu, 1981).

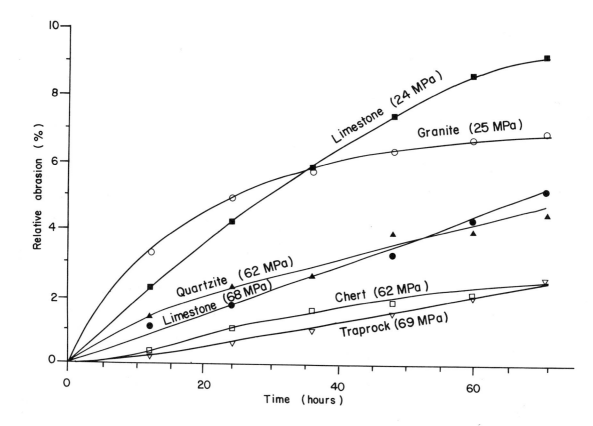

process is termed cavitation and is well known in hydraulic engineering where it affects outlet turbine runners, inlet pump blades, dam spillways, etc. (Knapp et al., 1970; Arndt, 1981). The mechanism was first proposed as a geologically important agent by Hjulström (1935) and later discussed by Barnes (1956).

Figure 5.6 illustrates some of the basic concepts involved in cavitation around an obstacle in a fluid flow. The fluid (water) accelerates around the protuberance reaching a maximum velocity and minimum pressure. If the pressure upstream of the obstacle is sufficiently low the minimum pressure will fall below the critical pressure, P_v. Cavitation occurs and spreads downstream. Cavity collapse is unstable giving rise to a pulsating tail as bubbles break loose and collapse.

5.4.1 Bernoulli equation

Cavitation becomes pronounced in localized regions of high velocity which have low pressures. The relation between velocity and pressure can be understood by recourse to the Bernoulli equation:

$$\frac{U_w^2}{2g} + \frac{P_w}{\rho_w g} + Z = H' \qquad (5.7)$$

where U_w = water velocity
P_w = water pressure
ρ_w = water density
Z = height above arbitrary datum.

The term H' is often called the total energy or total head and is the potential energy due to gravity of the fluid. The first term on the left-hand side represents kinetic energy, the second term the energy due to pressure in the fluid. The use of Equation (5.7) is limited by the fact that it assumes steady flow (no eddies) with no transfer of energy to or from the water, that flow is frictionless, and that the water is incompressible. Under most conditions in natural meltwater streams shear forces are produced in the direction of motion due to viscosity of the fluid. These consume energy which is not included in

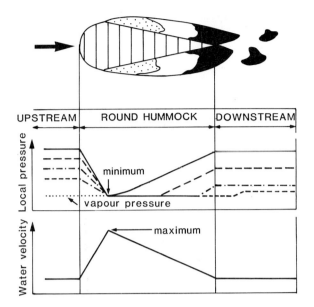

Figure 5.6 Cavitation round an obstacle in a fluid flowing from left to right. The fluid, accelerating past the restriction caused by the obstacle, reaches a point of maximum velocity and minimum pressure. If the pressure upstream is insufficient the pressure in the constriction caused by the obstacle will fall below the vapour pressure. As the upstream pressure is reduced cavitation will spread rearwards (from Pearsall, 1972).

Bernoulli's equation. The total head H' is thus diminished as energy is reduced by some small quantity; nevertheless Equation (5.7) can be used to show that any change in velocity is accompanied by variation in pressure or height or both.

5.4.2 Cavitation condition

The propensity for cavitation in a stream can be evaluated using a modified form of the **Euler number** (the Euler number is a dimensionless pressure coefficient: $(\rho_w U_w^2/2P_w)$ the ratio of the inertia force in the water to the pressure force):

$$\sigma_{cav} = [P_o - P_v] / [\rho_w U_w^2/2P_w] \qquad (5.8)$$

where P_v = vapour pressure or critical pressure at which cavitation occurs
P_o = local or reference water pressure.

Cavitation is less likely to occur under the condition $P_o \gg P_v$ than when $P_o \sim P_v$ or $\sigma_{cav} \sim 0$. Cavitation will definitely occur if $P_o < P_v$. P_v, it should be noted, varies with temperature (618 Pa at 0°C and 1228 Pa at 10°C) and dissolved air content. The critical value of σ_{cav}, therefore, at which cavitation commences is termed the *inception point* (σ_{crit}). The *desinent point* is defined as the point where cavitation ceases with increasing pressure. Hjulström (1935) used a form of Equation (5.8) to calculate the water velocity at which cavitation would commence [$(U_w)_{crit}$]. If P_o is assumed to be sea level atmospheric pressure (101 kPa), P_v at 0°C = 618 Pa and water density is 999.00 kg m^{-3} (at 0°C) $(U_w)_{crit}$ = 14.2 m s^{-1}.

5.4.3 Factors affecting cavitation

Because a non uniform pressure field may exist around channel obstacles and other factors alter the properties of the fluid, σ_{crit} is not a unique value and will vary with geometry, scale effects, air content, etc. allowing cavitation to take place at pressures above P_v.

Dissolved gas
The influence of dissolved gases can be quite profound even though concentrations in meltwater are usually small (see Subsection 5.6.4 below). During the growth of cavitation bubbles there may be diffusion of very small quantities of gas into the bubbles which act as a cushion during collapse and reduce the magnitude of the resulting shock wave. The collapse pressure of a cavity (P_c) is defined in terms of the gas content:

$$P_c/P_\infty = 0.157 \exp (P_\infty/P_{go}) \tag{5.9}$$

where P_∞ = constant external pressure
$\quad\quad P_{go}$ = partial pressure of noncondensible gas at a specific bubble radius

Arndt (1981) indicates that although there is not yet an exact understanding of the role of dissolved gas, its effect on attenuating cavitation erosion is well known. For the dissolved gas concentrations found in glacial waters (a few mg l^{-1}) it is likely that there will be significant reduction in the damage caused by cavitation.

Pressure
The reference pressure P_o is usually the pressure in the water upstream of the zone of cavitation. Other factors being equal the value taken by P_o will be related to atmospheric pressure. Since pressure falls with altitude, σ_{crit} will diminish accordingly so that at high altitudes streams issuing from glaciers will be more likely to attain critical cavitation numbers.

Channel roughness
The degree of surface roughness is very important to cavitation. Arndt (1981) gives a full discussion of the influence of roughness for both single bed protuberance and a roughened surface. Although the problem is highly complex the effect of roughness is to alter the inception characteristics of cavitation. The cavitation number thus depends on the relative roughness and thickness of the fluid boundary layer. In the case of a uniformly roughened surface an increase in relative roughness has a much smaller effect on the cavitation index than for an isolated roughness element.

5.4.4 Cavitation damage

As cavitation bubbles collapse high magnitude, but transient, pressure and thermal shock waves are created which cause rock material close to the imploding cavity to disintegrate. The maximum pressure created during the few microseconds of bubble collapse may reach values in the order of GPa and will typically be 10^8 Pa. The number of collapsing bubbles which actually damage a surface may only be a very small fraction of the total number of identifiable cavitating bubbles. That a certain number are 'erosive' depends upon a combination of initial cavity size, flow velocity, the pressure gradients, collapse pressure and distance to the boundary surface. The manner in which such pressure waves damage the channel wall may take several forms. The cavitation bubble may collapse to a minimum volume (dependant upon gas content and compressibility of the fluid). The bubble then commences to rebound and pressure waves are propagated outwards from the point of collapse. Dissipation of the energy flux of the shock wave obeys a characteristic **inverse square law** ($1/r_b^2$ where r_b = the radius of the bubble prior to collapse). In non-symmetrical collapse which may be typical, close to a boundary surface or in a pressure gradient, the bubble is distorted and forms a re-entrant microject possessing a very high velocity. The jet moves through the cavity interior and upon impact with the rock surface creates very high stresses. A third mechanism invokes an initial collapse similar to that described above but the jet is of insufficient magnitude to cause surface damage. The cavity collapses further in the form of a torus followed by rebound and shock-wave radiation.

Failure of the rock surface takes place by fatigue associated with the impinging shock waves. Individual mineral grains are loosened by crack propagation along intercrystalline zones of weakness. Loosened particles are removed by the force of high velocity water flow. The surface thus becomes pitted or fretted from the loss of mineral elements. The process of surface mass loss may be accelerated by a 'wave-guide' effect once a certain degree of roughness has been achieved (Knapp *et al.* 1970). Figure 5.7 shows a pit created by the collapse of a bubble with its

centre at C_1. A second bubble now collapses at C_2 but, due to the reduction in energy, proportional to $1/r_b^2$, it is insufficient to create further damage. However, the pit traps the propagating shock wave. As the wave shell travels down into the pit it is focused so that its diameter decreases whilst its intensity increases. In Figure 5.7 the shock wave has reached a diameter of half of that subtended at the pit rim and will have a pressure equal to that imparted by bubble collapse at C_1.

Knapp *et al.* (1970) suggest that such wave-guiding and stress amplification may explain another characteristic of cavitation erosion – the incubation period. After cavitation commences there is often a period during which little material is eroded from a surface, but after which the rate of erosion rises dramatically (accumulation period).

After the peak erosion rate is attained two hypotheses predict the continuation of erosion. According to one theory erosion rate drops off quickly to a steady value while the other model sustains the high erosion rate for a much longer period. Experimental evidence of cavitation erosion in metals suggests that material properties will determine the nature of the erosion rate curve.

The cavitation erosion of a meltwater channel surface will be controlled principally by fluid velocity and material properties. For the former parameter it is well established from engineering studies that the rate of erosion scales with U_w^6 while the mean size of pits scales with U_w^5. Figure 5.8 shows the relationship with U_w where the measure of erosion is given by the number of pits formed per unit area per unit time. Material properties control erosion rate and there is widespread agreement that strength and hardness are the best measures of erosion resistance (Pearsall, 1972; Arndt, 1981). High strength (tensile or compressive) and high hardness will inhibit erosion by cavitation. Although no experimental data are known for natural rock materials, work on metals shows a clear relationship between erosion rate and hardness. Examples and studies of unambiguous cavitation damage in open conduits approximating natural stream channels are principally restricted to the

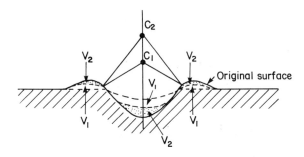

Figure 5.7 'Wave-guide' process for enhancing cavitation damage in a surface. C_1 is the point of collapse of a bubble generating the initial crater; C_2 is the point of collapse of a second bubble capable of enlarging the crater by the 'wave-guide' effect; V_1 is the volume of material displaced by the initial collapse at C_1; V_2 is the volume of material displaced by the later collapse at C_2 (after Knapp *et al.*, 1970).

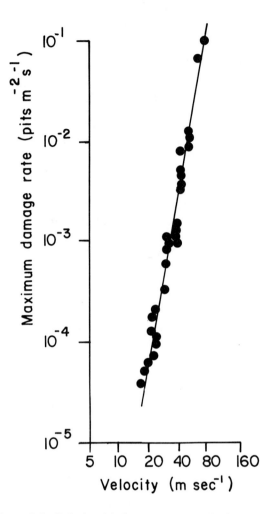

~0.5 m³ hr⁻¹ for a mean water velocity of ~7m s⁻¹. Once pits appear they may sufficiently disturb water flow so as to enhance the cavitation effect. Experiments with concrete surfaces by Kenn (1966) demonstrate that significant damage can occur in a period of hours. During a 72-hour test a maximum depth of erosion of 6 mm was observed with a water velocity of ~17.5 m s⁻¹ and a temperature of 20°C. Detailed studies by Houghton *et al*. (1978) attempted to evaluate cavitation resistance of several types of concrete. Experiments were undertaken with velocities of 37 m s⁻¹ in a 6.1 m long rectangular conduit. Although flow rates were considerably higher than in glacial streams the pattern of erosion was instructive. Initially discrete areas, downstream of the two experimental cavitations, became pitted. They later expanded and joined. A second area of erosion developed downstream from the first, followed by a third area. Finally all three areas coalesced and continued to enlarge and deepen into bowl-shaped depressions. Figure 5.9 shows the depth of erosion for various concrete test slabs. Although some of the concretes tested are difficult to relate to real rock surface the conventional concrete had a compressive strength of 41 MPa and the fibrous concrete 54 MPa – making them rock-compatible (see Section 3.2).

Figure 5.8 Relationship between water velocity and cavitation erosion (defined by number of pits generated m⁻² s⁻¹). Note that in most bedrock channels velocities will only be typical of the lower part of the curve (from Arndt, 1981).

erosion of concrete hydraulic structures such as spillways. Barnes (1956) quotes an erosion rate of ~20 mm hr⁻¹ and Price *et al*. (1955) report cavitation pitting of the Parker and Grand Coulee Dam spillways and conduits at velocities as low as 7.6 m s⁻¹. At Boulder Dam, Price reports the erosion of ≳1500 m³ of concrete in a spillway tunnel discharging on average 380 m³ s⁻¹ over a period of 4 months – a rate of

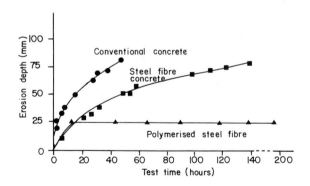

Figure 5.9 Experimental determinations of cavitation erosion for various concrete slabs. Water velocity = 37 m s⁻¹. The conventional concrete, similar to some rock surfaces, has a compressive strength of 41 MPa (after Houghton *et al*., 1978).

5.5 Chemical erosion by glacial meltwater

The chemical analysis of glacial meltwaters collected either directly at the glacier bed or upon discharge at the glacier terminus, show them to be rich in numerous dissolved chemical species. The occurrence of such constituents indicates effective hydrochemical reaction within the glacial environment, principally at the bed of ice masses. The quantity and variety of constituents reflect the geological setting traversed by meltwater and physical and chemical factors involved in the erosive process. Although detailed studies of the geochemistry of glacial meltwaters are limited, sufficient determinations exist to indicate some general trends and that the apparent rate of denudation by glacial waters is highly significant – in the order of 50–70% of the world average chemical denudation (i.e. 27 g m^{-2} a^{-1}) as deduced by Livingstone (1963). Chemical weathering would thus appear an active geological agent in glacial regions.

5.5.1 Chemical composition of glacial meltwater

Meltwater contains a large number of inorganic chemical species whose concentrations vary considerably from glacier to glacier and through time. Lists of the major, minor and trace constituents in natural water can be found in water quality and similar texts (e.g. Todd, 1980; Freeze and Cherry, 1979). Table 5.1 shows some typical concentrations for meltwaters from a variety of glaciers around the world. Since the six principal ions (Na^+, Mg^{2+}, Ca^{2+}, Cl^-, HCO_3^- and SO_4^{2-}) usually make up >90% of the total dissolved solids, whether dilute or highly saline, it is not unusual to make only a partial chemical analysis of water samples concentrating on these six ions with the addition of one or two other of the minor components such as potassium (K^+) and carbonate (CO_3^{2-}). Dissolved gases (e.g. O_2, N_2, CO_2, CH_4, H_2S, and N_2O) are also present in meltwater and, as observed in Subsection 5.4.3, important in the cavitation process, as well as

Table 5.1 Chemical composition of glacial meltwater (μg l^{-1})

Glacier	Ref.	Na	K	Ca	Mg	SiO$_2$	HCO$_3$	SO$_4$	Cl
Chamberlin Creek, Alaska	1	0.15	0.10	4.00	0.80	0.35	10.5	8.20	0.45
Gorner, Swizerland	2								
Sup6glacial		0.20	0.24	0.46	0.08				
Gornera		0.34	0.61	4.37	1.00				
Berendon, BC, Canada	3								
Supraglacial		0.01	0.02	1.97	0.04	0.15			0.91
Subglacial		0.10	0.11	8.55	0.13	0.21			0.93
Matanuska, Alaska	4	1.30	0.40	9.20	0.75	1.20			
Castner, Alaska		1.25	1.10	20.00	5.10	0.80			
Norris, Alaska		2.40	0.75	0.65	0.80	0.55			
Emmons, Washington, US	5	1.51	0.85	2.10	0.40	8.00			
Nisqually, Washington US	5	1.54	0.57	2.07	0.74	4.28			
Rhone, Switzerland	5	0.34	0.95	4.10	2.56	1.14			
Boverbreen, Norway	5	0.16	0.38	0.70	0.08	0.48			
Taylor, Antarctica	5	6.90	1.14	17.90	0.94	0.20			
Wright, Antarctica	5	2.14	0.40	0.00	0.09	0.20			
Grindelwald, Switzerland	5	0.22	0.62	17.50	0.53	0.66			
S. Cascade, Washington, US	6	0.29	0.88	3.00	0.39	1.50	9.20	2.40	0.25

Refs.
1 Rainwater and Guy (1961)
2 Collins (1979b)
3 Eyles et al. (1982)
4 Slatt (1972)
5 Keller and Reesman (1963)
6 Reynolds and Johnson (1972)

organic substances. Their concentrations are usually low, especially organic content, and rarely measured. CO_2 content is particularly crucial to carbonate dissolution.

5.6 Solute sources

Several sources of solutes in glacial meltwater may be recognized. The proportions contributed by each may change radically on several timescales. If measurements of the concentration of chemical species is to be used to estimate chemical erosion of glacierized terrain these various sources require identification and separation. The principal contributions come from precipitation at the ice surface (either snow or rain), ice and firn, groundwater and surface runoff, exchanges between meltwater and the

atmosphere and geological materials, Table 5.2 lists potential sources of chemical species in glacier ice.

5.6.1 Precipitation

Rainfall within a glacierized basin may add significant quantities of chemical species directly into meltwater streams or as rain and snowfall contributions to the glacier (Table 5.3). In most of these cases the ionic signature from precipitation will become modified as additional chemical alteration takes place during the passage of water through and beneath a glacier. Rainwater and Guy (1961) and Eyles et al. (1982) have analysed summer liquid precipitation on glaciers in North America. Their results, shown in Table 5.3, are compared with rainfall in non-

Table 5.2 Potential sources for chemical species found in glacier ice (after Lyons and Mayewski, 1984)

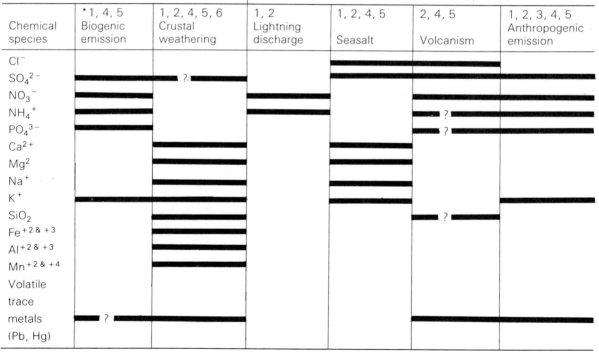

Chemical species	*1, 4, 5 Biogenic emission	1, 2, 4, 5, 6 Crustal weathering	1, 2 Lightning discharge	1, 2, 4, 5 Seasalt	2, 4, 5 Volcanism	1, 2, 3, 4, 5 Anthropogenic emission
Cl^-				▬▬▬		
SO_4^{2-}	▬▬▬	?▬		▬▬▬		▬▬▬
NO_3^-	▬▬▬		▬▬▬			▬▬▬
NH_4^+	▬▬▬		▬▬▬		? ▬▬	▬▬▬
PO_4^{3-}	▬▬▬				▬ ?	▬▬▬
Ca^{2+}		▬▬▬		▬▬▬		
Mg^2		▬▬▬		▬▬▬		
Na^+		▬▬▬		▬▬▬		
K^+	▬▬▬	▬▬▬		▬▬▬		▬▬▬
SiO_2		▬▬▬			▬ ? ▬	
$Fe^{+2 \& +3}$		▬▬▬				
$Al^{+2 \& +3}$		▬▬▬				
$Mn^{+2 \& +4}$		▬▬▬				
Volatile trace metals (Pb, Hg)	▬ ? ▬				▬▬▬	▬▬▬

* *Source characteristics*
Temporal distribution
  1 – cyclic (seasonal)
  2 – non-cylic (inter-annual and/or intra-annual)
  3 – significant only as of post-AD 1850

? – species production from this source uncertain.
Spatial distribution and magnitude of species
  4 – distance and/or elevation source to site
  5 – atmospheric circulation pattern source to site
  6 – aerial distribution of local ice-free terrain

(increasing importance of factors such as 5 (i.e., monsoonal flow) and 6 increased likelihood of 1 compared to 2)

Table 5.3 Chemical composition of rain and snow on ice sheets and glaciers

Glacier	Ref.	Na	K	Ca	Mg	SiO_2	HCO_3	SO_4	Cl
1 Rainfall ($\mu g\, l^{-1}$)									
Chamberlain, Alaska	1	300	100	700	100		5000	400	0
Berendan, BC	2	65	52	600	24				950
S. Cascade, Washington, US	3	90	39	60	12				
2 Snowfall ($\mu g\, l^{-1}$)									
Byrd Station, Antarctica	4	31.0	1.5	1.5	4.0				60.0
Greenland	4	20.0	2.4	5.4	5.0				31.0
Mizuho Plateau, Antarctica	4	14.0	1.2	2.8	2.1				
South Pole	4	3.0	8.0	8.0	6.0				1.0
S. Cascade, Washington	3	51.0	0	100	0				
D-10 Terre Adèlie Antarctica	5	439	14	12	16			92	646
3 Rainfall (non-glacial)									
N. Europe ($mg\, l^{-1}$)	6	2.05	0.35	1.42	0.39			2.19	3.47
SE Australia ($mg\, l^{-1}$)	6	2.46	0.37	1.20	0.50				4.43

1 Rainwater and Guy (1961)   4 Morozumi *et al.* (1978)
2 Eyles *et al.* (1982)   5 Delmas *et al.* (1982)
3 Reynolds and Johnson (1972)   6 Lerman (1979)

glaciated regions and snowfall on a variety of ice masses including the Greenland and Antarctic ice sheets. In the latter cases concentration of marine-derived components (e.g. Na) is a function of distance from the coast reflecting rainfall scavenging and dry-deposition of ions. Gas-derived ions such as SO_4, HO_3, NH_4 and Cl are the principal impurities deposited in central Antarctica (Delmas *et al.*, 1982). Meltwater from Antarctic snow is thus a very dilute mix of natural acids (H_2SO_4 and HO_3) and contains a small quantity of neutral salts.

5.6.2 Firn and ice

Glacier ice is essentially composed of frozen atmospheric precipitation so that its chemical composition should nearly approach that of snow or rainfall. Table 5.4 lists chemical constituents from a variety of glaciers and ice sheets and although comparison with precipitation is not easy some conclusions may be possible. For a glacier in which there is little advection of ice from up-flow line, ionic concentration may increase with depth as **enrichment** takes place due

to contact with solid impurities and migration of saline pore water. Berner *et al.* (1977) point out that pore water, saturated with gases at a given depth, can dissolve more gas and soluble constituents when reaching a deeper layer due to the decrease in partial pressure. With increasing depth in a glacier or ice sheet, the source of the ice lies at greater distances as illustrated by the flowline patterns of Figures 7.10 and 7.11. If the ionic content diminishes with distance (and

Table 5.4 Chemical composition of non-basal glacier ice ($\mu g\, l^{-1}$)

Glacier	Ref.	Na	K	Ca	Mg
Chamberlin	1	750	1300	600	0
Berendon	2	8.0	18.0	10.0	18.0
Tsidjiore Nouve	3	18.4	35.2	84.2	31.6
Gruben	3	32.2	39.1	72.1	27.9
G1, Terre Adèlie ~20m depth	4	800	150	360	80
G1, Terre Adèlie ~57m depth	4	40	<100	<100	10
South Cascade	5	80	78	60	5

1 Rainwater and Guy (1961)   2 Eyles *et al.* (1982)
3 Souchez and Lorrain (1978)   4 Lorius *et al.* (1969)
5 Reynolds and Johnson (1972)

altitude) from the ice margin there will be a drop in chemical concentration with depth. It is possible that this pattern is being shown by measurements from boreholes in Antarctica (Lorius *et al.*, 1969).

Basal ice

Water draining through a glacier's internal vein network may become enriched by uptake of gases and solutes. As such water approaches the bed where debris is entrained in basal ice the uptake of chemical species will be rapid and significant. Souchez *et al.* (1973) and Souchez and Lorrain (1975, 1978) have investigated this process in a series of detailed papers as a result of a careful analysis of basal ice beneath glaciers in the European Alps. They consider that the basal debris layer, rich in clay-size particles, acts as an **ion-exchanger**. The result is **electrolyte** filtration and the relative retardation of cations in water squeezed through the bottom layers. The basal debris layer constitutes particles (especially colloids) whose surface will possess adsorbed cations in a diffuse layer (see Section 5.9). When water is forced through this 'membrane' the cations **desorb** more rapidly and in greater quantity than alkaline salts. The divalent ions adsorbed on the particles are closer to the pore walls and are not detached as easily as monovalent ions. Thus the water, squeezed under a pressure gradient of, say, 30 Pa m^{-1}, will separate the cation species preferentially enriching the excluded water in Na and K. Table 5.5. shows the mean concentration of principal cations for glacier ice above the debris horizon and ice accretions – refrozen from squeezed interstitial water, from a cavity beneath Glacier d'Argentière. There is thus chemical enrichment and sorting at the bottom of glaciers under the

influence of basal debris. Souchez and Lorrain show that these effects may be used to distinguish the origin of basal ice, especially that found in regelation layers.

5.6.3 Surface runoff and groundwater

The runoff from mountain sides within a glacier basin may be added to the total meltwater discharge by entering the glacier drainage system at its margins. Eyles *et al.* (1982) measured the concentration of ions in valley side streams and their results give (in μg l^{-1}): Na = 130; K = 34; Ca = 4 750 and Mg = 150.

Groundwater may result in an important solute contribution, as there can often be a connection between the glacier drainage system and groundwater flow. It is likely that this component will be significantly different in chemical signature from the ice mass and also highly variable from glacier to glacier depending upon the composition of subjacent bedrock and sediments, and groundwater flow conditions. The proportion of solutes added by groundwater may also vary throughout the year as suggested by Lutschg *et al.* (1950) and Stenborg (1969). During the winter, when melting and glacial water flow is at a minimum, the groundwater input may be of increased importance.

5.6.4 Exchange between meltwater and the atmosphere

The contribution made to the chemical content of glacial meltwaters by exchange with the atmosphere is usually considered a minor factor. Nevertheless the highly turbulent nature of many superglacial streams and those descending steep mountain sides enables the efficient uptake of

Table 5.5 Chemical composition of basal ice (mg l^{-1}) (from Souchez *et al.*, 1973)

Basal ice type	Na	K	Ca	Mg	(Na + K)/(Ca + Mg)*
Glacier ice	0.80–0.27	0.19–0.26	0.15–0.40	0.07–0.08	0.4–1.1 (mean = 0.7)
Refrozen squeezed water	0.80–1.84	1.38–3.20	0.13–1.29	0.16–0.40	1.5–4.6 (mean = 2.92)

* in meq l^{-1}

certain chemical species. This is particularly true of gases such as CO_2. The continuous saturation and resaturation of turbulent well-aerated meltwater with CO_2 produces weak carbonic acid:

$$CO_2 + H_2O \rightleftharpoons H_2CO_3 \qquad (5.10)$$

The presence of H_2CO_3 is most important to the **carbonation** reaction of meltwater with bedrock materials and the subsequent uptake of Ca and Mg.

5.6.5 Solute enrichment of meltwater at the glacier bed

Solute enrichment in meltwater takes place by two primary processes – dissolution and ion exchange as water flows over freshly scoured subglacial bedrock or through fine-grained sediments containing a significant clay-size fraction. These are considered by many workers to represent the most important source of solutes in meltwater.

5.7 Dissolution

Water acts as a solvent for many inorganic and some organic materials. The effectiveness of water in the dissolution process arises from its high dielectric constant and the fact that its molecules, joining with ions, create a solution of hydrated ions. Positively charged ions (cations) create a stable group of molecules by attracting the negative end of the water molecule dipole. Anions (negatively charged) are attracted to the positive end of the water molecule but the association is much weaker.

By this process solid bedrock and loose particles are dissolved: an especially vigorous process along fractures and joints of newly exposed rock (from crushing and abrasion) which may become pitted and etched. Dissolution proceeds until an equilibrium concentration is achieved which depends upon temperature and pressure. The solubility of a material is thus the mass dissolved per unit volume solution at a given temperature and pressure. Lerman (1979) has shown that the rate of dissolution (in a closed system) is:

$$\frac{dJ_s}{dt} = N_r(C_{es} - J_s)^{ff} \qquad (5.11)$$

where J_s = solution concentration
C_{es} = solubility or equilibrium concentration
N_r = reaction rate parameter
ff = reaction order exponent.

The total quantity of matter dissolving from a bedrock or particle surface (i.e. the mass loss) is:

$$\frac{dM_s}{dt} = \frac{d^* S_{ar}(C_{es} - J_s)}{T_l} \qquad (5.12)$$

where d^* = diffusion coefficient for dissolved material
S_{ar} = reactive area of solid surface
T_l = thickness of the reactive laminar layer.

A linear rate of dissolution can be defined as:

$$D = \frac{d^*(C_{es} - J_s)}{\rho_r T_l} \qquad (5.13)$$

where ρ_r = density of the rock

It is possible to use Equation (5.13) above to estimate the linear dissolution rate anticipated beneath a glacier by meltwater. Taking silica as typical the following quantities may be defined: C_{es} = 6–60 mg l^{-1}; J_s as measured in typical meltwater streams in mg l^{-1} (e.g. Slatt ≤ 12; Eyles < 21; Rainwater and Guy ≤ 1.3);

d^* = 3.15×10^{-2} a^{-1};
ρ_r = 2 700 kg m^{-3};
T_l = 3×10^{-5} m.

If the solution is close to equilibrium with the solid, as might be expected from water expelled through a debris-charged basal ice layer, $C_{es} - J_s$ will be very small. At other times $C_{es} - J_s \approx C_s$. Taking two concentration differences, 1 mg l^{-1} and 50 mg l^{-1} as typical of glacial meltwaters Equation 5.13 will give:

D = 0.4 mm a^{-1} for   1 mg l^{-1} difference
D = 19.0 mm a^{-1} for   50 mg l^{-1} difference

It would appear that these values are

unacceptably high for subglacial dissolution. The mean global rate of surface lowering by chemical weathering is 0.01 mm a^{-1} – based upon the dissolved load of rivers (Livingstone, 1963) and is identical to that calculated by Rapp (1960) for the mean annual denudation rate for Karkevagge in Lappland, Sweden. The reason for the difference by at least an order of magnitude may be due to the fact that only a small proportion of a surface is usually exposed to water and hence solution, and to the development of protective coatings on surfaces (such as hydrous iron oxide which is well known from glaciated terrain). Until more quantitative measurements are performed the results of such theoretical calculations remain merely speculative. Nevertheless some surface chemical weathering rates achieve the values given by Equation (5.13) (see Brunsden, 1979).

5.8   Carbonation

Chemical analyses of glacial meltwaters invariably show that carbonate minerals are the dominant constituent. This is not surprising as 99% of the carbon on planet Earth is locked up in carbonate minerals, principally calcite ($CaCO_3$) and dolomite ($CaMg(CO_3)_2$) and constitutes a common component of most sedimentary, igneous and metamorphic rocks at the surface of the earth. Reynolds and Johnson (1972) suggest, for instance, that for the South Cascade Glacier watershed in Washington State, USA carbonation is the principal mechanism governing the composition of meltwater. Carbonation takes place from the reaction between dissolved CO_2 in meltwater and rocks containing carbonate minerals. CO_2 combines to form carbonic acid (H_2CO_3), and results in the production of H^+ and HCO_3 ions which are removed in solution:

$$CO_2 + H_2O \rightleftharpoons H_2CO_3$$
$$CaCO_3 + H_2CO_3 \leftarrow Ca^{2+} + 2HCO_3^-$$

$$(5.14)$$

It has already been shown that glacial meltwaters may contain very high quantities of dissolved CO_2 due to high partial pressures, the release of entrapped CO_2 from ice, extreme turbulence which continuously recharges water with CO_2 and low temperatures. According to Lemmens and Roger (1978) some of the H^+ ions produced during carbonation may be exchanged at the surface of clay minerals for K and Na ions.

5.8.1   Subglacial dissolution – precipitation dynamics

Hallet (1976, 1979a) has undertaken detailed studies of subglacial carbonate deposits (i.e. subglacially precipitated calcite). These comprise columnar spicules and coatings with pronounced furrows which occur predominantly in small subglacial cavities. Similar concentrations have been reported by Ford et al. (1970). The deposits are often finely laminated. Their origin is ascribed to precipitation from chemically enriched subglacial meltwaters associated with regelation processes involved in glacier sliding, especially over limestone terrain.

Under high-pressure zones on the upstream flank of bedrock protuberances basal ice melts and water is also squeezed out of capillaries and the vein system in the ice. Such squeezing may preferentially enrich the water in alkalis as suggested by Souchez and Lorrain (1975). The water then flows along the pressure gradient at the glacier bed. During its passage over carbonate rocks the water will become enriched in Ca and Mg. Water flow is towards low-pressure areas in the lee of bed hummocks where the water refreezes. Hallet (1979a) and Hallet et al. (1978) indicate that solutes in the water are preferentially expelled during refreezing and accumulate. If the concentration of solutes in the lee-side zones reaches saturation (0.2–0.02 mg 1^{-1}) they are precipitated giving rise to a thin veneer of $CaCO_3$. Hallet points out that such subglacial precipitation is also related to the amount of through-flow in the thin water film as the high discharges common in channels tend to flush out the solutes and reduce the concentration below saturation level.

Precipitates of silica and ferro-manganese have also been reported in similar lee-side locations on the glacier bed (Hallet, 1975; Andersen and Sollid, 1971). On the gneiss terrain of the forefield of Nigardsbreen, Norway brown pigmentation occurs in depressions on the lee of bedrock knobs (Figure 5.10). It is thought to consist of iron oxides derived from oxidation of magnetite and pyrite in bedrock and later subglacially precipitated.

5.9 Cation exchange

Solute enrichment also takes place by the exchange of ions, adsorbed on the surface of particles, during flow over and through materials that contain a high proportion of **colloid**al-sized material (i.e. clay minerals in the size range 1 000–1 μm). Colloids, resulting from weathering, usually coat the surface of larger particles.

Exchange of ions takes place principally on the surface of clay minerals which usually possess negative electrical charges resulting from isomorphous substitution of higher-**valence** for lower-valence atoms. The edges of clay minerals may also possess negative charges. Water molecules, with **dipole** characteristics and stray cations are attracted to the colloid surface to balance the negative electrical charges and are held in an electrostatic double layer. The retention of cations is termed *adsorption*. The ions adsorbed on particles or mineral surfaces can be readily exchanged for other cations in solution in meltwater (as long as the overall valence balance is retained). As an example, there is exchange of Ca^{2+} for Na^+ on a suspended clay mineral upon addition of calcium chloride to meltwater. The Na^+ ions are subsequently released into the water. The most important cation reactions in meltwater involve monovalent and divalent cations: $Na^+ — Ca^{2+}$, $Na^+ — Mg^{2+}$, $K^+ — Ca^{2+}$ and $K^+ — Mg^{2+}$. The main factors that determine subglacial ion exchange are time of contact between water and basal sediments and area of contact. The former is well demonstrated by the experiments of Lemmens and Roger (1978). Samples of fine-grained till were placed in dilute superglacial water. Very fast enrichment was observed as shown in Figure 5.11. **Diffusion** of the different cations was in the order: Mg < Na < K. Ca was not considered.

Figure 5.10 Iron oxide pigmentation on the lee of bedrock undulations exposed after recent ice retreat on the forefield of Nigardsbreen, Norway. Palaeo-ice flow direction right to left; Susan S. Wickham as scale.

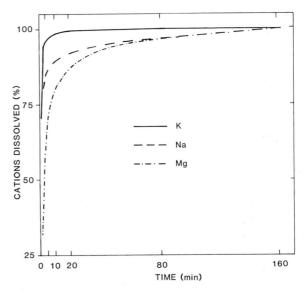

Figure 5.11 Experimental determinations of the rapid chemical enrichment of glacial meltwaters after contact with fine-grained till materials (from Lemmens and Roger, 1978).

Many other workers report the rapid ionic enrichment of samples from contact with sediments or rock. (Reynolds and Johnson, 1972; Tamm, 1924; Rainwater and Guy, 1961; Collins, 1977; Eyles *et al.*, 1982; Slatt, 1972). In addition,

Lemmens and Roger (1978) report the rapid uptake of Na and K in very dilute superglacial meltwater at the Tsidjiore Nuove Glacier when it flows over proglacial moraine. Only 30 m of flow over sediments was necessary for meltwater to achieve approximate chemical equilibrium.

The role of contact area is shown by observations by Lemmens and Roger (1978) at Tsansfleuron Glacier. Under fast melting conditions proglacial water spreads out into a thin sheet with a decrease in channel hydraulic radius (cross-sectional area/wetted perimeter) with a consequent increase in water–sediment contact. Pronounced enrichment of Na and K was observed. An opposite trend was found where hydraulic radius diminished – as occurs under more gradually increasing discharge conditions. This latter result is the principal explanation of the marked decrease in cation concentration with water discharge reported by Souchez *et al.* (1978).

5.9.1 Exchange between meltwater and suspended particles

Lorrain and Souchez (1972) have shown that a significant quantity of the principal cations in meltwater may be adsorbed on the surface of sediment particles in suspension. The ratio of adsorbed cations to dissolved cations was found, for water discharging from the Moiry Glacier, Switzerland, to vary from a few percent to 33%. These cations remain adsorbed on particle surfaces, however, and are not readily released to meltwater during transportation – a conclusion also arrived at by Collins (1979b) from studies of the electrical conductivity of meltwater from the Gornergletscher.

Eyles *et al.* (1982), working on meltwaters of Berendon Glacier, Alaska, found similar trends with adsorption being of greater importance for monovalent than divalent cations. Their results showed Na^+ to be the most adsorbed cation. The water from one subglacial sample suggested that adsorption is not an important mechanism for cation transport in water discharging beneath the ice. The results of Eyles *et al.* provide a relative

breakdown for the total material transported by Berendon Glacier water, as dissolved, adsorbed and suspended solids. It is clear that in all cases the contribution to cation flux from **suspensoids** is overwhelmingly greatest, and adsorption usually relatively minor.

5.10 Discharge of solutes in glacial meltwater

If the solute content of meltwater streams is to be used to gauge the rate of chemical erosion within the glacial environment it is necessary to continuously monitor the total amount of dissolved solids (**TDS**) as water chemistry varies throughout the year. This involves using indirect methods to measure TDS and a long-term sampling programme.

5.10.1 Measurement of TDS

Meltwater may be considered as an electrolyte solution since all the major and minor constituents are present in ionic form. This allows **electrical conductance** (or conductivity) of the water to be used as a measure of TDS. Conductance, whilst not simply related to TDS due to the presence of a variety of ionic and undissociated species, can nevertheless provide a rapid estimate, particularly in the range of 100–5 000 S cm^{-1}.

The technique has been widely used and several sets of measurements on glacial meltwater have been summarized by Collins (1979b). Comparisons are not, however, easy due to the lack of standardization in the experimental procedures but ranges of conductivity (excluding Alaskan measurements) lie between 6.0 and 60.0 S cm^{-1}. Nevertheless, high variability is a real keynote of the chemical composition of meltwater and reflects diverse sources of solutes as well as complex flow routeings (Collins 1979b).

There are considerable variations in TDS through time. Collins (1977) notes that diurnal cycles are present as shown in Figure 5.12 and TDS is often out of phase with the concentration of suspended sediments. For an increase in discharge TDS is shown to decrease which

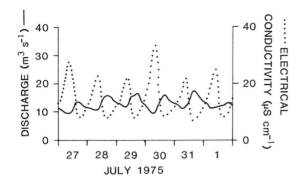

Figure 5.12 Variations in electrical conductivity of meltwater from the Gornera, Switzerland and discharge for the same period (27 July–1 August 1975) (from Collins, 1977).

suggests that the total input of the solutes to the stream is reduced rather than being the result of dilution.

5.10.2 Temporal variations

Several studies have demonstrated that there are distinctive variations in the discharge of solutes in meltwater streams. Collins (1977, 1979b, 1981) has made a number of investigations in the European Alps to evaluate the seasonal fluctuations of solute concentration. Many of the variations result from the mixing of water from the diverse sources of solutes within the glacier system (as outlined in Section 5.6) as well as complex and changing flow routeings.

In an early study to model temporal variability, Reynolds and Johnson (1972) attempted to fit a simple annual sinusoidal cycle to the annual chemical discharge of the South Cascade Glacier of the form:

$$C_{sol} = a_m \sin (2\pi t/366) + \overline{C}_{sol}$$
$$= a_m \sin (0.0172\,t) + \overline{C}_{sol} \qquad (5.15)$$

where C_{sol} = concentration of solutes
  a_m = amplitude
  t = cumulative time in days after 1 December.

The high concentrations in winter and low concentrations during the summer reflect enrichment and dilution of ions in meltwater

respectively in relation to water discharge. Data for this model, however, are limited to observations over the summer ablation period and one in February. Collins (1981) has attempted to provide more data over the annual cycle especially for the winter period by monitoring electrical conductivity of the Gornera, discharging from Gornergletscher, Switzerland. Figure 5.13 shows the seasonal variation in discharge and electrical conductivity for the year 1978–79. Flow during the winter is very low (<0.2 m³ s⁻¹) rising slowly at first and then more rapidly in May. Collins notes that the regular daily cycle was established by the beginning of June (Figure 5.13). Numerous specific events may be identified in the discharge record related to the draining of ice-dammed lakes, intraglacial water reservoirs or from periods of high ablation, thunderstorms, etc. The end of September marked the end of the principal

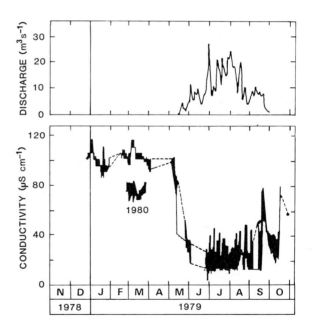

Figure 5.13 Seasonal fluctuations in discharge and electrical conductivity of meltwater in the Gornera, Switzerland. The discharge curve was recorded at a distance of ~1 km from the snout of Gornergletscher. Note the significant drop in conductivity with the onset of the melt season and streamflow (from Collins, 1981).

(summer) high discharge period. The electrical conductivity measurements, representing TDS, show periodic daily variations in the summer which are approximately out of phase with water discharge (Figure 5.12). Collins (1979b) has suggested that the total quantity of solutes input to meltwater may well diminish but there are no simple relationships between conductivity and discharge. Winter solute content was 2–10 times greater than during the summer and had only a limited diurnal variation. Collins (1981) has attempted to summarize and explain the relationship between solute concentration and discharge as shown in Figure 5.14 based upon hourly averages. The line E–F is equivalent to the lowest recurrent conductivity, independent of discharge and is controlled by the concentration of atmospherically derived solutes in precipitation and drainage through the glacier vein network. The peak conductivity, H, is associated with maximum winter solute concentration at minimum discharge and is due to the uptake of chemical species in contact with bed materials during these low-flow stages. The equivalent peak for summer ablation is shown at D. The line E–G–H relates conductivities during recession flow in spring and summer. The section CD locates the lowest observed recession flows resulting from varying proportions of surface runoff and basal discharge. Further aspects of these relationships are discussed by Collins (1981).

The total solute load discharged for a given period of time t can be obtained by integrating the product of the instantaneous electrical conductivity (Ω_w) and discharge (Q_w):

$$Q_s = \int_{t=0}^{t=n} Q_w \cdot \Omega_w \, dt \qquad (5.16)$$

Collins (1981) found that the bulk of the annual solute load is discharged during the summer. During this period the high meltwater discharge efficiently reworks chemical components from subglacial locations. For the winter period low discharge is responsible for only a minor flux of solutes despite very high conductivities.

The tapping of hydrologically independent meltwater bodies at the glacier bed has been

considered by Collins (1981) to account for some of the variations in meltwater chemistry. The water may be found in cavities or within basal sediments where prolonged contact with rock particles may give rise to considerable equilibrium enrichment.

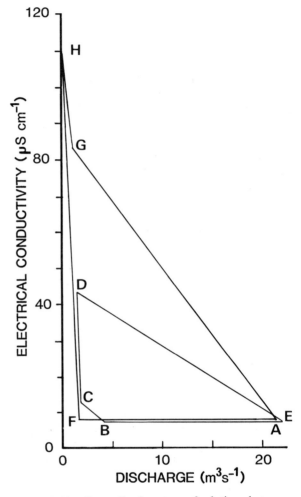

Figure 5.14 Generalized pattern of relations between solute concentrations (as measured by electrical conductivity) and meltwater discharge for the Gornera, Switzerland. Details are given in the text (from Collins, 1981).

6 Glacial erosion: process rates and rankings

6.1 Introduction

Brief summaries have been given in previous chapters of the principal aspects of abrasion, fracturing and meltwater processes (both chemical and mechanical) contributing to glacial erosion. Much remains to be learnt about the basic physics of these processes including controlling mechanisms, rates of operation and more particularly the relative importance of each process in the denudation of the Earth's surface under cold conditions.

There are few studies in which the different mechanisms of erosion have been examined and compared and relative magnitudes assessed. It is questionable whether any attempt should or could be made in view of the uncertainties in measurements, the need to make wide, often ill-considered generalizations, the pitfalls in extrapolation and the marked spatial and temporal variability known to occur in nature. Indeed there is considerable controversy among glaciologists on the efficacy of the various processes (see 'General Discussion', Symposium on Glacier Beds: the Ice–Rock Interface, *Journal of Glaciology* 1979). Nevertheless students of glacial geology have sought and will continue to enquire how important are abrasion, crushing and meltwater processes. For them this chapter has been written with circumspection. It highlights serious shortcomings in our knowledge of the glacial regime and methods of measurement and study which they may wish to remedy by their own research, rather than illuminate, per se, the question of rates and rankings.

6.1.1 Effects of temperature and geotechnical properties

It is quite clear that basal temperature plays a key role in determining erosional activity. Where ice close to the bed is below the pressure melting point numerous processes are inhibited, such as sliding, and hence most abrasion, meltwater production at the base and consequently both removal of crushed and abraded material and direct fluvial erosion of bedrock. Under cold basal conditions low velocities in turn reduce the outward flux of entrained sediments in traction at the bed or in higher level transport. On the other hand where pressure melting is widespread all the erosional processes appear to operate and often high rates of erosion are achieved.

Absolute rates of erosion and the relative rank of different mechanisms are also affected by the geotechnical properties of the bedrock traversed by ice. It is clear that bedrock control is considerably more complex than at one time considered. Besides the effects introduced by large-scale structural features more small-scale lithological or even microscopic factors are found to be critical in determining rock strength and resistance. Hard, brittle rocks such as granite and certain metamorphic rocks are particularly susceptible to crushing, fracture and some abrasion. Softer strata, exemplified by shales, thin-bedded sandstones, and some volcanic rocks, are readily abraded and also fractured and crushed. Very soft, **argillaceous** rocks like clays, marls and mudstones, are eroded dominantly by the ploughing component of abrasion and hardly affected by fracture. Meltwater erosion may show an almost independent relationship to lithology in contrast to the mechanical action indicated above – relative hardness (of suspensoids and channel materials) being but one factor in erosion (see Chapter 5).

6.2 The erosion budget

The total mass of material eroded per unit area from the glacier bed may be expressed in the form:

$$\Psi = \iint_a [M_a + M_d + M_m + M_q] da \qquad (6.1)$$

where M_a = mass of material eroded by abrasion

M_d = mass of crushed and fractured material

M_m = mass of material removed by meltwater erosion

M_q = mass of chemical substances resulting from meltwater erosion

a = area of glacier bed.

A total denudation rate may be expressed as mass per unit area per unit time.

The question for glacial geology is to ascribe typical quantities to M_a, M_d, M_m and M_q and to indicate their variability.

The *direct* measurement of bedrock erosion by ice is notably difficult. Many of the problems arise from factors discussed in previous chapters, principally those related to the task of placing intruments or markers at several subglacial locations and the long period required for gathering an adequate time series. In view of the highly variable nature of processes at the bed the representativeness of one or two sets of measurements must be highly questionable – although these are the very sources we shall use later in this chapter! *Indirect* methods may prove more satisfactory for obtaining total denudation rates, especially those that integrate loss of rock mass over a whole glacierized region. The separation of individual erosion components, in this case, however, may prove difficult and equivocal. These two philosophies for estimating erosion have not necessarily been applied explicitly to previous work. Most direct studies have concentrated on calculating abrasion, while indirect investigations of total erosion are based upon the long-term measurement of sediment and solute transport by meltwater streams issuing from glaciers. Few studies have paid attention to estimates of crushing and fracture or mechanical meltwater erosion.

6.3 Direct estimates of erosion rates

6.3.1 Measured abrasion rates

In Chapter 4 it was suggested that abrasional grooving of bedrock could account for a fraction of a cubic metre per year along the zone of asperity contact, depending upon rock strength and sliding velocity. Few direct measurements of abrasion have been made with which to check the typicalness of such theoretical considerations.

Embleton and King (1975, pp. 309–10) describe some of the early experiments undertaken to determine abrasion rates directly. The most interesting were those studies in the European Alps by Alfred de Quervain and O. Lutschg between 1919 and 1925 on the Upper Grindelwaldgletscher and Allalingletscher. Close to the front of the glaciers vertical holes were drilled into bedrock. Where ice advanced over the markers they were later able to determine the amount of surface lowering. Although the method was somewhat limited in application de Quervain and Lutschg derived rates in the order of 1.0–6.0 mm a^{-1} in rocks ranging from vein quartz to gneiss.

Boulton (1974) has reported the results of a series of experiments to determine the rate of abrasion beneath Glacier d'Argentière near Chamonix and at Breidamerkurjökull in Iceland. Smooth metal (aluminium) and rock (marble and basalt) plattens were fixed to bedrock beneath the ice. The edges of the plates were abraded preferentially when raised slightly above the surrounding surface. Details of the mean erosion rates recorded by Boulton are given in Table 6.1. They varied between 0.9 and 36 mm a^{-1}.

Direct measurements such as those described above are often difficult to extrapolate to either

Table 6.1 Measured subglacial erosion rates on rock plattens (from Boulton, 1974)

Locality	Mean abrasion rate [mm a^{-1}]	Ice velocity [m a^{-1}]	Platten charac- teristics
Glacier d'Argentière, Alps	⩽36.00	250	Marble
Breidamerkurjökull, Iceland	3.75	15	Marble
'' ''	3.40	20	Marble
'' ''	0.90	20	Basalt
'' ''	3.00	10	Marble
'' ''	1.00	10	Basalt

other parts of a single glacier bed or to ice masses in general. The ability to access only a limited and often marginal section of the glacier bed may reduce the value of such determinations. Abrasion produces grooved bedrock. The volume of rock removed by any groove can be estimated but what is often difficult to determine is the rate of grooving and the density of grooves over the whole glacier bed. Such estimates are needed in order to calculate M_a in Equation (6.1). It would appear from an inspection of recently deglaciated terrain that the density of grooving is highly variable depending upon bedrock hardness, large-scale bed geometry and palaeo-ice flow conditions. It may be unsatisfactory, therefore, to depend upon the scaling-up of relatively localized and short-term abrasion studies.

The abrasional process creates very small particles and more commonly a fine mineral flour of sub-millimetre to micron size (Figure 6.1). Rarely does abrasion produce clasts larger than a few millimetres. Some abrasional wear debris will be entrained into the base of the ice by a suite of processes further discussed in Chapter 7. A further quantity will be accessed by basal meltwater (either in films or distinctive channels (Figure 6.1.)) and flushed through the glacier to be discharged as suspended sediment at the terminus. During transit both in ice and meltwater modification of abrasional debris may take place (see Chapter 8): mechanical reduction or comminution is probable and chemical alteration by contact with meltwater highly significant.

6.3.2 Measured rates of bed crushing and fracture

Boulton (1979) argues (but not in regard to absolute rates) that the removal of fractured rock, which he terms 'plucking', is volumetrically important, although extremely specific to areas

Figure 6.1 Size distribution of fine-grained clastics in layers of subglacially deposited carbonate indicating the typical sizes transported in subglacial water films. The bar scale (upper left) shows the number of particles present in each size bin along the vertical scale (from Hallet, 1979b).

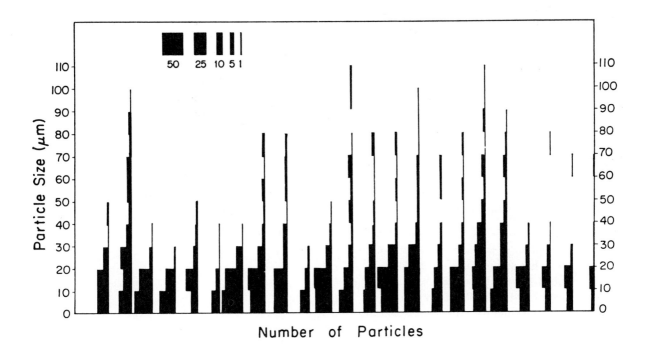

of high stresses on the downglacier side of bed undulations. This importance is disputed by several other glaciologists (e.g. Vivian, 1979; Röthlisberger, 1979).

Unfortunately there are virtually no direct quantitative measurements of the rate of erosion by the removal of crushed and fractured bedrock. In a recent study on processes in a cavity beneath the Grinnell Glacier in Montana, USA Anderson *et al.* (1982) observed the removal of several large blocks from bedrock ledges. Between 1977 and 1978 a block 0.5 m³ moved downglacier along a fracture by some 40 mm. The movement of this rock mass caused the failure of a further portion of the bed.

Crushed and fractured debris products tend to be much coarser than those from abrasional wear. Some clasts even up to metre-size may be transported by meltwater whilst comminution during ice transport may induce fragmentation and reduce clast size sufficiently for meltwater removal. A whole range of sizes of material from large blocks to small clasts will be entrained in the ice and transported towards the glacier snout or subglacially lodged. How much fractured bedrock is comminuted to a very fine grade during transportation (details of communition processes are discussed in Chapter 8) and therefore 'appearing' as a primary *abrasion* product is uncertain. Meltwater will sometimes remove considerable quantities of material from the ice, especially during periods of increasing discharge in the spring–early summer when tunnels are being enlarged. Estimates of this loss can only be speculative but could be calculated if tunnel expansion was known.

6.3.3 Measured rates of mechanical and chemical meltwater erosion

Very few direct observations have been made of erosion by meltwater beneath glaciers. Robert Vivian (1975) has summarized several observations from the European Alps. He reports the observations of J.-P. Perreten on the formation of potholes to 2.0 m depth and of 2.3 m width in ophiolites by meltwater

discharging from Gornergletscher over a period of 3 years. Assuming the pothole has a conical shape this is equivalent to the remarkable rate of 2 500 kg a^{-1}. To obtain an average value for the role of meltwater in this way requires knowledge of the density of potholes per unit area of the glacier bed – which is unknown. Further measurements indicate 100 mm of erosion in quartzite in five years. The products of mechanical meltwater erosion depend upon bedrock composition and are usually in the range from clay–silt-size particles to fine–medium-grained sand. Such debris may be kept in suspension until discharge at the portal or deposited during periods of recessional flow.

6.4 Indirect estimates of glacial erosion rates

The discharge of sediments in ice and meltwater can be used as a measure of the rate of erosion beneath glaciers. The success of these methods relies on the careful estimation of the terms of a sediment balance equation for the glacierized area. A typical budget is:

$$SQ_w + SQ_m = \iint_b E'\, db + \iint_a P_s\, da + D_r \pm S_i \pm S_b \tag{6.2}$$

where SQ_w = discharge of sediment in meltwater
SQ_m = discharge of sediment in ice
E' = sediment yield from bed erosion
P_s = sediment precipitated onto ice surface
D_r = sediment added by rockfall
S_i = sediment entrained in ice
S_b = sediment in storage at glacier bed
a = area of glacier bed
b = area of glacier surface.

In the following sections SQ_w and SQ_m are examined.

6.4.1 SQ$_w$: discharge of sediments in meltwater

Early estimates of the rate of glacial erosion, including those of H.F. Reid in Alaska and S. Thorarinsson in Iceland are summarized by Embleton and King (1975). Significant problems are encountered in estimating total sediment flux (suspended, bed and solution loads), and are described in Chapter 9. Unless an adequate frequency of regular sampling is undertaken, short-duration, high-magnitude flood events which evacuate considerable quantities of subglacial sediments may go unobserved. Estimates will, therefore, be reduced by a factor of two or three at best and an order of magnitude at worst. In addition not all erosional products are necessarily removed by water – some are entrained in the ice and may never communicate with the glacier hydraulic system. Collins (1979a) has also indicated that whilst erosion may occur over much of the ice–rock interface, evacuation of debris in meltwater is related to hydrological conditions which are restricted to only certain parts of the bed and vary through time.

Some of the most useful estimates are those reported by Østrem (1975, 1982). He and his colleagues have measured the annual sediment yield (SQ$_w$) from a number of glacier basins in Norway deriving average erosion rates (SQ$_w$/area of glacier). These vary between 0.07 and 0.61 mm a^{-1}. Although SQ$_w$ does not appear related to glacier size, even where climate and geological factors are comparable, relations between different glaciers seem to remain similar from year to year, although only records of a few years are available. Østrem's results are summarized in Table 6.2 and other, less rigorous estimates of average glacial erosion by similar methods, in Table 6.3

It is difficult to separate out the relative importance of the various erosive mechanisms from the meltwater discharge of sediments. The suspended load will sample only the finer fractions produced by a complex mixture of abrasion, crushing and comminution during

Table 6.3 Estimated subglacial erosion rates for selected glaciers based upon measurement of suspended sediment transport

Glacier	Mean erosion rate [mm a^{-1}]	Source
Muir, Alaska	19.0	Reid (1892)
Muir, Alaska	5.0	Corbel (1962)
Hidden, Alaska	30.0	Corbel (1962)
Engabreen, Norway	5.5	Rekstad (1911–12)
Storbreen, Norway	0.1	Liestøl (1967)
Heilstugubreen, Norway	1.4	Corbel, (1962)
Hoffellsjökull, Iceland	2.8–5.6 [max]	Thorarinsson (1939)
Kongsvegen, Svalbard	1.0	Elverhoi et al. (1980)
St Sorlin, France	2.2	Corbel (1962)
Imat, USSR	0.9	Chernova (1981)
Ajutor-3, USSR	0.7	Chernova (1981)
Fedchenko, USSR	2.9	Chernova (1981)
RGO, USSR	2.5	Chernova (1981)

Table 6.2 Estimated subglacial erosion rates for Norwegian glaciers based upon suspended and bed load sediment transport (from Kjeldsen, 1981)

Glacier	Years of measurement	Mean annual sediment transport [10^3 kg]	Sediment yield [Tonnes km^{-2} a^{-1}]	Mean erosion rate [mm a^{-1}]
Nigardsbreen	13	21 400*	447	0.165
Engabreen	12	22 560*	594	0.218
Erdalsbreen	7	8 130	645	0.610
Austre-Memurubre	6	7 540	842	0.313
Vesledalsbreen	6	775	191	0.073

It should be noted that the individual yearly summaries given by Kjeldsen (1981) upon which the table is based are different from those published by Østrem (1975)
* Figures in column 2 are based upon measurements of both suspended sediments and incremental coarse bed load material on a delta close to the glacier front.

transport, and the products of direct bedrock erosion by debris-charged meltwater. Bedload, however, may be almost exclusively confined to material produced by fracturing. These associations are indicated in Figure 6.2. It might seem from this scheme that long-term measurements of bedload may be a rather good guide to the minimum rate of crushing and fracture of the glacier bed. The very largest blocks produced by these processes (and hence the most significant volumetrically) are probably not moved by water but by the ice. This is deduced from the size cut-off in the coarsest grades of material observed in subglacial channels. Indeed it would need to be demonstrated that the large meltwater channels capable of discharging the coarsest sediment fraction access statistically typical areas of the glacier bed and hence their sediment discharge could be appropriately scaled. The quantity of crushed material will also be reduced by the comminution process which generates the size of debris likely to be suspended.

6.4.2 SQ_m: Discharge of sediments in ice

Few studies have been undertaken to estimate the flux of sediments in glacier ice but such work is clearly necessary in order to derive a full sediment budget and hence net erosion.

Consider the front of a glacier flowing in steady-state. Sediment is located in the ice at a high level and in the basal transport zone. Assuming that only the basal stratified facies (see

Chapter 7) is derived from entrainment of eroded bed materials (i.e. the diffused facies represents P_s and D_r in Equation 6.2) the flux of sediments per metre width of glacier is given by:

$$SQ_m = U_s h_d \overline{C_o} \qquad (6.3)$$

where h_d = thickness of basal stratified horizon
     U_s = basal sliding velocity (taken as applicable through the entire depth h_d)
     C_o = average debris concentration in h_d.

Barnett and Holdsworth (1974) calculated SQ_m for a section of the Barnes Ice Cap which feeds Generator Lake in north-central Baffin Island, NWT, Canada. U_s was estimated at $\sim 17 \pm 3$ m a^{-1}, h_d as 8 m and C_o at $8 \pm 2\%$ by volume. The debris discharge was thus calculated at 11 ± 5 m^3 m^{-1} a^{-1}. The average annual erosion rate may then be computed by dividing this figure by the drainage basin area. A similar type of study was undertaken at Bondhusbreen in Norway by Hagen et al. (1983). From their reports C_o is ~ 10 kg m^{-3}, estimated from measurements of basal ice in tunnels. Debris was concentrated in the lower ~ 5 m of the glacier and the average basal ice velocity 30 m a^{-1}. These data yield a flux of $\sim 1\,500$ kg m^{-1} a^{-1}. By

Figure 6.2   Flow chart indicating the relative sediment contributions to bed and suspended load in meltwater streams from the products of the four principal glaci-erosional processes (abrasion, crushing/fracture, mechanical and chemical meltwater activity).

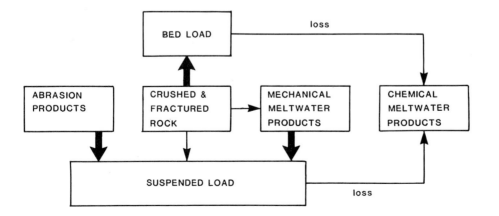

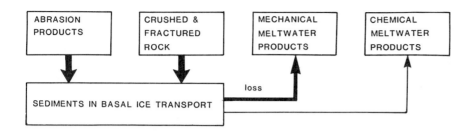

Figure 6.3 Flowchart indicating the relative sediment contributions to basal ice transport from the products of the four principal glaci-erosional processes.

extending this analysis to only certain size grades of material it may prove possible to provide some estimates of the relative contributions from different mechanisms. A primary problem, however, lies in deciding what size fraction may be characteristic of abrasional products. It is quite clear that the large clasts (upwards of several mm) must result from crushing and fracture (Figure 6.3).

6.4.3 Average chemical erosion by meltwater

It would appear from the discussion in Chapter 5 that the role of chemical denudation in the glacial environment is substantially more potent and quantitatively greater that at one time considered. Rapp (1960) working in Arctic Sweden calculated that the transport of dissolved salts was *the* most important factor in mass loss in Karkevagge,

yielding some 26 tonnes $km^{-2} a^{-1}$, equivalent to a surface lowering of 0.01 mm a^{-1}. Given a mean density of surface rocks of 2 700 kg m^{-3} this gives a loss of 0.027 kg $m^{-2} a^{-1}$. Similar studies of the solute content of meltwater streams issuing from glaciers have estimated subglacial chemical erosion. Eyles *et al.* (1982) calculated the total geochemical denudation (including adsorption) for the Berendon Glacier basin in British Columbia, Canada at some 1 050 mequiv $m^{-2} a^{-1}$ – about 2.5 times the approximate world average denudation rate (i.e. equivalent to 0.025 mm of surface lowering per year). A major problem arises from the interpretation of such figures for estimating subglacial erosion. In Chapter 5 it was suggested that the bulk of the uptake of chemical species occurs at the surfaces of very fine-grained sediments rather than

Figure 6.4 Flow chart indicating the multiple sources of and interactions between the sources contributing chemical products to glacial meltwater.

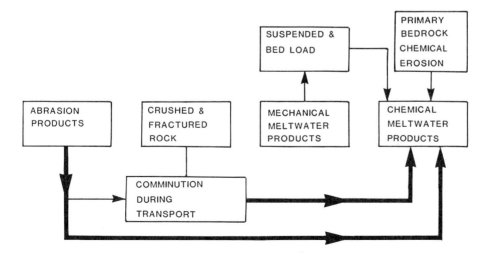

homogeneous bedrock. If this inference is correct the bulk of the solutes will be derived from exchange between subglacial meltwater and material originating from abrasion and comminution (and which form the bulk of the suspensoids). Such a conclusion would seriously diminish the absolute rate of surface mass loss due to chemical processes on bedrock, and give rise to complex solute transport paths as indicated in Figure 6.4

6.5 Conclusions

The state-of-the-art in separating the various glacial erosive mechanisms is insufficient to allow an adequate assessment of relative rates. Indeed even to assess accurately total mass loss by glacial activity is still an area of much needed research. It is to be hoped that by careful consideration of these tasks future workers will be able to design experiments and observational campaigns to deduce, by either direct or indirect means, these important components. If I were to be asked to provide a crude ranking of the importance of each mechanisms under certain conditions the lists in Table 6.4 would be a first attempt.

Table 6.4 Ranking of glacial erosional mechanisms

Erosion mode	Cold base	Melting base Hard rock	Soft rock
Abrasion	2	2	1
Crushing/ fracture	1	1	2
Mechanical meltwater		3	3
Chemical meltwater		4	4

1 = Highest
4 = Lowest

7 Entrainment and characteristics of sediments in ice

7.1 Introduction

Glaciers and ice sheets act as slow-moving conveying systems for eroded rock waste or that transferred to the ice by sub-aerial processes. Because the transit time for most materials is extremely long ($10^2 - 10^3$ a) the ice acts as a major store of sediments. This is especially true of large ice sheets where minimal subglacial melting combined with sub-zero ice/bedrock interface temperatures in their outer margin may prevent the easy escape of sediment in meltwater. If the average concentration of debris, in a basal layer of 10 m thickness of the Antarctic ice sheet, is 30 kg m^{-3} ($\sim 1\%$ by volume) the ice sheet would store some 3.6×10^{15} tonnes of debris. In temperate glaciers it is likely that a substantial amount of the finer sediment fractions is removed by water flowing through and away from the glacier bed.

Chernova (1981) has reported the results of measurements on glaciers in the middle Asian belt of the USSR (Table 7.1). It can be seen, especially for some of the larger glaciers, that the ratio of sediment subject to fluvial transport away from the glaciers to that brought to the snout in ice may be $>50\%$. It is, however, difficult to make generalizations from such few measurements or to determine the cause of the considerable range of results shown in Table 7.1. In this chapter the mechanisms of entrainment of glacial sediments into the ice which occur at the glacier bed and at the surface of an ice mass are outlined and later the characteristics of the entrained sediments are discussed.

7.2 Subglacial entrainment mechanisms

Crushed and fractured bedrock and wear debris produced by abrasion at the bed must be removed for erosion to proceed. It is possible for some of the very finest materials (usually < 200 μm) to be transported in thin subglacial water films as suggested by Vivian (1975). Hallet (1979a) found that fine rock grains were

Table 7.1 Annual sediment budget for glaciers in central Asia (from Chernova, 1981)

Glacier	(1) Glacier area km^2	(2) Unconsolidated sediments transported to glacier terminus 10^6 kg a^{-1}	(3) Increase in moraine volume 10^6 kg a^{-1}	(4) Sediments removed by glacial river 10^6 kg a^{-1}	(5) Balance between 2 and 3 (%)	(6) Glacial river discharge 10^6 m^3 a^{-1}
Fedchenko (1925–1959 average)	662	8 300	500	7 800	94	1 000
Zaravshanskiy (1850–1932)	134	2 200	400	1 800	82	476
RGO (1911–1935)	109	1 100	260	940	87	306
IMAT (1963–1968)	3.8	15.4	6.9	8.5	55	12
Ajutor-3 (1935–1970)	3.4	9.6	7.4	2.2	22	10
Karabatkak (1935–1949)	4.7	13.6	8.8	4.8	35	8

incorporated in subglacial calcite precipitations as a product of water migration associated with regelation. The predominant size was < 30 μm, which must represent the typical thickness of the water layer (see Figure 6.1). Larger particles may be transported in major subglacial water channels. These are opened up during the spring with increasing water flow and pressure. Sediment at the bed is tapped by these waters and flushed out through the hydraulic system.

There are few useful measurements to assess the relative contributions of subglacial water and ice in transporting debris. In a recent study of Bondhusbreen, an outlet glacier of Folgefonni, Norway Hagen *et al.* (1983) made a comparison between crude estimates of the englacial debris flux and the measured transport of sediment in subglacial streams. Their results indicate that 90% of the debris is transported by the river.

7.2.1 Regelation

Ice, close to the melting point, deforming over an irregular bed may induce pressure fluctuations and melting on the upstream side of hummocks – a basic mechanism in sliding. On the downstream side of the bed protuberance

Figure 7.1 Entrainment by refreezing of meltwater and incorporation of small particles at the top and in the lee of bedrock undulations. Flowlines are picked out by this process. A graph of the accretion rate is shown below.

refreezing of the liberated water may occur adding a thin layer, a few centimetres thick, of regelation ice to the glacier sole. Loose debris may be readily incorporated into this regelation layer as the ice in the regelation zone has a velocity component which is directed upwards (Figure 7.1). Nye (1970) has calculated the thickness of this regelation layer (h_r) due to sliding as:

$$(h_r - \bar{h}_r)^2 = \left(\frac{a'}{2\pi \, k_*^2} \right) \qquad (7.1)$$

where $\bar{h}_r$ = average thickness of the regelation layer
 a' = roughness constant (Eq. 4.22)
 k_* = material constant (Eq. 4.22)
 ≈ ($L/4c_d\overline{K}\eta_i$)
 L = latent heat of fusion
 c_d = constant related to melting point depression
 $\overline{K}$ = mean thermal conductivity of ice and rock
 η_i = ice viscosity.

The thickness, h_r, is usually a few tens of millimetres which corresponds closely to subglacial observations such as those of Kamb and La Chapelle (1964). It should be remembered that Equation (7.1) is a formulation for 'clean ice'. Morris (1979) has demonstrated the present inadequacy of classical regelation theory in accommodating a single, idealized clast in pure

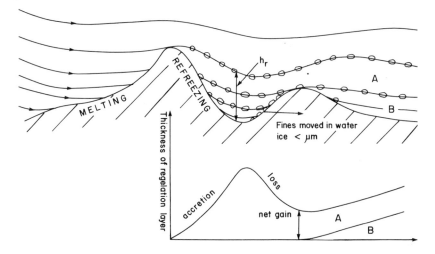

ice adjacent to a flat bed. It is highly unlikely, therefore, that calculation of the thickness of the regulation layer in dirty basal ice will be anything more than an order-of-magnitude.

Regelation, which need not be a continuous process, results in the incorporation of small rock fragments into the freezing water layer (Figure 7.2). It has already been shown that subglacial water films may transport small particles but the freezing process may incorporate fresh material from the substrate by accretion. Boulton (1972) has suggested that the most favourable particles for such incorporation are those from argillaceous strata. Isotopic, ice fabric and textural studies of basal ice layers can indicate active regelation.

Lliboutry (1964) has suggested the entrainment of debris may take place in subglacial cavities. Cavities, it should be remembered, are usually found on the downstream side of bedrock protuberances when ice sliding velocities are high and the overburden pressure insufficient to close them off. The normal pressure experienced at the glacier sole often fluctuates and if this variation is distributed around the critical value to suppress

cavities it is possible that gaps at the interface may periodically open and close. As cavities develop, the ice in contact with the bed is pulled away and any debris contained close to the surface due to regelation is moved out, and has a greater chance of remaining within the ice mass. This mechanism, similar to the hydraulic jack process (Subsection 3.5.2), allows periodic sediment concentrations to be incorporated into glacier ice.

7.2.2 Ice-debris accretion

A further, large-scale, mechanism for entrainment of debris from the bed has been proposed by Weertman (1961). This is essentially an ice-debris accretion model which may operate under non-steady-state flow and is dependent upon changing thermal conditions at the glacier bed.

Under relatively thick ice the temperature gradient in the glacier may be insufficient to drain away all the heat from the bed (i.e. geothermal heat and that produced by sliding motion – see Equation 1.27). The excess heat melts basal ice producing liquid water which flows, as previously discussed, outwards under a generalized pressure gradient proportional to the ice thickness. As the ice thins, the temperature gradient steepens and melting from excess heat ceases: freezing conditions take over at the base. While this process may incorporate material in a similar way to water and debris in the regelation film Weertman argues that changes through time of the location of the 0°C isotherm (in response to changing ice thickness for instance) will lead to large-scale rafting of the refrozen zone into the moving glacier.

7.2.3 Block incorporation

Moran (1971) and Moran et al. (1980) propose another mode of incorporation of subglacial materials to explain the presence of thrust blocks and source depressions in the prairie region of North America.

Figure 7.2 Core section of regelation ice taken from basal layers of Taylor Glacier, Antarctica. Note the small mm and sub-millimetre size particles and cloudy bands of clay-grade fractions. The hand of E.J. Jankowski for scale.

Following Weertman (1961), inland ice is of sufficient thickness to raise basal temperatures to pressure melting point, while towards the front ice is frozen to bedrock. Meltwater from the ice/water interface escapes into and flows through unfrozen and porous substrata. The pressure gradient induced by the ice overburden drives the water towards the glacier margin where it encounters frozen and impermeable ground beneath colder ice. Pore water pressures are significantly raised, decreasing the shear strength of subglacial sediments. This allows the glacier and water to easily lift large blocks or slices of sedimentary material (which may comprise till or bedrock) into the moving ice. Such large-scale behaviour clearly disrupts the pattern of debris already entrained and generates a suite of **glacitectonic** structures. It is easy to see that the mechanism proposed by Moran is merely a variant of the Weertman process (7.2.2).

7.2.4 Overriding

Shaw (1977a), following the work of Hooke (1973), has suggested that advancing cold-based glaciers may entrain debris by the overriding and incorporation of frontal aprons. Most cold glaciers possess steep marginal cliffs against which fallen ice blocks, refrozen meltwater, superglacial and intraglacial debris (washed or fallen), and wind-blown material may accumulate as aprons or ramparts. These characteristics, it is argued, give rise to the bubbly nature, distinctive foliation and debris content of basal ice in cold glaciers.

7.3 Super- and englacial entrainment mechanisms

In addition to entrainment at the bed debris is also incorporated at the glacier or ice sheet surface. Several processes are operative although there has been considerably less study of individual mechanisms and their quantitative contribution to the glacier load.

It is important to separate those materials which become englacial from those that remain on the glacier surface. In general if sediments are added in the accumulation area burial and some englacial travel, although at a high level, will ensue. Material added to the glacier in the ablation zone will tend to remain at the surface. However, debris falling into crevasses, carried into moulins by streams, or rocks of sufficient weight may also enter the ice at or progress to deeper levels.

7.3.1 Rockfall

Where rocks crop out on the flanks of a valley or outlet glacier, or where nunataks pierce the ice cover, material may be added to the ice by the fall of loosened rock. This section will not concern itself with the mechanisms involved in the weathering of mountain walls in cold climates: several pertinent articles deal with such phenomena and readers are referred to Rapp (1960), Mellor (1973) and Selby (1982). Stress-released blocks fall towards the glacier surface and may shatter from impact during travel. The total amount of debris added is difficult to estimate and no measurements have been made on glaciers. By taking Rapp's (1960) calculations of the mass of rockfalls for an Arctic environment in Karkevagge, North Sweden, it is possible to arrive at an order of magnitude estimate of between 10 and 100 tonnes $km^{-2} a^{-1}$ of rock wall surface.

7.3.2 Avalanches

Mountain sides and nunatak slopes above glaciers are often subject to avalanche activity. Such phenomena include the effects of snow, ice, soil and rock, often in various combinations. The mechanics of avalanching need not be of direct concern here, good reviews are readily available (e.g. Voight, 1978; Perla, 1980).

In glaciated country it is likely that most avalanches will be principally of snow, with some of ice, the debris content varying from location to location. Occasionally pure rock avalanches or landslides may occur such as on Sherman Glacier, Alaska in 1964. The effect of all varieties of avalanche is to transfer large quantities of ice, snow and rock debris onto a glacier surface. Small to medium-scale avalanches descending tens or hundreds of metres may carry 10^2–10^5 m^3 of mixed material at speeds of up to 50–60 m s^{-1}. Snow and debris may be transferred a considerable distance including travel right across a valley glacier as is frequent in high mountain country such as the Himalayas and Andes. Both velocity and runout distance of avalanches depend on slope angle, coefficient of friction and entrainment of additional material.

Frank *et al.* (1975) describe periodic debris avalanching off the crater of Mount Baker (a large stratovolcano in Utah) onto several flanking glaciers. In 1973 material travelled 2.6 km down Boulder Glacier before terminating about 1.4 km from the snout. Six major avalanches occurred between 1958 and 1973. An avalanche last century (1890) is reported to have run out 11 km beyond the snout of Rainbow Glacier. Hindcasting from the wedge-shaped scar on Sherman Peak of Mount Baker suggests a volume of snow, ice and rock of 35 × 10^3 m^3. The debris resulting from this event was detected as thin beds of mud exposed at the glacier terminus.

7.3.3 Atmospheric precipitation

Material is added to the surface of glaciers and ice sheets from the atmosphere. Fallout of volcanic ash, condensation nuclei in snowflakes or the deflation of sediment from adjacent dry, ice-free terrain are examples of such contributions.

There appear to be four principal sources of such substances:

- continental (particles and dust, resulting from weathering of the earth's surface, volcanic material, etc.)
- marine (salts derived from sea spray)
- extraterrestrial (interplanetary dust, meteorites, meteorite ablation debris)
- anthropogenic (pollutants and contaminants)

Aerosols (microparticles) and gases

In general two primary groups of microscopic substances contributed from the atmosphere may be recognized, **aerosols** and trace gases. The former are particulate matter referred to in glaciological literature as microparticles. They range in size from small clusters of molecules (0.001 μm radius) to large salt and dust particles with radii up to about 10 μm (0.01 mm). The latter constitute dissolved and adsorbed materials.

The chemical concentrations of primary substances in snow from Greenland and Antarctica in comparison with pure water and rainfall in urban areas has been discussed in Chapter 5 and is shown in Table 7.2. On average the chemical concentration in Antarctic snow is 10^{-2}–10^{-3} lower than in rainfall from mid-latitude coastal areas. Interesting differences occur between the very low concentration found at Mizuho Plateau in East Antarctica (some 200 km from coast), Byrd Station in West Antarctica (over 500 km from coast) and Camp Century in N.W. Greenland (100 km from coast). In addition, the balance between silicate dust and components derived from salts in sea water is markedly different (Table 7.2).

At Camp Century the bulk of microparticles (75% or more) are silicates (especially clay minerals). Kumai (1977) detected over 40 different silicates in each snow crystal, ranging in size from 0.05–8 μm, but mostly <1 μm. Only 1% appeared to be contributed by sea salt. In Antarctica, at Mizuho Plateau, as well as Byrd Station, Morozumi *et al.* (1978) found that only 10% of the major components was made up of silicate dust, 90% being derived from sea salt! This difference may be due to the pronounced differences in land–sea distribution in the Northern and Southern Hemisphere, and to correspondingly complex atmospheric circulation patterns.

Table 7.2 Silicate dust in polar snow (μg kg^{-1}) (from Morozumi et al., 1978)

Sampling station		Silicate dusts	Elements in silicate dusts				Total amounts of elements in snow–ice			
			Na	K	Mg	Ca	Na	K	Mg	Ca
S122	⎫	8	0.2	0.2	0.2	0.3	53	2.1	13	3.18
Y200	⎪	6	0.1	0.1	0.1	0.21	14	1.0	2.3	3.76
Mizuho Camp	⎬ Mizuho Plateau, Antarctica	2	0.02	0.01	0.002	0.04	14	1.2	2.1	2.78
Y300	⎪	8	0.2	0.15	0.2	0.3	11	0.56	1.7	1.93
Y135	⎭	12	0.3	0.2	0.3	0.42	12	1.0	3.0	1.81
Byrd Station, Antarctica		4	0.1	0.1	0.1	0.1	31	1.5	4.0	1.3
Camp Century, Greenland		20	0.5	0.5	0.5	0.7	21	3.1	8.0	6.4
Tokachidake, Japan		90	2	2	2	3	750	50	200	100
Asahidake, Japan		100	2	2	2	4	590	40	100	130

Figure 7.3A Volcanic ash in the Byrd Station ice core, Antarctica. Left shows volcanic dust as observed in the core, right depicts a thin section showing fine grain crystals in the band. Depth of sample 1415 m (~1700 a BP). (Courtesy A.J. Gow, CRREL).

The annual fallout of silicate and sea salt components at Camp Century, Byrd Station and Mizuho are shown in Table 7.2. Windom (1969) estimated that 0.21 kg m^{-2} ka^{-1} of atmospheric dust was accumulating in Greenland snows. The annual contribution of such matter to the ice sheet from the atmosphere is 3.7×10^8 kg.

Large particles

Volcanic ash and wind-borne dust particles > 10 μm in size are also found in glacier ice (Figure 7.3). The latter are typically concentrated into areas where there is substantial ice-free terrain which acts as a local source for fine debris. Valley glaciers in mountainous terrain, small ice caps and ice sheet margins skirted by bare rock are thus most likely to act as repositories for fine materials. Dort (1967) reports considerable accumulation of wind-blown material in Sandy Glacier, Victoria Land, Antarctica. Southerly prevailing winds with velocities of 10–16 m s^{-1} pick up sand from the floor of adjacent ice-free Wright Valley and carry it several kilometres horizontally and about 1 km vertically to be released in the vicinity of Sandy Glacier. The material has a narrow size range (0.125–1.0 mm with >50% in the 0.25–0.5 mm range) and forms layers in the ice 20–250 mm in thickness with an average of ~100 mm.

Figure 7.3B Debris band (originally wind-transported clays) cropping out on the surface of Bersaerkerbrae, East Greenland.

Volcanic deposits are well known from glaciers in Iceland, Alaska, Greenland and Antarctica. Vinogradov (1981) reports that most of the recent glaciers in regions of active volcanism in Kamchatka are overlain by a variable cover of pyroclastic material. Average thickness of surface volcanic debris layers is 0.3–0.4 m. The thickness of scoria produced by the 1945 Avachinskiy eruptions in the accumulation area of Kozełskiy Glacier was 1.6 m. In Iceland Steinthorsson (1977) discovered 30 ash layers during drilling operations at Bardarbunga, a small dome on the northwest side of Vatnajökull. The 2 164 m core drilled to bedrock at Byrd Station in Antarctica revealed 25 distinct **tephra** layers and some 2 000 additional dust bands (grain size < 5 μm) and probably also of volcanic origin (Gow and Williamson, 1971; Kyle *et al.*, 1981). The tephra layers contain glass sherds and lithic fragments within the size range 20–100 μm. The core location is some 300–500 km from the Cenozoic volcanic province in northern Marie Byrd Land and Mount Takahe has been identified as the most likely source (Kyle and Jezek, 1978). Kyle *et al.* (1981) also report tephra (up to 70 μm in size) in a core from the 906 m drill hole at Dome 'C' in central East Antarctica. They consider the glass sherds chemically similar to those from the Byrd Station Core. Dome 'C', however, is $\sim$ 2 400 km distant from Mount Takahe.

In recent years some 6 000 meteorites and meteorite fragments have been discovered on the Antarctic ice sheet. The meteorites are carried within the ice and in certain favourable locations become concentrated at the surface where ice flow is restricted (such as close to mountain ranges) and where there is also net ablation by sublimation giving rise to blue ice fields. The annual infall rate of meteorites larger than 3 kg to Antarctica is about 48.5 so that it is possible to calculate the approximate steady-state population contained within the ice sheet – 760 000 meteorites or 1.5×10^8 fragments (Olsen, 1981).

7.4 Character of sediments: basal

7.4.1 General disposition

Observations and measurements of the sediment properties of glaciers rich in basal debris are few. Several field reports describe basal sediments in the front, sides and margins of current ice masses (e.g. Boulton, 1970; Shaw, 1977a; Goldthwait, 1960, 1971). Subglacial tunnels, although they access more restricted portions of the glacier bed, nevertheless provide highly diagnostic information on the characteristics of basal layers (e.g. McCall, 1960; Kamb and La Chapelle, 1964; Peterson, 1970; Holdsworth, 1974). One of the most recent and detailed studies of basal sediments was conducted by Lawson (1979) on the Matanuska Glacier in Southern Alaska, and his careful results are used to illustrate points throughout the following sections of this chapter. The review by Goldthwait (1971) also contains useful additional material.

Glacial sediments are transported in closely spaced basal layers which may, in places, be discontinuous (Figure 7.4). Lawson (1979), following earlier workers, recognized two principal **sedimentary facies** in the basal debris zone: a lower *stratified* and an upper *dispersed* region (Table 7.3 and Figure 7.5). The two may be separated by a distinctive contact. The stratified facies comprises conformable and

Figure 7.4A A dipping sequence of step faulted debris-rich layers toward the base of Taylor Glacier, Antarctica.

Table 7.3 Characteristics of basal debris layers, Matanuska Glacier, Alaska (from Lawson, 1979)

FACIES/ subfacies	Thickness (m)	Principal features	Ice — Grain size [diameter, mm]	Ice — Debris content by volume %	Debris — Principal features	Debris — Grain size [max − 5ϕ] M = mean σ = RMS	Roundness	Fabric
DISPERSED	0.2–8	Uniform in appearance; upper contact distinct and planar; horizontal to upglacier dip	10–40	0.04–8.4	Uniform, massive	Sandy gravel to gravelly sand M: −1.1 σ: 1.5–3	Subangular to angular with minor sub-rounded to round	
STRATIFIED	3–15	Stratification, often showing internal deformation; downglacier to steep upglacier dip; irregular, sharp upper contact; subfacies interlayered	<4 (exceptionally 10)	0.02–74	Layers, lenses, zones compose strata; thin and thicken rapidly; lateral extent limited	Silt to sand to gravel	Rounded to subrounded	Strong single mode; mean axis (V_1) parallels local ice flow; $S_1 > 0.8$
discontinuous	0.05–2			Variable (0.02–36); dependent on sampling technique	Lenses, layers, platelets, aggregates aligned subparallel to stratification; debris streaming	Silt to sandy silt M: 3.5–6.5 σ: 1.3–1.9		
suspended	0.001–1.2			0.02–60	Suspended particles and aggregates without orientation	Silt to silty sand M: 4.3–6 σ: 1.3–2.5		
solid	0.01–1.7			>60	Well-defined layers; may show sedimentary structures	Silt to gravel M: −2.5–5 σ: 1–2.6		

S_1 is a normalized eigenvalue which gives the strength of the cluster of long axes about the mean axis (see Mark, 1973)

unconformable intercalations of sediment. Layers may be lenses or lumps which sometimes comprise frozen till, outwash or homogeneous bedrock. Such layers may dip upglacier and towards the glacier central axis. Many investigators have reported large- and small-scale deformation structures in basal ice layers including open, isoclinal, overturned and recumbent folding; as well as normal, reverse and thrust faults (see Figure 8.3). These features usually result from strong longitudinal compression and are arguably responsible for

Figure 7.4B   Basal sequence, Taylor Glacier, Antarctica. Note the very sharp break between the lowermost, heavily debris-charged horizon (>25% by volume) and cleaner ice above. Large clasts (25 cm diameter) are embedded in the basal layers. A series of debris-rich bands occurs several metres higher.

moving debris to higher levels in the glacier. In such deformed ice Boulton (1970) following Gripp (1929) recognizes debris layers which parallel the foliation and those that cut it.

Above the stratified zone there is a fine suspension of debris particles which constitute the dispersed facies. This layer has often been referred to as **'amber ice'**. It may also contain thin bands with higher sediment concentrations.

It is important in further discussion of the characteristics of basal sediments, to recognize the difference between the nature of materials *in the ice* and the same debris *after release* in till or outwash. We shall see that it is often impossible to unequivocally deduce or understand sediment transport mechanisms in the glacier from study of deposits alone.

7.4.2 Texture

Basal sediments are typical polymodal and exhibit a considerable size range. Lawson (1979) notes, however, that each basal facies is usually texturally distinctive (Figures 7.6 and 7.7). This conclusion, if correct, is important since it demonstrates for that period of time in which sediments remain in the glacier gross homogenization, diagnostic of *deposited* till, does not occur. The processes of entrainment may thus produce several textural sub-populations; *deposition* mixes the sub-populations so that the textural similarities of any one till result not from transportational phenomena but from the depositional process. In Chapter 8 transportational factors are considered in detail and this notion further discussed.

A further point brought out by Figure 7.6 is that the dispersed facies, on average, carries coarser material than the lower stratified facies which is dominated by clay and silt-sized fractions – the results of comminution by strong shear forces in the zone of primary traction.

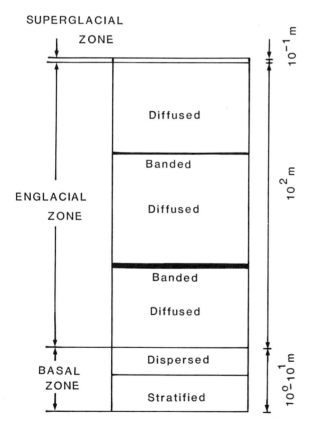

Figure 7.5   Schematic representation of ice facies (from Lawson, 1979).

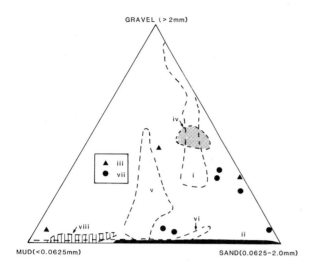

Figure 7.6 Grain size characteristics of various ice facies from Matanuska Glacier, Alaska. Roman numerals refer to the following Lawson categories: i superglacial debris, ii diffused facies, iii banded facies, iv dispersed facies, v composite of stratified facies, vi discontinuous facies, vii solid subfacies, viii suspended subfacies. The solid subfacies has no preferred size distribution. There is poor textural diversity and samples exhibit poor–very poor sorting (from Lawson, 1979).

7.4.3 Debris concentration

There are remarkably few systematic studies of the concentration of sediments in glaciers. Many generalizations from ice masses appear in the glacial literature but their representativeness may be disputed in view of the extreme variation in sediment facies suggested above. Any 'average' or unqualified estimate should be used only with extreme caution. Goldthwait (1971) quotes determinations which range from 0.03 to 64% by weight. Lawson has given esimates from 0.002 to 75%. Pessl and Frederick (1981) have made a

Figure 7.7 Typical grain size curves for basal debris from the solid sub-facies of Matanuska Glacier, Alaska (after Lawson, 1979).

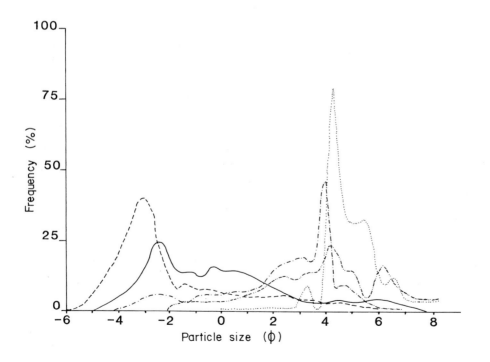

literature search of such data and suggest an average value for most temperate glaciers is 25% by volume reaching 30–90% by volume in marginal zones.

Real problems attend an adequate understanding of variations in debris concentration in both space and time. In general, concentration might be expected to rise with distance down flowline as the processes of entrainment become increasingly effective. Changing basal conditions, especially temperature regime and water flow, will also govern concentration fluctuations. That pronounced variation exists is clearly demonstrated by strong disagreements between glaciologists (e.g. *Journal of Glaciology* (1971, 1972) between Boulton and Andrews). It is unfortunate, therefore, that over the last 100 years, and after the investigation of innumerable glaciers only gross generalizations on sediment concentration can be made.

Observations from ice cliffs, ice cores and subglacial tunnels indicate that there is usually a distinctive basal horizon of relatively uniform debris content above which the concentration shows a marked exponential decrease. Bands and rafted-up material give rise to major discontinuities in concentration. Yevteyev (1959) and Hagen *et al.* (1983) have presented detailed studies from ice cliffs near Mirnyy, Antarctica, and the base of Bondhusbreen in Norway which support such relationships.

Drewry and Cooper (1981) have formalized this type of concentration variation by the use of Heaviside function of the form:

$$C_o(y)/C_o(0) = U_o(0) - U_o(y1)(1 - \exp(-k_1(y - k_2))) \tag{7.2}$$

where $C_o(y)$ = concentration at any height, y, above the bed

$C_o(0)$ = initial debris concentration of basal layer

k_1 = constant

k_2 = constant proportional to thickness of basal layer of constant debris content

$U_o(0), U_o(y1)$ = Heaviside step functions.

Figure 7.8 shows the relationship between Equation (7.2) and debris concentration in a cold ice sheet base (Yevteyev's data from near Mirnyy, and the average concentration for the Byrd Station ice core). Although Equation (7.2) appears to satisfactorily give the background diminution of sediment concentration in these cases it should be tested with additional observations.

From the available studies there does not seem to be clear separation between debris concentrations in cold, sub-polar or temperate glaciers despite discussion in the literature. Concentrations in some debris bands of Antarctic glaciers for instance yield values in excess of 25% by weight quite comparable to temperate glaciers.

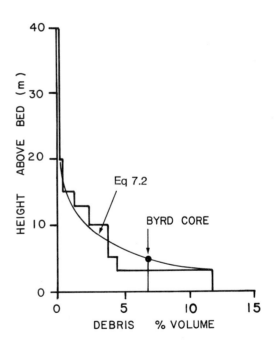

Figure 7.8 Vertical variation in debris concentration in basal ice from cliffs near Mirnyy, Antarctica as reported by Yevteyev (1959) and point measurement of average concentration in the base of the ice core from Byrd Station, Antarctica. The Heaviside function (Equation 7.2) is shown for comparison (from Drewry and Cooper, 1981).

7.4.4 Debris layer thickness

Observations in the Alps, Norway and Iceland led Boulton (1971) to suggest that the debris layer of temperate glaciers is typically very thin – possibly no more than 0.05–0.1 m in thickness and only exceptionally reaching 1 m. In contrast cold ice caps and ice sheets appear to possess much more substantial debris horizons. At Taylor Glacier in Southern Victoria Land, for instance, a layer 5 m thick is evident, while in the Camp Century core debris extended 17 m above the bed. The data from ice cliffs around the Greenland and Antarctic Ice Sheets show that several metres to tens of metres is typical. Bentley (1971) suggests that one interpretation of an early, low amplitude **'P' wave** arrival on seismograms from West Antarctica is moraine-charged basal ice with a thickness of up to 600 m!

Basal layers are not likely to be of constant thickness, especially towards the margins of glaciers where complicated flow over bed irregularities may thicken and thin debris horizons. For large ice sheets it would seem appropriate to assume the thickness of the basal layer must thicken only slowly from the centre where horizontal velocities are low and basal melting predominates. Towards the margins, however, layer thickness may increase rapidly if there is **add-freezing** of water driven outwards by pressure gradients. Weertman (1966) has calculated the thickness of the debris layer under conditions of net accumulation in an ice sheet and also where an outer ablation zone might exist to align **particle paths** upwards and hence thicken the debris zone (Figure 7.9). In the accumulation area the thickness of the debris-rich zone at any distance (x) from the centre is given by:

$$h_d = (3h\,U_u/2\dot{A})[1 - (x_m/x)^{1/\gamma*}] \qquad (7.3)$$

where γ_* = the ratio of horizontal to vertical

velocity to a maximum value of $\dfrac{3h\,U_u}{2\dot{A}}$

where U_u = vertical ice velocity
$\dot{A}$ = accumulation rate.

Taking U_u as 0.01 m a^{-1}, $\dot{A}$ as 0.30 m a^{-1} and h = 2 000 m the basal debris layer would be 100 m in thickness.

In the ablation zone, beyond 'x_c' the layer thickness (h_d) is:

$$h_d = (3h\,U_u/2\dot{A})\{[(\dot{A}/\dot{A}_{bb})H - (x_m/x_c)^{1/\gamma*}]$$
$$[(L' - x_c)/(L' - x)]^{1/\gamma*} - (\dot{A}/\dot{A}_{bb})\} \qquad (7.4)$$

where $\dot{A}_{bb}$ = ablation rate
L' = ice sheet half width.

When the layer thickness reaches the value of 3h/2 it will intersect the ice sheet surface, The distance (x_c) at which this occurs is:

$$(L' - x_c) = (L' - x_e)\{(\dot{A}/\dot{A}_{bb}) + 1 - (x_m/x_e)\}/$$
$$\{(2\dot{A}_{bb}/3U_u) + (\dot{A}/\dot{A}_{bb})\}^{\gamma*} \qquad (7.5)$$

Given values for $\dot{A}$ and $\dot{A}_{bb}$ in the order of tens of centimetres per year and 1 m a^{-1} respectively, a few cm a^{-1} for U_u and γ_* between 0.5 and 1 Weertman shows that the proportion of the glacier surface covered by transported debris may be up to 50%, but is usually 0.2 to 10%.

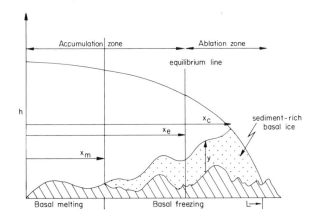

Figure 7.9 Variations in the thickness of the basal debris zone (h_d) in a typical section through a glacier or ice sheet margin with an ablation zone. x_m is the distance to the boundary between melting and freezing conditions at the glacier bed; x_e is the distance to the equilibrium line; x_c is the distance to the point on the glacier surface where basal debris crops out (after Weertman, 1966).

7.5 Character of sediments: super- and englacial

7.5.1 General disposition and comparison with basal sediment

Superglacial materials are readily observed in the ablation areas of ice masses, especially during the summer melt season (see Figure 8.5). Englacial sediments can usually be examined only in situ from cores extracted from boreholes, in tunnels and crevasse/cliff sections (see Figure 7.3).

In general superglacial debris is subject to rapid mechanical and chemical breakdown resulting in distinctive sediment facies. It must also be remembered that englacial material may become superglacial where ice flow lines intersect the ice surface due either to ablation or glacitectonic mechanisms (which may also emplace basal debris at a high level in the glacier, even reaching the surface as further discussed in Chapter 8). Figure 7.10 and 7.11 attempt to summarize the distinctive locations and location changes that are exhibited by debris in ice sheets and glaciers.

In glaciers the presence of debris is governed strongly by initial entrainment and subsequent flow patterns. In ice sheets where the superglacial contribution is decidedly minor it is the vertical flow component which is dominant and most materials will approach the bed to form a distinctive zone above any horizon resulting from basal entrainment (Figure 7.11).

A general distinguishing characteristic between super- and englacial debris and basal materials will be lack of evidence for strong crushing and fracture typical of basal traction. There will, as a consequence, be few striated clasts and a predominance of angular particles, often weathered and oxidized with little preferred orientation and lack of substantial consolidation. Boulton and Eyles (1979) also point to a coarser mean size, high angle of friction and low density (with frequent voids). The composition of superglacial debris is usually independent of subjacent bedrock, may comprise far-travelled rock types with pronounced survival of weak lithologies and may also contain certain organic materials including spores and pollen and occasionally large fauna and flora (Figure 7.12).

In many cases the distribution of size, sorting and lithology will vary remarkably from place to place on the glacier surface, once again emphasizing the diversity of sediment sources and the difficulties of generalization.

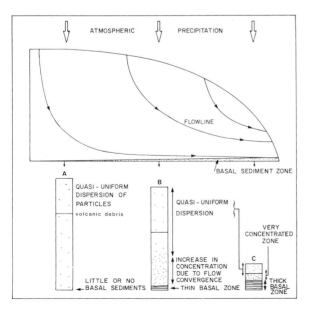

Figure 7.10 General model for the disposition of sediment in an ice cap or ice sheet. Three flowlines are shown. **A** in the centre of the ice sheet the bulk of debris is supplied from atmospheric precipitation. Dominant downward vertical motion generates a quasi-uniform dispersion of particles (excepting distinctive volcanic ash layers), but there is little basal debris as entrainment is inhibited by low horizontal basal velocities. **B** further out vertical strain and increased horizontal motion thins the deeper ice increasing the apparent concentration of particles. Thick ice at the pressure melting point close to the bed and a significant horizontal velocity may assist basal entrainment. **C** towards the margin ice thins concentrating debris further. The basal debris zone also attains its thickest development.

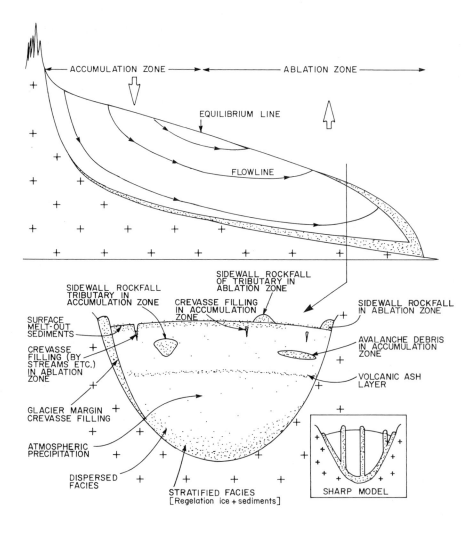

Figure 7.11 General model for the disposition of sediments in a valley glacier. The location and quantity of debris varies markedly down the glacier and is dependent upon the relative size of the accumulation and ablation zones (i.e. critical to the position of the debris-carrying flowlines). A transverse section through the ablation zone is illustrated. An earlier model proposed by Sharp (1960) is also illustrated but considered less realistic.

7.5.2 Texture

Figure 7.5 shows the location of superglacial, diffused and banded facies as defined by Lawson (1979) for samples from the Matanuska Glacier in Alaska. It is important to note the difference in this example between the very fine-grained diffused facies and superglacial debris suggesting different origins. Lawson remarks that 40–80%, by area, of coarse material consists of particles larger than 32 mm in diameter. Silt–clay content is usually well below 20%. Typical grain size characteristics are given in Figure 7.6. Similar distributions biased towards the coarser fractions

Figure 7.12   Carcass of a mountain sheep in the ablation zone of Lyell Glacier, Yosemite National Park, California. When the carcass was discovered sheep had been extinct for 50 years in Yosemite. Estimates suggest it had been entrained in the glacier for 200–300 years (from Dyson, 1962).

are reported by several other workers (e.g. Boulton, 1978).

   In general, sizes range from clay to blocks of several tens of metres. The paucity of finer grades in some superglacial deposits may result from the absence of pronounced abrasional processes which are shown (in Subsection 8.2.2) to be important in producing fine grades in the basal debris zone. Localized transport from meltwater released by ablation may also carry away fine material at the surface. Such processes may be responsible for the pronounced shaping of particles giving rise to small deposits of well rounded and sorted detritus. This contrasts with weathering at the surface which usually assists in

development of angular fragments. Table 7.4 shows the roundness of some pebbles in three principal facies recognized by Lawson (1979). Figure 7.13 provides a plot of particle shape for Breidamerkurjökull, Søre Buchananisen and Grinnel ice cap.

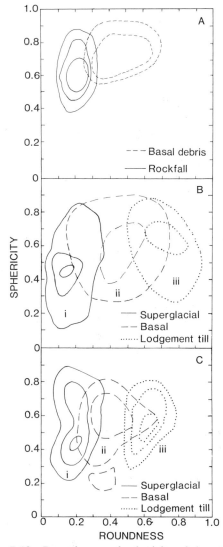

Figure 7.13   Roundness and sphericity of clasts (based upon Krumbein's index) for material in transport in basal layers, in high level transport and in lodgement till. **A** Samples from Breidamerkurjökull, Iceland (Boulton, 1978); **B** Søre Buchananisen, Svalbad (Boulton, 1978); **C** Grinnel Ice Cap, Baffin Island, Canada (Dowdeswell *et al.*, 1985). Contours are at 5, 10 and 15% of data per 1% area.

Table 7.4 Frequency (per cent) of roundness groups of clasts in three principal facies of Matanuska Glacier, Alaska (from Lawson, 1979)

	Very angular	Angular	Sub-angular	Sub-rounded	Rounded	Well rounded
Superglacial	49	39	12			
Basal, dispersed facies	4	43	43	8	2	
Basal		3	18	38	32	9

7.5.3 Debris concentration

The concentration of debris in the superglacial zone is usually very low ranging from a sparse sprinkling of individual particles to a more dense but nevertheless loose aggregate. Boulton (1978) suggests a volume concentration of <1% for medial moraines in Iceland. High concentrations may occur where sediment has been collected by meltwater streams into superglacial lakes.

The bulk of englacial sediments have concentrations typically in the order of 10^{-2} to 10^{-3}% by volume. There may be occasional zones where the concentration is higher (giving rise to weak stratification) produced by periodic events such as avalanche activity, volcanic fallout or from streams carrying substantial debris. Higher concentrations are usually in the order of 0.05–1.0% by volume.

7.5.4 Debris layer thickness

In an early detailed study of Wolfcreek Glacier in the Yukon, Sharp (1949) reported that superglacial debris gave rise to a continuous cover of 0.3–0.6 m in thickness on stagnant ice. Local variations in thickness are controlled by the topography of the ice surface as well as the presence of debris-rich bands in the ice. Tarr (1908) reported thicknesses of between 5 and 6 m on glaciers in the Alaska Coastal ranges.

In areas of high ablation rate, and where sublimation may play an important role, the thickness of debris may reach very great values, in the order of tens of metres. In addition the lower parts of such glaciers may be entirely covered by a continuous layer of debris, broken only by stream cuttings and thermokarst features.

Such phenomena are characteristic of many high-altitude, low-latitude glaciers (Figure 7.14).

Nakawo (1979) investigated superglacial debris on G2 Glacier in the Mukut Himal of Nepal and found that debris thickness increased downglacier to 2 m at the snout, over a distance of about 3.8 km and with an elevation decrease of 220 m. At the same time he observed that particle size decreased.

Glazyrin (1971) has examined the thickness change with elevation down a glacier flowline which is dependent upon the ablation rate (i.e. under the debris) and the debris concentration (C_o). During a time (dt) the mass per unit area of superglacial debris (M_o) will increase by an amount:

Figure 7.14 Stream-cut section through the terminal (ablation) zone of Glaciar Vizcacha, Patagonia. Note the pronounced flowlines carrying debris, picked out by differential ablation, which raise debris to the glacier surface and where melt-out concentrates sediments into a thick superglacial horizon (or lag) (from Bertone, 1972).

$$dM_o = C_o (A_{bb}k)\, dt \qquad (7.6)$$

where A_{bb} = mean annual ice ablation

   k = constant proportional to the reduction in ice melt due to the presence of debris layer.

During the same period the ice flows to lower elevation in the ablation zone by an amount $dz = -U \sin \alpha_s\, dt$, and the surface debris layer thickness (h_d) increases by:

$$\frac{dh_d}{dz} = -\frac{C_o A_{bb}\, k}{U \sin \alpha_s\, \rho_d} - h_d \frac{d\rho_d}{dz} \qquad (7.7)$$

where U = ice velocity

   α_s = surface slope of glacier

   ρ_d = dry bulk density of the rock debris.

Figure 7.15 shows the estimated debris thickness for the Ajutor-2 Glacier in western Tien Shan Province, using Equation (7.7).

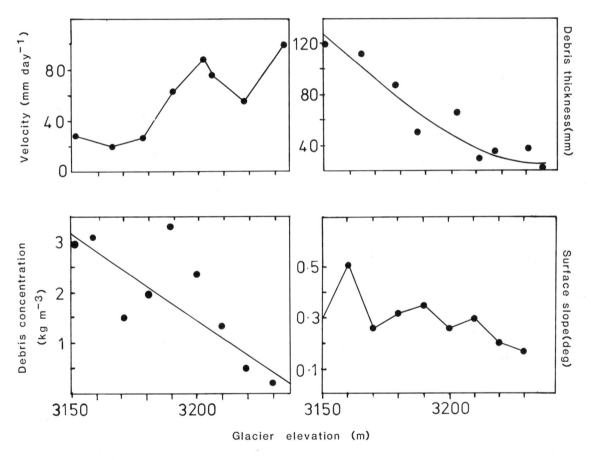

Figure 7.15   Superglacial moraine layer thickness (h_d) calculated from Equation (7.7) for Ajutor–2 Glacier (western Tien Shan, USSR). The other curves show · mean annual velocity, surface slope and debris concentration in the ice; all are functions of elevation above sea level (from Glazyrin, 1975).

8 Transport and modification of sediments in ice

Basal sediments and those in high-level transport are advected through the glacier system. During transportation processes operate which may alter the position of sediments. This process, termed migration, may involve only a few metres of lateral and vertical migration, while on a larger scale ice and sediments may be imprinted with structures resulting from glaci-tectonic activity. In addition, the size and shape of clasts may be altered and distinctive orientations imposed.

8.1 Migration of debris

8.1.1 Vertical migration

It is often observed that close to, but above the basal debris-rich zone there is a finer suspension of solids. The upward migration of particles in the dispersed facies (sometimes called 'amber ice') may result from a variety of mechanisms.

Weertman (1968) applied a simple shear model (i.e. no vertical strain) based upon the interaction of clasts of an ideal spherical shape. Under steady-state conditions and during an infinitely long time it is possible to determine a characteristic upward **diffusion** distance:

$$\lambda^* = (2\,C_*\,h\,/\,\dot{A})^{1/2} \qquad (8.1)$$

where C_* = sediment concentration-dependent diffusion constant
$\dot{A}$ = accumulation rate
h = ice thickness.

According to Equation (8.1) the distance diffused is only an order of magnitude greater than the size of the particles involved. Thus for fine silt-sized material λ^* may be only a few tens of millimetres. In most cases the dispersed facies is characterized by a significant variety of clast sizes and the layer often possesses a relatively uniform height so that in these circumstances the Weertman mechanism cannot play a major role.

Holdsworth (1974) has presented a simpler diffusion model related to vertical strain in basal ice. The upward velocity of debris is given by:

$$U_{vd} = \dot{\epsilon}_z\,h \qquad (8.2)$$

where $\dot{\epsilon}_z$ = vertical strain rate.

The time required to change the thickness of the debris-charged layer is:

$$dt = (h_{d(o)} - h_{d(f)})\,h\,/\,z_f\,z_i\,\dot{\epsilon}_{z(o)} \qquad (8.3)$$

where $h_{d(o)}$ = initial debris layer thickness
$h_{d(f)}$ = final thickness
$\dot{\epsilon}_{z(o)}$ = vertical strain rate at zero depth (i.e. the ice surface).

The time to achieve a corresponding change in the concentration in debris volume is thus:

$$dt^* = C_{o(o)} - C_{o(f)}\,/\,z_i\,C_{o(o)}\,\dot{\epsilon}_{z(o)} \qquad (8.4)$$

where $C_{o(o)}$ = initial debris concentration
$C_{o(f)}$ = final debris concentration.

The important result of Equation (8.4) is the characteristic *diffusion* time which is in the order of 10^4 a.

8.1.2 Lateral migration

Migration occurs in the horizontal plane, moving sediment laterally, but also has the effect of changing debris layer thickness and concentration. Lateral dispersion occurs near the bed due to flow over and around obstacles. The process has been elucidated by Boulton (1975, 1978) and an example is shown in Figure 8.1. As a uniform debris-rich basal ice layer approaches a bed hummock its flow may accelerate due to enhanced plastic deformation around the sides of the protuberance (Equation 1.11 and 1.12) with little debris carried over their tops. Basal ice is thus initially separated into narrow but thick zones between obstacles of high sediment concentration, while ice overlying the crests of **bedforms** is thin with low sediment concentration. It is possible, therefore, by ice streaming around a field of such obstacles to develop a complex and laterally dispersed basal sediment layer.

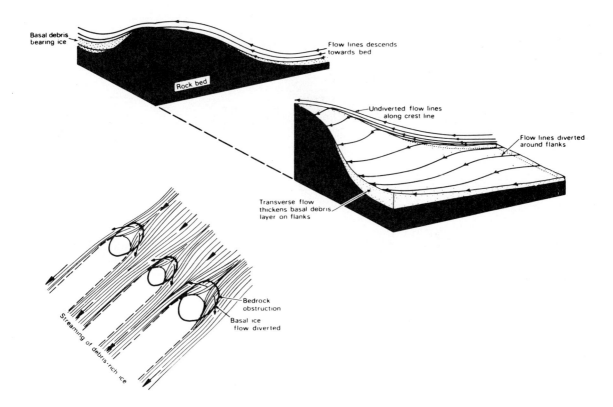

Figure 8.1 Lateral and vertical debris migration induced by the streaming of basal ice around bed undulations (from Boulton, 1976).

Figure 8.1 also shows the effects of basal layer streaming which may give rise to a vertical component of migration emplacing basal sediment at a higher transport level. Further complicating factors arise if there are strong shear strains in the ice or if some or all of the various streaming zones travel at different velocities due to physical conditions encountered as they pass bed obstacles. Such a process may result in the complete mixing or **homogenization** of ice and sediment in basal layers. This conclusion is quite different from the model suggested by Lawson (1979) and discussed in Subsection 7.4.2 in which separate sediment populations within a glacier are considered to remain distinctive during transportation.

The notion of homogenization of basal ice layers and the break-up of an inherited stratigraphic continuity at depth is given support from results of radio echo-sounding. Internal radar reflecting layers, which represent depositional horizons, within the Antarctic ice sheet are *not* observed in a zone above bedrock despite adequate system performance (Figure 8.2). The most plausible explanation is that the continuity of layers has been destroyed by complex flow over rough subglacial terrain (Robin *et al.*, 1977; Robin and Millar, 1982). It would appear that the relative degree of mixing of basal sediments will depend upon bed roughness and basal flow conditions so that pronounced variation may arise between one ice mass and another.

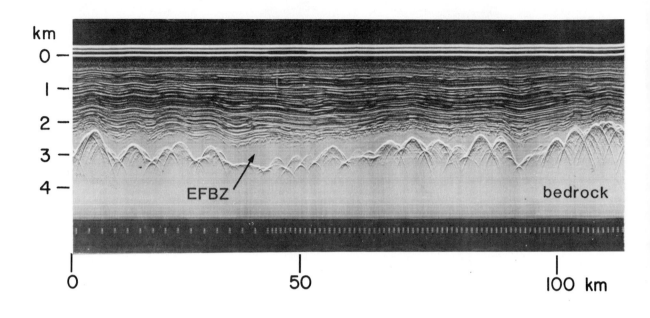

Figure 8.2   Radio echo-sounding profile from East Antarctica showing a zone above bedrock in which internal reflections (layers) failed to be recorded despite adequate radar system performance. This 'echo-free basal zone' (EFBZ) is thought to be due to high shear strain in the lowermost ice layers which combined with complex flow and ice intermixing over bed undulations destroys the contiguity of layers.

8.1.3 Glaci-tectonic processes and structures

The deformation of ice may produce major and minor structures within glaciers and ice sheets such as folds, foliation and faulting. These in turn give rise to the transposition of entrained debris. Figure 8.3 shows various structures exhibited by glaciers in Svalbard and Antarctica. The development of structural features usually results from a combination of longitudinal stress (compressional or tensional) at specific temperature and bed roughness boundary conditions. Figure 8.4 shows large-scale features produced by ice flow from a region where basal sliding is dominant to one in which the ice is frozen to the bed. This situation is typical of glaciers with bed hummocks which give rise to cold patches in the manner described by Robin (1976). It is likely that ice structures will develop where the strain rate is highest. This will be in the transitional zone which has horizontal dimensions in the order of the ice thickness and is characterized by a longitudinal strain rate (Weertman, 1976) given by:

$$\dot{\epsilon}_x = \pm \, [\overline{U} - U_s]/h$$
$$= \pm \, U_s/kh \qquad (8.5)$$

where $\overline{U}$ = mean ice velocity
  U_s = sliding velocity
  k = a constant between 1–3.

The strain rate is obviously compressive in the lower part of the ice mass, across the transition, and extensional in the upper part.

The effects of any roughness possessed by the glacier bed will be superimposed on those of the thermal and **rheological** transition shown in Figure 8.4. From studies in folded rocks it is also

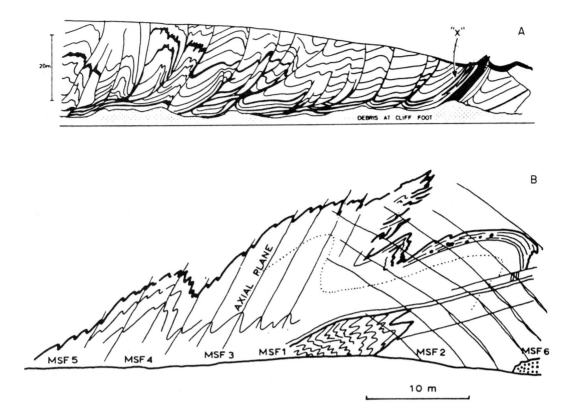

apparent that structural style and geometrical transposition during folding will depend upon the relative competence of the ice matrix and individual debris layers as well as the initial wavelengths of the layers.

Several possible patterns may develop: angular versus rounded folds, thickening of layer **hinges** and thinning on **limbs**, maintenance of a uniform layer thickness or overall layer thickening without major folding. Small, low-amplitude perturbations may be present in debris layers in a glacier where flow is over an irregular bed or characterized by transient behaviour. As the ice undergoes deformation all the irregularities inherent in a real layer may respond, but one wavelength (λ_p) will usually become dominant according to the relationship (Sherwin and Chapple, 1968; Huddleston, 1973):

Figure 8.3 Complex structural features developed in ice during flow. **A** Traces of folds resulting from transverse compression on Borebreen, Svalbard (from Boulton, 1970). **B** Overturned folds in the margin of the Antarctic ice sheet near Mawson Station (from Kitzaki, 1969).

$$\lambda_p = 2\pi \, h_d \left[\frac{\eta_d \, (l_e - 1)}{6 \, \eta_i \, 2l_e^2} \right]^{1/3} \qquad (8.6)$$

where h_d = debris layer thickness
l_e = an elongation term
η_d = viscosity of debris
η_i = viscosity of ice.

Folds whose thickness–wavelength ratio increases become those most amplified.

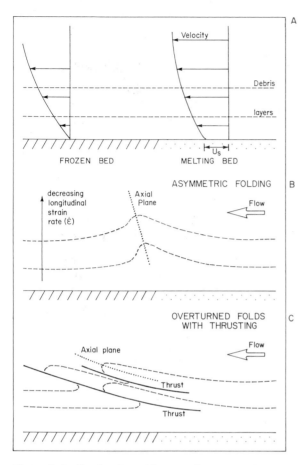

Figure 8.4 Production of large-scale structural features in ice and debris due to changes in bed conditions from melting (motion dominated by sliding) to frozen (motion dominated by internal creep). **A** illustrates the location of debris layers with generalized vertical velocity profiles for melting and frozen beds. **B** shows production of asymmetric folding due to ice deceleration and increased longitudinal compressive strain. Note that the most change in strain between melting and frozen bed is in the lowermost layers. Higher in the ice column the difference in longitudinal strain diminishes. **C** with an increase in the difference in ice flow between sliding and non-sliding fold structures may become overturned with the development of pronounced thrust planes (cf. Figure 8.3).

8.2 Changes in clast size: comminution

Debris transported by ice may be broken down or comminuted resulting in a decrease in size of individual clasts and a related increase in the total number of particles. This is probably one of the most significant and characteristic activities associated with glaciers. The relationships between these two factors will depend upon material properties and the processes involved. It is important to differentiate between two modes of size reduction. **Attrition** is the scratching, rubbing or breaking off of small fragments from a much larger parent clast due to high surface stresses while during *fracturing and crushing* rock particles are broken down into a range of fragments with dimensional ratios similar to, or smaller than, the parent. For glacial sediments it is possible to separate distinctive processes responsible for size change in clasts within the superglacial and basal environments.

8.2.1 Comminution of superglacial sediments

Superglacial sediments (Figure 8.5) are subject to various weathering processes such as thermal and chemical breakdown assisted by water at the glacier surface. The net result is usually the shattering and fracture of rocks, the degree of comminution depending upon the geotechnical properties of the rock.

Cryostatic fracture
Much has been written about the effectiveness of 'freeze–thaw' or 'frost-shattering' activity but few experiments have been undertaken and little adequate theory is available to judge qualitative assumptions. It has been commonly assumed that the wedging action of frost in rocks can be accounted for simply by the expansion of freezing water. The process is more complex, however, and depends in the first instance upon the character and distribution of discontinuities and voids in rock. If the rock is non-porous its susceptibility to breakdown by freezing strains is usually very low (Lautridou and Ozouf, 1982) which explains the scepticism about effective frost wedging of some workers (Selby, 1982).

Figure 8.5 Blocky medial moraine on Bersaerkerbrae, East Greenland.

Connell and Tombs (1971) have shown that the pressure exerted against an obstacle such as the side of a macroscopic fissure in a rock, by the unconfined growth of polycrystalline ice can exceed 20 kPa. More recently Davidson and Nye (1985) have shown from experiments of freezing water in simple slots that an ice plug, which commences to extrude from a real rock crack, might exert pressures in the range 10^1–10^3 MPa. As indicated in Chapter 3 the tensile strength of most rocks lies within the range 10^6–10^7 MPa. It would appear likely, therefore, that **freezing strains** are sufficient to fracture homogeneous rock. If the crack in which ice wedging takes place lies along a line of weakness the bulk tensile strength of the rock may be significantly reduced so that a considerable degree of fissuring may be possible. Besides the pressure exerted by growing

ice some investigators consider that free water in a fissure is forced under a hydrodynamic pressure, due to the growth of an ice lid, into the crack tip where it may cause wedging. The effectiveness of this process is again limited to the maximum stress that can be exerted by the ice. The net result of these processes is the flaking of extremely susceptible patches of a rock surface and will account for a certain proportion of rock comminution.

Pore-water effects

Most rocks contain pore spaces capable of containing water that can freeze. Internal hydraulic pressure is created during freezing. The magnitude of this pressure depends upon a number of factors principally porosity (and pore size distribution), degree of water saturation, freezing rate, permeability and the distance that water must travel to find a significant pressure reduction (Gordon, 1968). If the level of water saturation is such that the pores are filled by exactly 91% water the voids will be exactly filled by ice upon refreezing and no change in pressure will occur (remembering that 1 mm³ of water occupies 1.09 mm³ of space after freezing). This condition may be expressed as:

$$w_c = (n_* / 1.09\rho_d) \qquad (8.7)$$

where w_c = water content
n_* = porosity
ρ_d = dry bulk density.

If water saturation is >91% the rock pore will experience stress and dilate unless the excess water can be expelled during freezing. Given a low permeability the strain resulting from both void dilation and the hydraulic pressure exerted by unfrozen water trapped in the rock may be sufficient to break the rock. Figure 8.6 shows the expansion caused during freezing. Lautridou and Ozouf (1982) consider the rate of freezing is critical as at low rates internal water has a greater chance of migrating into larger intergranular voids or to the free surface of the rock – thus lowering the freezing strain.

The volumetric freezing strain (ϵ_f) is given by rewriting Equation (8.7).

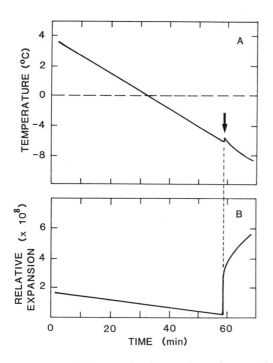

Figure 8.6   Initial expansion in a rock specimen at the commencement of freezing (shown by arrow).
A Regular decrease in temperature of specimen,
B corresponding relative expansion of specimen.

$$\epsilon_f = 1.09 \, w_c \rho_d - n_* \qquad (8.8)$$

Mellor (1973) found that the values of freezing strains measured in the laboratory were usually below ϵ_f by up to an order of magnitude when calculated from Equation (8.8). However, the strains were close to the tensile strength of the rocks involved (sandstone and limestone). They exceeded the strain associated with the initiation of microcracking in the rock during tensile tests (see Chapter 3). Both the limestone and sandstone had high porosities, in the order of 15-20%. For a granite with a porosity of $\sim 1\%$ the volumetric strain was only about 25% of the tensile strength indicating the importance of pores in the freeze–failure process.

Particle size characteristics
The particles released by the frost shattering process depend, in part, upon the frost susceptibility of the rock. Lautridou and Ozouf

(1982) have studied rock debris from numerous experiments. They found that dense, lithographic limestones preferentially release larger particles (10–20 mm). Chalk, on the other hand, was found to yield considerable quantities of fine material.

In their investigation of the fine grades from frost shattering Lautridou found that many siliceous rocks produced few particles $<0.5 \, \mu m$ in size and believes that this represents a minimum size limit for comminution by this mechanism. Certain rocks containing significant proportions of clay produced finer end products. Lautridou also notes that the size limit of fragmentation is considerably smaller than the size reduction limit – the smallest size for a given rock type below which frost shattering does not occur. For limestone the former is 0.2 μm while the latter is 1–5 mm.

The comminution of rocks exposed at the surface of a glacier will take place if the rocks are porous and there is effective uptake of water. This restricts the effectiveness of freeze–thaw reduction to summer ablation areas in temperate and sub-polar glaciers or where solar radiation may locally melt ice or snow around rocks in accumulation zones.

Salt weathering
The crystallization of salts in rock pores also acts as a powerful comminution process for superglacial debris. The process is particularly effective where sea spray can be blown inland, but salts may also be derived from the chemical breakdown of the rock itself. The growth of salt crystals in pores from percolating saline water has a similar effect as freezing of pore water – volumetric strains are created which may crack the rock. The process has been discussed by Cooke and Smalley (1968), Evans (1970) and Goudie et al. (1974).

8.2.2   Comminution in the basal zone

The processes and resulting particle comminution that occur in the superglacial environment are distinctive from those in basal sediment layers. Very large tractive forces at the base of a glacier result in crushing and grinding of sediments. We have seen that the sediments found in basal layers comprise a mixture of particle sizes (Figure 7.7). It is important to recognize that comminution can only take place if particles are either in contact with each other or with the bed. It is unlikely, for instance, that clasts dispersed englacially will become significantly broken since the forces acting on the particles will be small.

Two processes appear to dominate comminution of clasts at the glacier bed – abrasion and crushing (in a similar manner to the erosion of bedrock discussed in Chapters 3 and 4). Phenomenologically, abrasion will give rise, exclusively, to fine-grained material or 'rock flour' with consequent smoothing and striating of clasts. Crushing, on the other hand, will produce a variety of particle sizes. Separation of the two components may be difficult (Haldorsen, 1978). Boulton (1978) suggests that crushed material is usually coarser than 1ϕ (0.5 mm) while abrasion products are predominantly finer than 1ϕ.

It is difficult to make direct observations and measurements of the processes of crushing and abrasion of sediments beneath glaciers – a problem typical of much glacial activity. One method of investigation which has proved useful is to apply the results of comminution and grinding of industrial aggregates in milling machines. Although there are numerous types of comminuting equipment, the closest approach to the glacier environment are ball and autogenous tumbling mills for simulating crushing and abrasion respectively.

Considerable theory and numerous empirical relationships have been developed in the use of such machines for size-reduction of rocks. While the traction zone of glaciers may be significantly different from such engineering practices to make any transfer of results merely qualitative the experience gained from milling experiments, in the absence of measured glacier data, are nevertheless instructive.

Theoretical consideration of grinding
The crushing or abrasion of sediments in basal transport will depend upon factors similar to those important for the crushing and abrasion of bedrock, namely, hardness or strength of rock materials, the nature and magnitude of the imposed forces (i.e. tension, compression, **torsion**, flexure, and shear) as well as the duration of the process. A successful approach to understanding comminution, particularly crushing, is to assess the energy required to break rocks and create new surface areas. Such ideas were first proposed by von Rittinger (1867) and later by Kick (1885) and have received substantiation in numerous carefully controlled experiments (Kelly and Spottiswood, 1982).

Hukki (1961) has proposed a general form of the energy equation proposed by these earlier works:

$$dE_{com} = -k_*' \, d\left(\frac{2r_c}{2r_c^{m'}}\right) \qquad (8.9)$$

where E_{com} = energy required for comminution
k_*' = material constant
$2r_c$ = particle diameter
m' = exponent.

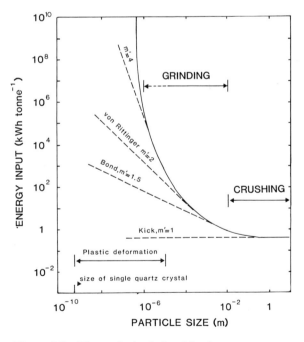

Figure 8.7 Theoretical relationships between energy consumed in comminution and particle size according to various models (as given in text) (from Hukki (1961) after Kelly and Spottiswood, 1982).

Equation (8.9) is shown in Figure 8.7 for different values of the exponent m'.

If the tensile strength of the rock (σ_t) is taken as a representative material property, Equation (8.9) becomes:

$$E_{com} = \sigma_t \Delta A_f \qquad (8.10)$$

where ΔA_f = new surface area created.

Most rocks are brittle materials and such theory fails to take account of the energy of elastic deformation necessary to bring a clast to the point of fracture. Tanaka (1962) has shown that this energy requirement is proportional to the volume of the material but independent of the size of the particles:

$$E_{com} = V (\sigma_t^2/2E) \qquad (8.11)$$

where V = volume of material
  E = Young's modulus of elasticity.

Bond (1952) developed a further relationship for the energy of comminution in which the energy possessed by an assemblage of clasts is based upon Griffith crack theory (see Subsection 3.5.3) — the length of the crack required to break a clast of specified size. Bond finds the energy proportional to the reciprocal of the square root of the sieve or screen size through which 80% of the ground particles pass. In grinding from a size S_{c1} to S_{c2}, the energy is proportional to the difference:

$$E_{com} = 10B_i \left(\frac{1}{\sqrt{S_{c2}}} - \frac{1}{\sqrt{S_{c1}}} \right) \qquad (8.12)$$

where B_i = material constant called the Bond Index
  S_c = grade attained by 80% of the ground particles.

The Bond Index (B_i) in Equation (8.12) provides an empirical measure of how easily clasts of a given material can be crushed or comminuted. Table 8.1 gives typical values for a selection of rocks.

Figure 8.7 shows the theoretical relationships between energy consumed in comminution and particle size, and the results of experimental grinding. In order to comminute coarse material

Table 8.1 Bond Index of grinding (see text for further discussion)

Rock/mineral type	Bond index [B_i]
Basalt	19
Dolomite	13
Granite	11
Limestone	14
Marble	4–12
Quartzite	11
Sandstone	11
Slate	16
Flint	29
Silica sand	16
Fluospar	10

Equation (8.12) suggests that only a small amount of energy is required. Below about 1 mm (0ϕ), however, Bond–Rittinger theory indicates exponentially increasing quantities of energy are needed to further comminute the material. Figure 8.7 further suggests that for a given supply of energy at the glacier sole sediments will not be ground finer than a given mode. This, in engineering terms, is called the 'limit of grindability', defined as the smallest particle size that can be fractured for a given material. For limestone particles, for instance, the limit of grindability is ~3–5 μm and for quartz particles 1 μm (Rumpf, 1973). This goes some way to explain the **'terminal grades'** identified by Dreimanis and Vagners (1969, 1971) in tills. The energy expended in crushing tends to exploit all lines of weakness within particles and clasts. These are made up of microcracks located at crystal boundaries as well as cleavage planes, concentrations of dislocations and inclusions. Many of these are governed by the evolution of the rocks involved (crystallization in igneous rocks, pressure–temperature history and depositional processes for metamorphic and sedimentary rocks respectively) (Slatt and Eyles, 1981).

Rock weaknesses have been placed in two groups termed intergranular and transgranular by Kelly and Spottiswood (1982) and intercrystalline and intracrystalline (Slatt and Eyles, 1981). The relative strength of these two sets of incipient failure planes in clasts will determine whether monomineral or lithic fragments are produced as a result of crushing. At the base of a glacier, Slatt and Eyles (1981) favour, but do not prove, the creation of lithic fragments of varying size and independent of the crystal configuration of the original clast. If this is true, rates of comminution will probably remain constant as the reduced fragments are smaller but compositionally similar to their parent. If, on the other hand, comminution results in smaller, monomineral constituents, grinding rates may well be altered as these new, individual constituents may behave mechanically as entirely different substances to the lithic fragments.

Results of grinding experiments

The rate of grinding and the resulting size distribution of products in milling machines depend in large measure on characteristics of the specific equipment used. For ball mills (Lowrison, 1974) these will include ratio of particle size to ball diameter, ratio of particle to ball density, ratio of particle to ball hardness, ball loading, speed of rotation and volume of material in the mill. Figure 8.8 illustrates the effect of progressive grinding on samples of limestone. Several features of these curves are important. Although there is a progressive fining of the material in both cases, the coarser particles disappear more rapidly than the finer. This indicates that larger particles are subject to preferential comminution due, perhaps, to the rapid exploitation of the microfractures. If this were the case then rate of grinding may be related to the size distribution of the planes of weakness in the rock. A second feature is that different materials possess characteristic grinding rates – related in most cases to their hardness. The changes that take place to the grain size distribution during grinding are also instructive for the same two materials when crushed separately and then together. The latter case more nearly approximates conditions at the glacier bed where several lithologies may be present. The grain size distribution in this case will depend upon the relative proportions of the rock type and their relative grindabilities. In

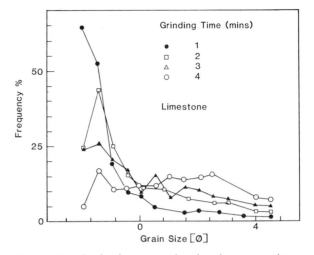

Figure 8.8 Grain-size curves showing the progressive evolution of size fractions with grinding in a rod mill for 4 minutes. Note the manner in which the initially dominant coarse fraction is rapidly reduced and an increasingly fine mode progressively develops (data from Remenyi, 1974).

general softer rock will grind finer than the harder, but both will comminute less rapidly than if ground separately. In the examples of silica and limestone, the rate of comminution of silica, the harder, less grindable material was found to rise as that of limestone decreased showing a transfer of energy to the coarser grades. Typical results for *crushing* suggest that the reduction in particle size tends towards a characteristic size fraction (Figure 8.9). This conclusion accords with the suggestion of Dreimanis and Vagners (1969) from studies of tills in S.E. Canada, that there are typical 'terminal grades' for different lithologies – sizes which appear resistant to further comminution.

Figure 8.9 also demonstrates the characteristic grain size distribution resulting from *abrasion*. For the process of reduction by abrasion autogenous machines are run full to avoid percussive impact effects due to the falling of particles during tumbling. Abrasion takes place at the contacts between grains and clasts as they continually slide over each other. Considerably finer fractions are generated by abrasion than from crushing.

It would appear from Figure 8.9 that the

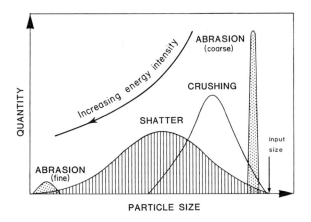
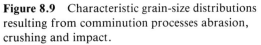

Figure 8.9 Characteristic grain-size distributions resulting from comminution processes abrasion, crushing and impact.

comminution processes operating at the glacier bed, particularly abrasion and crushing, may give rise to distinctive and independent grain size modes. Haldorsen (1981), in a study which included grinding experiments, came to similar conclusions for the interpretation of till samples. She suggests that the bulk of the fine fraction found in tills is the product of clast–clast attrition while the coarser modes represent the products of crushing. If both these processes were operative during transport of basal sediments it might be expected that particle size curves would mimic Figure 8.9 and display a bimodal distribution. Dreimanis and Vagners (1971) show that tills may, in fact, display such bimodal form, and argue that it results from the characteristic changes that are shown to occur in ball milling experiments.

8.2.3 Bimodal or multimodal grain-size distributions?

For sediments constituted from materials of a single or dominant mineralogy, Dreimanis and Vagners (1971) suggest that crushing activity by tractive forces at the glacier bed may give rise to distinctive grain size distributions. There is a characteristic mode representing rock fragments, usually coarser than 0.1 mm, and another for finer mineral particles. As crushing is continued

the mineral mode becomes dominant. Dreimanis and Vagners equate duration of crushing found in ball mill experiments, as illustrated in the previous section, with distance of transport in glaciers. Till grain-size distributions from the Hamilton–Niagara area of southeastern Canada suggest that close to its source till will be dominated by the rock-fragment component but at greater distance down the (palaeo)-flowline the mineral mode becomes more significant (Figure 8.10). Drake (1972) also used tills in New England as proxy data for processes within basal ice and found a significant decrease in size of clasts (i.e. Dreimanis and Vagners's rock fragment proportion) with distance from its source.

Several points from these investigations are worthy of attention. Dreimanis and Vagners (1971) used only the results of ball mill crushing

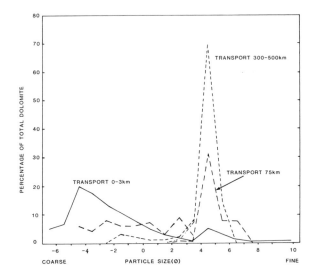

Figure 8.10 Frequency distributions of a dolostone dolomite from three till samples in the Hamilton–Niagara region of SE Canada illustrating the effects of progressive transport and comminution (distance of travel is here transposed for time of grinding) (from Dreimanis and Vagners, 1971). Reprinted by permission from A. Dreimanis and U.J. Vagners, 'Bimodal Distribution of Rock and Mineral Fragments in Basal Tills,' in *Till: A Symposium*, ed. Richard P. Goldthwait; copyright © 1971 by the Ohio State University Press.

to develop their theory. It has been shown that such experiments can only provide part of the explanation of bimodal distributions. The crushing literature indicates that where materials are crushed together and have significantly different grindabilities, the coarser grains of harder materials become surrounded by the finer component. Under such circumstances the bulk of the grinding energy will be transferred by the harder to the softer and finer material – with a *continued* increase in its grinding rate. The net result is shown in Figure 8.11. A bimodal distribution may merely represent material that has not proceeded to the tertiary stage of comminution and whose 'terminal grades' are possibly difficult to recognize.

It is also interesting to inspect the detailed size distributions for basal solid subfacies given by Lawson (1979) for Matanuska Glacier, Alaska. Five grain size curves show pronounced variability with both the rock-fragment and the mineral modes dominant in different analyses (Figure 7.7). Lawson's data are for entrained

sediments in transit and the pronounced separation of grain size characteristics does raise problems for the results of Dreimanis and Vagners which are for deposited till. Although progressive comminution will naturally occur the instantaneous grain size distribution will be complex and depend upon the characteristics of newly entrained (and principally coarse debris), recycled debris of mixed size as well as far-travelled finer fractions. Meltwater flow, especially in Weertman films at the bed, may selectively scavenge fine-grained components

Figure 8.11 Evolution of grain-size curves during progressive comminution in a reducing mill. Note how the coarse-size fractions persist until half-way through the crushing process giving a 'transient' bimodal distribution (after Lowrison, 1974). The inset provides a model for comminution of debris in the basal zone of a glacier. There is rapid initial size reduction followed by a period of slow comminution giving rise to apparent terminal grades. Thereafter there is continued size reduction.

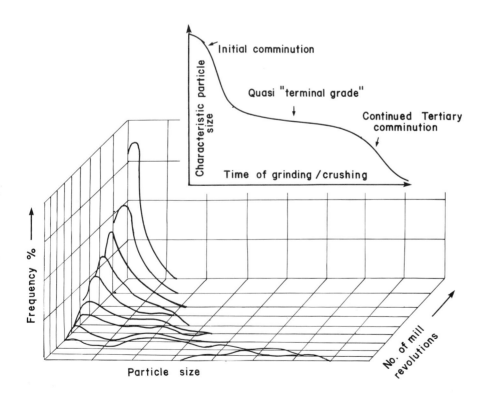

(<0.2 mm). Furthermore, while several distinctive grain-size distributions may be recognized within a glacier or ice sheet, it is unlikely they will be retained during deposition except under extremely favourable and probably rare occurrences (see Section 9.7).

8.3 Changes in clast shape

During transport a number of processes operate to change the shape and surface characteristics of materials in transit; several of these are related to crushing and comminution.

8.3.1 Striation and facetting

From some of the earliest studies of pebbles in tills it has been well established that clasts may become worn, striated and smoothed during basal transit. Holmes (1941) has given details of the relations between the direction of striations on pebbles and clast shape. The same author, from an exhaustive investigation of several thousand pebbles in tills from central New York State, found that 28% of clasts showed evidence of striation. Flint (1971) reports that this proportion is usually no more than 5–10%. A study of some New England tills by Drake (1972) revealed that observed striations varied with clast lithology, being present in 3% of hard but only 0.1% of soft rocks. More recently, Humlum (1981) found 20% of pebbles larger than 25 mm from the forefield of Slettjökull, Iceland exhibited striations.

Striations will be produced only by interference between particles and between particles and the bed. Sediments in dispersed high-level transport in glaciers and ice sheets will remain unscratched. Particle–particle and particle–bed contacts are controlled by debris concentration, relative velocities, and clast/bed properties (e.g. hardness) in a manner similar to that outlined for bedrock abrasion (and the details given in Chapter 4 should be consulted). The variation in the properties of striated clasts from deposited till as reported above may reflect the relative influence of these physical factors in addition to

Figure 8.12   Facetted 'bearing surface' of a large limestone block exposed in the forefield of Mueller Glacier, New Zealand. Note the striations and sichelbrüche.

depositional mixing of several sedimentary facies in transport, only some of which may exhibit marked striation to pebbles.

Many clasts are distinctively facetted during transport, especially where there has been prolonged contact with the bed. Figure 8.12 shows a smooth 'bearing surface' on a limestone block with evidence of abrasional grooving.

8.3.2 Rounding

Many clasts when initially entrained in basal ice possess angular forms as a result of crushing and fracturing of bedrock. Drake (1972) reports that 70% of pebbles in New England tills exhibit a **Krumbein roundness** of 0.1 when incorporated. Significant quantities of rounded clasts, however, may be present if sediments in subglacial meltwater channels are entrained, or if glacifluvial outwash is overridden.

Abrasion during transport will round clasts. This process does not continue until all pebbles are perfectly rounded, however, as periodic crushing will tend to return clasts to angular shapes. Drake's study of some 1 800 pebbles demonstrates that a steady-state may be achieved between rounding and crushing processes during transport: the average clast survives for a time of

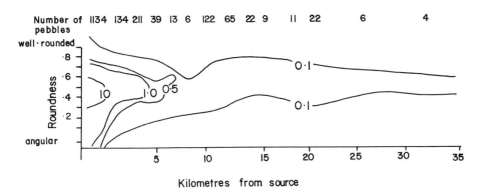

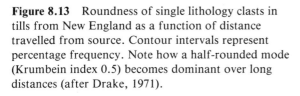

Figure 8.13 Roundness of single lithology clasts in tills from New England as a function of distance travelled from source. Contour intervals represent percentage frequency. Note how a half-rounded mode (Krumbein index 0.5) becomes dominant over long distances (after Drake, 1971).

sufficient length to abrade to a Krumbein roundness of 0.5 before being crushed (Figure 8.13). Drake's results are confirmed by more recent studies of basal debris which demonstrate a dominant semi-rounded character to particles (Lawson, 1979; Dowdeswell, personal communication). This state is achieved rapidly – within the first 2 km of travel and, thereafter, Drake supposes, pebbles go through several cycles of crushing and abrasion, gradually reducing the overall size of sedimentary materials.

8.3.3 Shape

It is likely that certain clast shapes are more susceptible to change – either by crushing or abrasion – so that during transport there may be selective evolution of pebble shape. Drake (1972) found that blade- and rod-shaped clasts (defined by the **Znigg fields**) were most easily destroyed by basal processes while spheres were least affected. After considerable transport a hierarchy of stable forms evolves: spheres, discs, rods and blades. Humlum (1981) reports an identical hierarchy based upon samples of till from the forefield of Slettjökull. Dowdeswell (personal communication) found a similar pattern for clasts

in basal ice of Watts Glacier, Baffin Island with 40–56% pebbles being spherical and 6–12% bladed. Drake also suggests that crushing and abrasion may be approximately equal in their effectiveness.

8.4 Fabric development

Much has been written during the last 100 years on the development and interpretation of fabrics in glacial sediments. Most studies have focused upon measurement and explanation of clast orientation in deposited till. Some workers contend that these provide a clue to the flow direction and flow behaviour of ice sheets and glaciers for palaeoenvironmental reconstruction purposes. Others believe that as significant orientations may be imparted during deposition and by post-depositional processes, the meaning and application of till fabrics is complex and equivocal. In this first section the mechanisms giving rise to observed fabrics in glacial sediments in transit only are examined; fabrics in deposited sediments are discussed in Chapter 9.

8.4.1 Englacial debris fabrics

In order to minimize the energy expended during ice deformation, particles will orient themelves so that the torque on a given clast is reduced to a low value. If a clast is held at a high level in a glacier or ice sheet it may be subject to little applied stress during transport: the vertical dimension over which significant stress gradients act due to ice flow may by very large compared to the size of the particle. No preferred orientation may ensue under such circumstances. Towards the base of an ice mass, however, significant stress regimes develop and particles are subject to shear in the ice. In this position, close to but not necessarily in contact with the bed, important englacial clast fabrics are generated.

Consider a **prolate** pebble whose dimensions in the long, intermediate and short axes are denoted by a_1, b_1 and c_1, and are large compared to the size of randomly oriented ice crystals (Figure 8.14). The clast is subject to the principal stresses σ_1 and σ_2, transmitted from the ice by a large number of crystal contacts with the surface of the pebble. σ_1 and σ_2 make an angle of contact with the mean slope of the bed of $\phi_1{}^*$ and $\phi_2{}^*$. If the ice is in pure shear $\phi_1{}^* = 45°$ and $\phi_2{}^* = 135°$. The torque or turning moment on the clast in response to σ_1 and σ_2 is, according to Holdsworth (1974a):

$$\Omega_t = b_1{}^2 a_1{}^2 (\sigma_2 \cos \Gamma + \sigma_1 \sin \Gamma) \qquad (8.13)$$

where $\Gamma = (\phi_1{}^* - \theta_x)$
       θ_x = angle of inclination of clast to x-direction.

The maximum moment is $\dfrac{d\Omega_t}{d\Gamma} = 0$

The results of Equation (8.13) are given in Figure 8.14 and indicate that a clast will rotate anticlockwise if $\phi_1{}^* > \theta_x$ and clockwise if $\phi_1{}^* < \theta_x$ in order to minimize the movement. Alignment will be favoured in the direction of the principal stress σ_1. Achievement of such a position is dependent upon clast shape. Symmetrical pebbles will do so easily but other clasts achieve equilibrium with some angle between the long axis (a_1) and direction of σ_1 (i.e. there is a minimum cross-sectional area normal to the direction of transport in the plane of shear).

Within a glacier or ice sheet the principal stress directions will change depending upon compressive or tensile regimes. Such changes will be reflected in the foliation and fabric development in the ice – in turn transmitted to

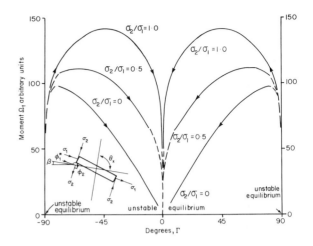

Figure 8.14 Rotation of a clast under the effects of two principal stresses (σ_1 and σ_2). The figure plots the moment (Ω_t) on the clast against Γ, the difference between the angle of the principal stress (referred to the x-direction) and θ_x, the angle of inclination of the clast to the x-direction (after Holdsworth, 1974a).

the clast. With a single principal shear component clasts will tend to align into the plane of shear and paralled to the ice flow. If two principal shear stresses are present a stable orientation may not be achieved, the clast changing alignment as one or other becomes dominant during transport.

Few measurements of englacial fabrics have been made. Some observation from ice sheet margins near Thule, Greenland, Meserve Glacier, Antarctica and Casement Glacier, Alaska have been summarized by Lindsay (1970), while more recently orientations have been reported from Svalbard (Boulton, 1970), Matanuska Glacier, Alaska (Lawson, 1979) and Slettjökull, Iceland (Humlum, 1981). A selection of englacial fabrics is presented in Figure 8.15 which are equal-area projections, contoured using Kamb's (1959) method. Important elements of these diagrams are diversity of orientations with both parallel and transverse directional maxima, although long axes show a consistent upglacier dip of several degrees. Lindsay (1970) found that the most frequently occurring orientation was parallel to a

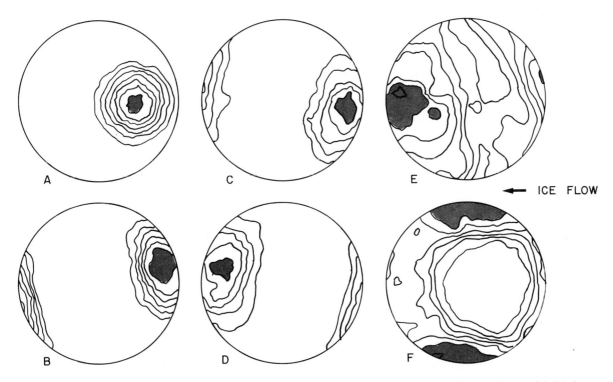

← ICE FLOW

Figure 8.15 A selection of typical englacial debris fabrics (Schmidt equal area projections). Contours are at 2 σ intervals, dark areas show areas of highest concentrations. **A–C** Matanuska Glacier, Alaska (from Lawson, 1979) **D–E** Clast long-axis (a_1) Casement Glacier, Alaska (after Lindsay, 1970) **F** Clast long-axis (a_1), Greenland ice sheet (after Lindsay, 1970).

single direction of shear, confirmed by subsequent studies. A secondary, less strongly developed mode is found when there is a secondary shear direction and here fabrics may develop with a steep downglacier dip.

8.4.2 Basal debris fabrics

Clasts held within the basal part of glaciers and ice sheets may come into contact with the bed. At the interface the stress regime changes and particles are subject to reorientation by being dragged or rolled. Englehardt *et al.* (1978) observed such motions in boreholes at the base of the Blue Glacier. A cobble that had remained in one overall position for three days had rotated gradually due to the drag of the bed beneath. For the purposes of elucidating orientation mechanisms it is assumed that clasts are neither broken, nor undergo significant shape changes during the period of fabric generation.

Consider an ellipsoidal clast resting upon a perfectly smooth bed. There is solid friction between clast and substratum (see Chapter 4).

The particle is subject to forces, proportional to the projected area, exerted by ice flow. The total moment on the clast about its pivotal point is:

$$\Omega_b = I\,(d^2\theta^* / dt^2) \qquad (8.14)$$

where $\quad$ I = moment of inertia
$d^2\theta^*/dt^2$ = angular acceleration (proportional to the ice velocity and the initial angle between clast long axis and flow direction).

If the force imposed by the ice is sufficient to overcome clast–bed friction the particle will also roll and slide.

In reality the glacier bed or sliding surface is irregular and ice flow unsteady so that a particle in traction will be subject to constant reorientation. It is also likely that vertical velocity gradients in basal ice layers will subject a clast to angular accelerations which tend to impart a dip to the particle in the upglacier direction.

Clasts in transit in basal ice and in contact with the bed will be expected to display orientations imposed by local ice flow. Indeed striations observed on clasts appear to indicate that many sliding particles are oriented more or less parallel to the local ice motion. Nevertheless, due to bed irregularites these orientations may diverge from the regional pattern of ice flow.

It is clear that orientation of particles in contact with the bed is highly complex and any simple model must take into account particle shape, relative velocity between clast and ice, bed roughness and sediment concentration which will influence particle–particle interactions. Drake (1974), for example, found that particle shape was an important factor in fabric development. Rod- and blade-shaped clasts were well aligned paralled to ice flow, spheres and discs possessed a much weaker alignment.

9 Process of till deposition and resedimentation

9.1 Introduction

Chapter 8 examined sediments in transport in glaciers and ice sheets: an environment where debris is dispersed within, yet confined by, a slowly deforming, viscous medium. Ultimately these materials are released from the ice during the process we understand as deposition. Their characteristics become dramatically modified and frequently experience episodic recycling. Deposition may take place in a variety of settings such as lakes, estuaries or the open ocean, creating what may appear initially to be a confusing ensemble of sedimentological process–facies relationships, and made more bewildering by a substantial body of literature. Nevertheless, the depositional spectrum may be structured and subsequently viewed through key environmental conditions. A starting point for the reader, no less for much glacial sediment, is the terrestrial deposition of till (Figure 9.1).

Much has been written about till genesis and till properties as the result of scientific and engineering studies in North America and Northern and Western Europe. It is not intended here to debate the niceties of **stratigraphic** interpretations of till

Figure 9.1 **A** Complex depositional processes at the terminus of surging Roslin Gletscher (behind the camera) in East Greenland. Note pronounced till ridges and confused topography resulting from decaying ice lobes, the extensive areas occupied by lakes due to inhibited surface drainage and the association of fluvial activity (braided streams) surrounding the zone of till deposition. **B** A typical till section showing a lower massive till deposit separated from an upper till unit by finer-grained glaci-fluvial sediments. Some deformation in the upper till horizon is apparent. Lake Tekapo, South Island, New Zealand.

units nor to explore their use in the development of glacial chronology. The focus of the chapter is exclusively the processes of deposition and remobilization of sediments, and resulting characteristics. It is important to make some comments on terminology. 'Till' has traditionally been used to describe poorly sorted sediments deposited directly by glacier ice. In the glacial environment, however, such sediments are commonly subject to mobilization and reworking by agents such as water and experience gravity flow. The genetic term 'till', therefore, is no longer applicable and new nomenclature is necessary. Till deposits exhibiting the effects of resedimentation are often referred to as **'flow-till'**, 'water-lain till', 'aqueous till', etc. Eyles *et al.* (1983) and Miall (1983) suggest non-genetic terminology with the use of 'diamict' applied to these glacial deposits following an earlier, though somewhat desultory usage (see Flint, 1971). **Diamict** may be defined simply as a poorly sorted clastic aggregate but without reference to genesis. Till would thus be a glacial diamict.

9.2 Deposition of basal sediments: lodgement

Two primary mechanisms are recognized for the deposition of sediments in transit in the lowermost layers of a glacier or ice sheet: lodgement and **melt-out**. Much of our knowledge of these processes has come from the exhaustive work of Boulton.

Particles are lodged at the base of an ice mass when the force imparted by ice flow is insufficient to maintain their forward motion. In Chapter 4 clast velocity was defined as (Equation 4.15):

$$U_p = . U_b - U_r \qquad (9.1)$$

where U_b = basal ice velocity = sliding velocity (U_s).

U_r = relative decrease in particle velocity due to frictional drag.

In the most simple terms the conditions for lodgement are: $U_p = 0$, $U_b = U_r$, or $U_b/U_r = 1$. For any value of the ice velocity an increase in frictional resistance may lodge particles, and for a given frictional force at the bed a reduction in

U_b may cause a moving particle to be arrested. To develop these ideas it is necessary to examine factors controlling clast velocity, which were discussed in Chapter 4.

9.2.1 Friction force

The force inhibiting clast motion is made up of two terms, adhesion (a_d) and ploughing (P):

$$\begin{aligned} F_d &= a_d + P \\ &= A_r \tau_{ci} + (2d^2/4) \cot \theta_a \, \sigma_y \qquad (9.2) \end{aligned}$$

where A_r = real area of contact between clast and bed

d = radius of indentation

θ_a = half-angle of conical asperity tip

σ_y = yield strength of substratum

τ_{ci} = shear strength of clast/bed interface.

For a clast moving over hard, smooth substrata the adhesional term will dominate, but where the glacier bed is composed of soft, deformable materials the ploughing component will be more significant.

For a specified set of clast and substratum material properties (i.e. τ_{ci} and σ_y) frictional resistance is dependent upon A_r and to a lesser extent upon θ_a. It has already been shown that A_r is proportional to the normal load (Subsection 4.4.1) so that a clast moving at the bed will be retarded if the load it exerts on the bed is increased by a critical amount. In the hypothesis enunciated by Boulton (1975) the normal load is viewed as the product of clast weight *and* overlying effective cryostatic pressure. Changes in this force are considered by Boulton as the primary factor in lodgement, ice velocity (U_b) taking a secondary role. A_r is also affected by overall clast size and shape. Boulton (1975) considers that particle shape is, in large measure, governed by lithological characteristics and small-scale structures in rock such as joints and foliation. Plate-shaped particles, for example, are more likely to lodge than those of spheroidal form.

Hallet (1979, 1981) regards the normal load acting upon basal clasts as made up of particle weight plus that component of ice flow directed against the bed (U_n), but independent of cryostatic pressure. In consequence, particles lodge either as U_n rises in magnitude or as U_b diminishes to a value where basal ice can be readily melted by geothermal heating.

9.2.2 Particle velocity: variable lodgement

Equations (9.1) and (9.2) indicate that, in general terms, clasts will be moving relative to the ice. An extension of this notion envisages relative motion between different sized clasts. Such behaviour leads to size-sorting of debris in basal traction and preferential lodgement. Boulton (1975) has elaborated this concept in detail.

Above the bed, in the dispersed facies, clasts are essentially in suspension in the ice and their velocity is the same as the basal ice velocity ($U_p = U_b$). When a clast impinges on the glacier bed its forward motion is retarded by an amount proportional to the frictional force between bed and particle. As outlined in Subsection 4.4.3, Boulton adapted Weertman's sliding theory to calculate clast velocity, and deduced that for small particles the same conditions might apply as for small bed obstacles where regelation will be favoured, and the term U_r in Equation (9.1) is dominant. For large clasts the result is the same as for large bed protuberances which favour enhanced creep, and U_r will again be high. By this argument, Boulton suggests that clasts both larger and smaller than a critical and intermediate size (between -3ϕ and 7ϕ) will lodge more readily. According to Boulton lodgement takes place when:

$$[2r_c \{BP_n' \mu_k (A_a)\} + \{P_n' \mu_k/A_a\} (\psi K_p/L)\}]/U_i \geqslant 1 \qquad (9.3)$$

where r_c = clast radius
      B = thermally activated ice hardness parameter in the flow law of ice
      P_n' = effective normal stress
      μ_k = coefficient of dynamic friction
      A_a = apparent area of clast
      ψ = coefficient related to pressure melting
      K_p = thermal conductivity of clast
      L = latent heat of fusion.

Boulton has simplified his analysis to yield a 'critical lodgement' parameter (L_{crit}), applicable to tills exhibiting a wide range in component sediment size and shape. Lodgement takes place when:

$$P_n'/U_i^m \geqslant L_{crit} \qquad (9.4)$$

where m = exponent dependent upon ice flow characteristics (~ 0.3).

Boulton's hypothesis that a certain fraction of the basal sediment assemblage will lodge preferentially and that size sorting by this mechanism takes place has been criticized by Hallet (1981). Hallet argues that the force moving particles in basal ice into contact with the bed, and the effective contact force resisting motion are both controlled by viscous drag and hence proportional to clast size. The ratio, therefore, of driving to resisting forces is independent of particle dimensions: all clasts should move at the same rate, no size sorting should be anticipated and preferential, non-random lodgement of certain fractions should not occur. According to Hallet, particles lodge when the sliding velocity drops and basal ice melts around clasts as the effects of geothermal heat become dominant.

9.2.3 Particle interference effects

Hallet's theoretical objections to differential motion between clasts of different size relate to isolated particles in similar contact with a smooth, homogeneous bed. Given that lodgement of some clasts does take place there will still be more particles moving in 'suspension' in the ice above the bed (Figure 9.2). Collisions will inevitably take place between particles moving at higher levels and those already at rest on the bed, and relative retardation and preferential lodgement must occur. Boulton (1975) and Boulton and Paul (1976) suggest that this process will give rise to clustering of clasts, with concentration of particles 40–50% by volume, and possessing dimensions of a few metres to several hundred metres.

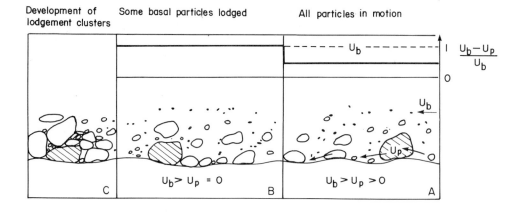

Figure 9.2   Motion and lodgement of entrained particles in the basal part of an ice mass. Ice flow is from right to left. Upper right graph shows normalized basal ice velocity (U_b) relative to particle velocity (U_p) for the shaded clast. When $U_p = U_b$ the scale is set to 0, when $U_p = 0$ the scale reads 1.0). **A** All particles are moving over the bed, although at different velocities according to size. **B** Conditions now favour lodgement of a certain size of particles (shaded clast). **C** Retardation of large clasts can cause the development of lodgement clusters.

9.3 Lee-side cavity deposition

Peterson (1970) described the expulsion of debris-laden basal ice layers into subglacial cavities on the lee-side of bed protuberances beneath Casement Glacier, Alaska. During flow over irregular bedrock the basal debris layer is thickened in response to lateral and vertical dispersive processes described in Chapter 8. As the basal ice moves over the cavity the support previously provided by the hummock disappears and basal debris is detached from the overlying and cleaner ice, slowly accumulating on the cavity floor (Figure 9.3). Peterson used the term **'till curls'** for certain such cavity deposits. More recently Boulton and Paul (1976) and Boulton (1982) have extended the discussion of similar phenomena which are illustrated in figure 9.3.

9.4 Some characteristics of lodgement and cavity tills

Diamicts, deposited at the glacier bed by lodgement and in cavities, are subject to continued ice flow which will shape and streamline the sediment body. Under typical sub-glacial conditions ice velocity, normal and shear stresses and the presence and flow of water will tend to vary with time, on seasonal, annual or longer timescales. Such fluctuations may either enhance or reduce the deposition rate or even suppress lodgement and the presence of cavities altogether. Under such conditions till will be eroded. It is thus possible to map out in time–space coordinates the sedimentation/erosion pattern at a glacier bed.

Boulton (1982) has discussed streamlined bedforms generated by lodgement around a nucleating centre and also by deformation of sub-glacial till. In the former case there is upglacier migration of the depositional hummock in response to velocity and normal stress (as evinced from Boulton theory, above). Migration occurs rapidly if ice velocities are high and effective cryostatic pressure low (Figure 9.4). If movement is also induced in the glacier sediment bed Boulton (1982) envisages a suite of till bedforms termed mobile, static and residual. The mobile features result from the accumulation of sediment around slowly deforming till where the strength of the sediment is an important factor. Boulton's static bedforms arise where there are

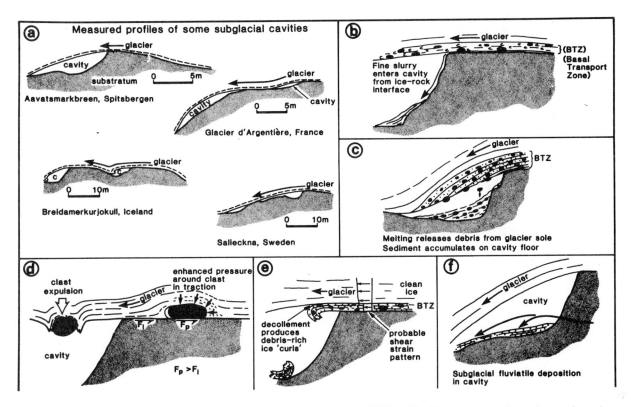

Figure 9.3 Modes of till lodgement in subglacial cavities (from Boulton, 1982).

irregularities (e.g. bedrock protuberance) in the plane of **decollement** (that is, the surface plane of decoupling between mobile sediments and 'true' bed); sediment will deform over the surface giving rise to stationary or static but asymmetric features lying above the irregularity. According to Boulton's theory the formation of residual bed features relies on differences in the strength of subglacial till which gives rise to discrete and resistant cores of sediment around which more deformable material is removed. Boulton (1982) has modelled 2-D bed forms of specified geometry and shows that the dominant bedform is likely to display a steep stoss side.

The composition, structure, grain-size and geotechnical properties of lodged till have been described extensively. It is worth noting that till will experience post-depositional disturbance which progressively alters its physical characteristics and reduces the applicability of the term 'till' itself. Good reviews have been given by Dreimanis (1976) and Boulton and Paul (1978).

Bulk grain-size characteristics of lodged glacial diamicts will display similarities with those of the sediments in transit in basal ice (see Chapters 7 and 8). Some clustering or nucleation of clasts may occur, however, with the creation of boulder or clast lags (Boulton, 1976). Meltwater flowing at the surface of and within the till horizon may remove finer fractions giving rise to some preferential sorting.

Lodgement tills are often subject to consolidation but the pattern is complex due to the time-variable effects of pore-water pressures in the subglacial till horizon. These aspects have been discussed by Boulton and Paul (1976) who suggest that till sequences can be subject to considerable remolding by changes to intra-till water pressures and subsequent deformation by basal ice.

Numerous structures are developed in response to consolidation and deformation. Small-scale, low-angle shear planes are often present alongside high-angle transverse joints. Massive till units may

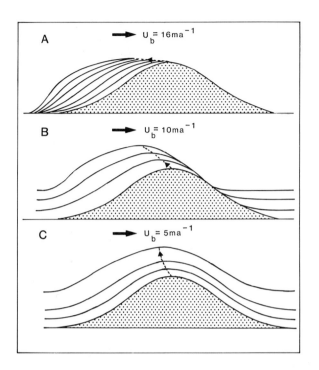

Figure 9.4   Accretion of lodgement till to a bed hummock (with amplitude/wavelength ratio of 0.25) for different basal ice velocities under a constant effective pressure of 2.3 MPa. Crest line migration is shown by the dashed line and is less pronounced at low velocities (from Boulton, 1982).

display pronounced plateyness or fissility, splitting into shear lenses (often along planes picked out by lithological changes such as bands of silt or coarse sand). Krüger (1976) has suggested that some of the above features constitute criteria that may be found useful to distinguish lodgement till from other glacial diamicts. His five diagnostic elements are: small lenses of sorted material, smudges, small-scale deformation of till matrix and smudge clasts, consistent striation of clasts and clasts possessing stoss- and lee-sides.

9.5 Basal melt-out

Sediments may be deposited by the *in situ* melting of debris-rich basal ice. Melting occurs if the heat generated by basal sliding (or from strain heating where there is a pronounced basal velocity shear)

and that contributed by geothermal heat is not fully conducted through the overlying ice. Deposition is thus proportional to the vertical temperature gradient within the ice sheet or glacier (dT/dh). Nobles and Weertman (1971) have examined the process in some detail and establish the initial inequality for basal melting (cf. Equation 1.27):

$$\Lambda_g + \Lambda_s > K_i (dT/dh)_i \tag{9.5}$$

where Λ_g = geothermal heat
        Λ_s = heat from sliding motion

If the basal ice layers have a debris concentration by volume, C_o, the sediment deposition rate is given by:

$$\dot{S}_d = [C_o/1 - C_o] [\Lambda_g + \Lambda_s - K_i (dT/dh)_i/L] \tag{9.6}$$

Values of Λ_g and Λ_s have been discussed in Chapter 1 (Equations 1.23 and 1.25). Their ice-equivalent melt per year can vary from a few μm to cm. Nobles and Weertman (1971) consider several other factors that influence basal sediment melt-out rates. The role of topography and ice thickness variations are shown in Figure 9.5. It is well known from heat flow studies that in an isotropic medium, regional heat flow is normal to the isotherms. Heat flowlines thus diverge beneath hills and converge beneath valleys. Where there are local irregularities in a glacier or ice sheet bed, $\dot{S}_d$ will, by this mechanism, be greater in the hollows than over hummocks.

In addition, major changes in ice thickness alter the temperature gradient, assuming that the overall boundary conditions remain the same – such as might be experienced over a few kilometres of an ice sheet or glacier. Where ice is thin (Figure 9.5) the temperature gradient is steeper and is more effective in draining away geothermal heat and that produced by sliding. The result is the amount of heat remaining to melt debris from basal ice is considerably reduced. In areas of thick ice, the temperature gradient being less steep, more heat is available to melt basal ice and $\dot{S}_d$ is greater, possibly by up to 20%. These variations are illustrated in Figure 9.5.

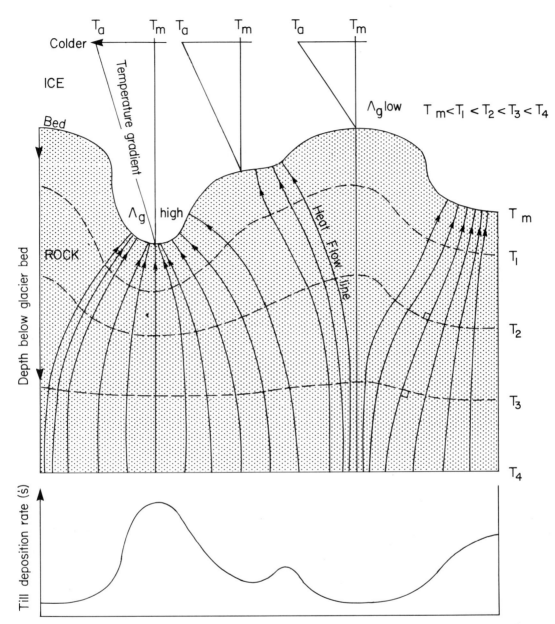

Figure 9.5 Processes resulting in melt-out deposition of till. The figure shows an irregular bed of a glacier or ice sheet. Temperature isolines in the subglacial rock are designated, in increasing magnitude, $T_1 - T_4$. Note how with depth the isolines damp out the influence of the irregular rock surface. The ice–rock interface is at the pressure melting point (T_m). Heat from the earth flows approximately perpendicular to the isolines. Where there are bedrock peaks the isolines diverge giving rise to a lower surface heat flux than in valleys where the heat flowlines converge.

The temperature at the glacier surface is T_a and the base is at the T_m. At the three locations shown, however, due to depth effects, the gradients will be significantly different. Both these factors give rise to enhanced meltout conditions in the valleys. A graph of till deposition rate is shown beneath.

Table 9.1 Characteristics of melt-out and lodgement till (after Lawson, 1979)

Process	Deposit	Texure type 1) mean (ϕ) 2) $\sigma(\phi)$	General features Clasts	Internal structure	Clast fabric
Melting of buried ice	Melt-out till	Gravel 1) 1 to 6 sand 2) 1 8 to silt; 3.5 silty sand; sandy silt;	Clasts randomly dispersed in matrix.	Massive; may preserve ice strata.	Strong; unimodal parallel to local ice flow; low angle of dip; $S_1 > 0.75$
Lodgement at glacier sole	Lodgement till	Gravel-sand-silt; silty sand	Clasts randomly dispersed to clustered in matrix.	Massive; shear foliation, other 'tectonic' features.	Strong; unimodal(?) pattern; orientation influenced by ice flow and substrate; low angle of dip.

9.5.1 Characteristics of basal melt-out tills

Sediments deposited by basal melt will retain their integrity so long as they are not subject to major disturbance. The most favourable conditions for the preservation of basal melt-out till are those beneath stagnant or relatively inactive ice, and where subsequent resedimentation is minimal. Nevertheless, minor changes will occur such as gradual consolidation under cryostatic pressure. The degree of compaction is related to water drainage in the till as discussed in detail by Boulton and Paul (1976). Consolidation gives rise to well-developed jointing and sometimes to low-angle shear planes. Lawson (1979) indicates that mixing of the sediments is secondary with some loss of structural features and strata boundaries originally present in the ice, but with preservation of bulk texture and fabric. The principal characteristics of melt-out till are summarized in Table 9.1. Figure 9.6 shows a section through basal ice and melt-out till close to the terminus of the Matanuska Glacier, Alaska. In general, structural, fabric and textural components of the melt-out till will be almost identical to those of the parent, debris-charged ice. Preservation of features will be particularly pronounced where there is little ice motion and slow melt rates.

Significant changes relate to the loss of some structural integrity and compaction of the sediments either by particle settling or overburden stress under **drained conditions**.

Lawson (1981b) has given a summary of the principal effects of melt-out on sedimentary properties of till (Table 9.1). Structureless pebbly silts and sandy silts originate by the slow even melting of relatively uniformly dispersed debris in the ice. Textural characteristics are retained, however, but with a marked increase in packing density, as smaller particles migrate into voids left as the ice melts. Where there are greater concentrations of sediment in the ice, melt-out gives rise to discrete and slightly deformed layers and lenses. Crude or weak stratification may

Figure 9.6 Characteristics of basal melt-out till beneath the Matanuska Glacier, Alaska. To the left is shown a layer of debris-rich ice (stratified facies) overlying melt-out till. Two fabric diagrams are shown for debris within the ice and for the till. To the right is an idealized melt-out till section depicting:
A structureless pebbly–sandy silt, B discontinuous laminae, stratified lenses of texturally distinctive sediments in massive pebbly silt, C bands and layers of texturally-, compositionally-, or colour-contrasted sediment. Layers and laminae appear to be draped over clasts (after Lawson, 1979, 1981).

Surface forms	Contacts– basal surface features	Pene– contemporaneous deformation	Geometry– maximum dimensions	Miscellaneous properties
Similar to ice surface; may be deformed.	Upper sharp, may be transitional; sub-ice probably sharp.	Possible; observable if structured sediments present.	Sheet to discontinuous sheet; km² to m² in area, m thick.	Internal contacts of strata are diffuse; loose.
Similar to base of ice.	Image of substrate.	Possible subglacial.	Discontinuous pockets or sheets of variable thickness and extent.	Usually dense, compact.

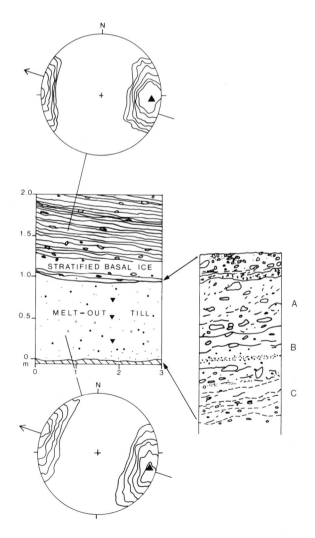

result from the melting of alternating ice-rich and debris-rich layers. Draping of laminae occurs over irregular zones of sediments or individual clasts. Fabrics will usually show only minor modifications during melt-out such as a small increase in dispersion (i.e. scatter of clast orientation) and a reduction in dip angle.

9.6 Autochthonous deposition of surface melt-out sediments

Englacial debris arrives at the surface of a glacier in areas of net ablation, where flowlines are directed upwards. In some cases tectonic structures within an ice mass may also bring sediments to the surface. Sediment melts out, *in situ*, either from scattered solids in the ice or from distinctive bands, and will accumulate on the surface forming a variety of features depending upon the configuration of the debris-charged zone and as described by Boulton and Paul (1976) and Boulton and Eyles (1979). Such sequences may be termed *autochthonous* since they result *in situ* and experience little disturbance during their initial deposition. Surface melt-out till is also favoured within stagnant ice and is best developed where little water is released during ablation – that is, say, where sublimation constitutes the primary ice loss mechanism such as at high altitudes. Under such conditions thick surface melt-out debris horizons may be produced (Figure 7.14). Continued ablation leads to the production of a layer or discrete zone of surface debris which will protect the underlying ice from ablation, thus slowing the melt-release process. These relationships are shown in Figure 9.7 and have been reported on a number of occasions (Østrem, 1959; Loomis, 1970; Drewry, 1972; Nakao and Young, 1981). Assuming that the temperature gradient through the upper till layer is linear, that all heat from the debris surface is conducted downwards to the ice, and that typical thermal conductivities are constant, the ablation rate for the buried ice ($\dot{A}_{bb}'$) is:

$$\dot{A}_{bb}' = [K_d T_o / h_d L \rho_i^*] \qquad (9.7)$$

where K_d = thermal conductivity of the debris layer

h_d = debris layer thickness

ρ_i^* = density of ice with dispersed debris

L = latent heat of fusion of ice

T_o = temperature of debris surface

Texture of melted-out material may tend to be coarser than that in the ice matrix immediately beneath, sometimes attributable to the effects of a pronounced downglacier dip of the ice foliation superimposed on a particle size decrease down-glacier. Some loss of structures present in the ice will occur. Fabrics, however, tend to retain direction but lose dip due to settlement as the effects of foliation in the ice are obliterated by melting. This pattern should be compared with Figure 9.6.

9.7 Flow till: Resedimentation processes

The processes of melt-out at both the lower and upper surfaces of a glacier, by definition, liberate water. If the drainage conditions within a developing sedimentary till sequence are good the water will have only a minor effect on structure, texture and orientation of the deposit. **Fines** will be washed into the voids increasing

particle–particle contact, boundaries between strata will become somewhat blurred and while overall clast orientations may be retained there is likely to be some increase in dispersion or scatter about the mean directional axis. If, however, there is no significant escape of water the physical attributes of the sediments may be fundamentally affected and again the term 'till' is no longer adequate. In the worst case, water will so saturate the sediment as to allow it to flow under its own weight down shallow slopes (i.e. **liquifaction, fluidization** and gravity flow) (Figure 9.8). In these circumstances, the diamict takes on entirely new characteristics in which the grain-size distribution becomes dominated by sorting and settling processes during flow, fabrics exhibit modes related to flow direction and particle–particle interactions; overall structure and morphology are related to bulk density, flow velocity and distance travelled. The flow process effectively destroys the sedimentological properties of the original till material and thus the new deposit must be given another name such as 'flow till' (Hartshorn, 1958; Boulton, 1970, 1971) or 'glaci-sediment flow'. Lawson (1979) notes that in the case of Matanuska Glacier, Alaska 95% of the sediments in the snout area are the result of resedimentation emphasizing the importance of the processes. The character of glaci-sediment flows is principally dependent upon water content and induced flow behaviour and gives rise to a process–facies continuum.

9.7.1 Flow mobilization

Sediments resting upon a glacier surface will commence to flow when applied forces exceed the internal strength of the materials. For a static sediment packet with a given water content, the stress acting on any element will be composed of the total stress, σ, less the effects of pore water pressure (p_w). Any pressurized water in the voids forces the particles apart and thus reduces the frictional strength of the debris. The effective stress is thus:

$$\sigma' = \sigma - p_w \qquad (9.8)$$

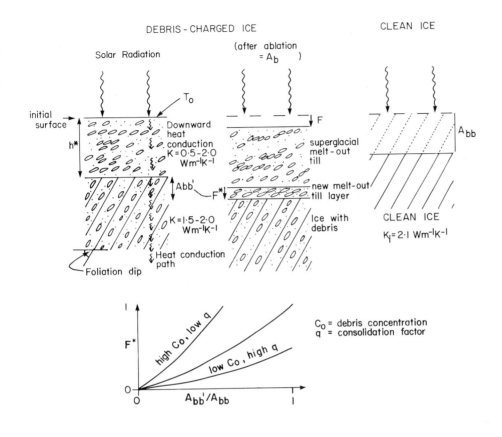

Figure 9.7   Processes involved in the melt-release of sediments from dirty ice at the surface of a glacier. Over clean ice melting results from incoming solar radiation. Over a period the surface is lowered by an amount A_{bb}, the clean-ice ablation rate.

Over debris-charged ice there is a surface layer of a previously melted-out till of thickness h^*. Heat is transferred downwards towards the dirty ice proportionally to the thermal conductivity of the upper debris layer ($K_d = 0.5$–2.0 W m^{-1} K^{-1}) and results in melting by an amount A_{bb}' (which is $< A_{bb}$). After ablation the ice surface is lowered by $F^* < A_{bb}'$, due to the presence and compaction of debris melted out and now added to the superglacial till horizon. If there is pronounced foliation and dip to the ice note how orientations of clasts may change during melt-out. The lower figure indicates the thickness of the melt-out layer F^* as a function of differential ablation for various values of debris concentration in ice (C_o) and till consolidation (q).

The strength of a cohesionless sediment assemblage (i.e. composed principally of bulky clasts) is due to friction between particles and their degree of interlocking. In cohesive sediments, where there is a significant clay component, account has also to be taken of relative compressibility, low permeability and electrochemical forces between particles. Thus the strength of wet sediment to shearing is:

$$\tau_s = c + \sigma' \tan \phi \qquad (9.9)$$

where c = a cohesion term
   ϕ = angle of internal friction of solids
   σ' = effective stress.

Equation (9.9) may be understood by reference to the Mohr–Coulomb failure criterion discussed in Chapter 3. ϕ is thus the slope of the Mohr envelope and c is its intercept on the shear stress axis. For cohesionless debris $c = 0$. Flow or failure of debris on the glacier surface will occur

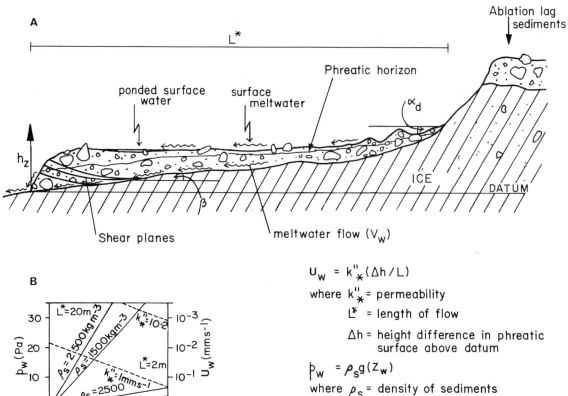

$$U_w = k''_*(\Delta h/L)$$

where k''_* = permeability

L^* = length of flow

Δh = height difference in phreatic surface above datum

$$p_w = \rho_s g(Z_w)$$

where ρ_s = density of sediments

Z_w = depth below phreatic horizon

when the shear strength of the sediment–water mix (slurry) is exceeded by the shear stress (driving moment) acting on the material; that is when:

$$\tau_s/\sigma' \sin \alpha_d \leqslant 1 \qquad (9.10)$$

where α_d = surface slope of the debris.

Examining each part of this safety factor expression (Equation 9.10) flow will occur if the *shear strength is reduced* beyond the critical amount. This situation is most commonly achieved by increasing pore water pressure due to seepage within the debris. If the water in the sediment is not moving its pressure is given by:

$$p_w = \rho_s g Z_w \qquad (9.11)$$

where ρ_s = density of the flow till

Z_w = depth below the **phreatic** surface (water table) – see Figure 9.8.

Figure 9.8   Characteristics of superglacial flow till processes.

A An ablation lag, saturated with meltwater, flows away over the ice surface of gradient β. The flow surface has a slope α_d. Towards the frontal lobe shear planes may be present. Melting continues and water saturates the flow. Water movement may occur as runoff at the surfaces or by throughflow; some is ponded. A phreatic horizon may be defined (the level of zero or atmospheric pressure in the water filling voids in the flow) h_z is the height above datum.

B Throughflow water velocity, U_w, is given for various values of bedslope (β) and flow permeability (k''_*). Water pressure in the flow (p_w) is also graphed for two lengths of flow (L^* = 2 and 20 m) and for two values of mean sediment density (P_s = 1 500 and 2 500 kg m^{-3}).

In most cases, however, water will be in motion through the sediment and the pressure is:

$$p_w = \rho_s g h_p \tag{9.12}$$

where h_p = pressure head
     $= (h_t - h_z)$
          where h_t = elevation of total head
               h_z = elevation above some arbitrary datum.

Failure and flow also occur when the *driving moment is increased* beyond a critical amount. This can arise from the introduction of debris to the sediment layer by the slumping of material from adjacent ice-cored areas – **allochthonous** addition, or from the continued melting of debris-charged ice beneath the sediment layer – autochthonous addition, thickening the flow proper by lowering its effective base (see Section 9.6). A further effect is produced by change to the angle of the slope upon which the debris lies. When the debris is at rest any increase in β results in an increase in the shear stress in Equation (9.10). These relationships are shown in Figure 9.8. Bagnold (1968) has considered the minimum slope condition (β_{min}) required for debris to flow:

$$\beta_{min} > C_o \tan\phi \, [C_o + (\rho_s/\rho_r - \rho_s)] \tag{9.13}$$

where C_o = debris concentration
     ρ_r = density of rock particles in the flow till.

If $C_o = 0.25$, $\tan\phi = 0.5$, $\rho_r = 2\,700$ kg m^{-3} and $\rho_s = 2\,000$ kg m^{-3} the limiting slope (β_{min}) upon which the debris will commence to flow is 2.3° – a typical value for temperate glaciers. It is thus clear that sediments resting upon a glacier surface, subject to water saturation from summer melt, are likely to be highly mobile.

9.7.2 Sediment flow behaviour and depositional facies

A continuum of sediment flow species may exist depending upon the mechanical properties of constituent materials, the degree of water saturation and surface conditions. Towards one end of this spectrum are 'stiff', high-strength flows that move only slowly and change their characteristics gradually while at the other extreme there are highly mobile, rapidly evolving 'fluidized' flows. Lawson (1979) has attempted to generalize some of the principal characteristics of such glaci-sediment flows (such as mean grain size, maximum flow thickness, shear strength, porosity, bulk density, etc.) primarily in relation to their water content. Some of these associations are shown in Figure 9.9.

Figure 9.9   *Characteristics of sediment flows as a function of water content. The boxes shown in the thickness vs. water content graph refer to the range of flow conditions found in Lawson Type I–IV flows (after Lawson, 1979).*

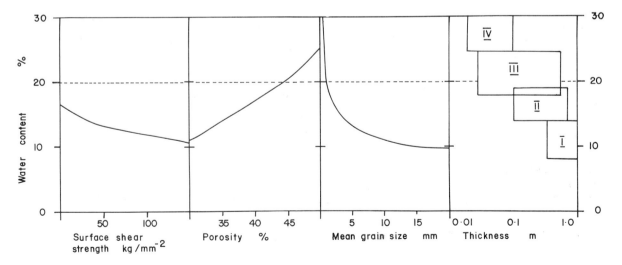

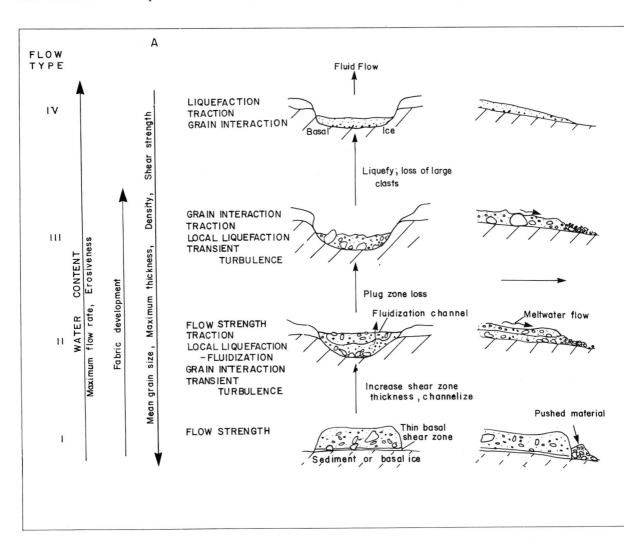

It should be noted that critical water content values lie usually below ~ 20% by weight.

Lawson (1979, 1981a) as a result of a pioneering study of resedimentation process on Matanuska Glacier has recognized four major flow types, and termed here 'Lawson Flows', based upon these characteristics (shown in Figure 9.10 and summarized in Tables 9.2 and 9.3).

Lawson Type I flows

Possessing a low water content (8–14% by weight) Type I flows are typified by a relatively uniform thickness, a high density and debris concentration, steep frontal slopes (> 45%) and low velocities

Figure 9.10   Characteristics of the pattern of flow till behaviour, depositional forms and facies (after Lawson, 1979).

Lawson flow types I–IV, characterized by increasing water content, are illustrated (see text for detailed descriptions). Note that forms are part of a fully transitional sequence. **A** cross- and longitudinal-profiles. Channels are usually ice-floored with walls of ice and sediment. Grain support and transport mechanisms are indicated.

B plan of flow systems and modes of deposition. Source areas are low angled (1–7°), mid-sections are steep and irregular. Stacking and coalescing of deposits occur frequently.

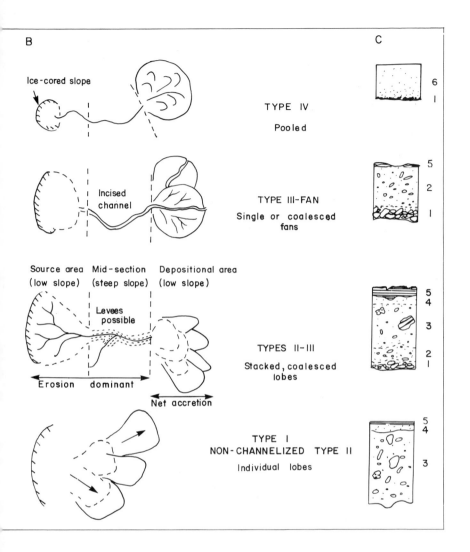

B

Ice-cored slope

TYPE IV

Pooled

6

Incised channel

TYPE III-FAN

Single or coalesced fans

5
2
1

Source area
(low slope)   Mid-section
(steep slope)   Depositional area
(low slope)

Levees possible

Erosion   dominant

Net accretion

TYPES II-III

Stacked, coalesced lobes

5
4
3
2
1

TYPE I
NON-CHANNELIZED   TYPE II

Individual lobes

5
4
3

C

C idealized sediment characteristics resulting from each of the flow types. Lawson recognizes six zones: (1) texturally homogeneous, increased gravel content of traction origin, massive to graded, pebble fabric weak to absent; (2) massive and texturally heterogeneous, absence of large grains (from settling), possible increase in silt and clay (from elutriation), weak pebble fabric; (3) massive and texturally distinctive, sometimes structured sediments, pebble fabrics absent, clasts commonly vertical; (4) massive and fine-grained (sand–clay) similar to matrix of zone 2, few coarse clasts due to settling during or after deposition; (5) stratified – diffusely laminated silts and sands (meltwater origin); (6) massive to partially or fully graded silty sand, fabric absent.

Basal contacts may be conformable or unconformable, sharp, transitional or planar.

(2–5 mm s^{-1} on slopes of between 1 and 10°). Motion is principally by failure and shear along a basal plane, although a small amount of shear will take place within the body of the flow and more pronounced shear in the frontal snout. Shear is expressed in terms of the velocity gradient. Such flows have a limited run-out distance, in the order of a few hundred metres.

Type I flows are usually texturally homogeneous except in the snout region and sometimes along the flow margins which are distinctively coarse. Such size sorting is more apparent in flows possessing a higher water content (Types II and III) where there is pronounced vertical shear throughout the whole flow, rather than at the snout alone. Bagnold (1954, 1968) and Johnson (1970) have investigated

Table 9.2 Principal characteristics of Lawson-type sediment flows (after Lawson, 1979)

| | Lawson flow type | | | |
	I	II	III	IV
Morphology	Lobate with marginal ridges; non-channelized.	Lobate to channelized.	Channelized.	Channelized.
Channel-wise profile	Body constant in thickness with planar surface; head stands above body; tail thins abruptly upslope	Body constant in thickness with ridged to planar surface; head stands above body (less than Type I); tail thins upslope.	Mass thins from head to tail; irregular surface.	Thin continuous 'stream'; planar surface.
Thickness (m)	0.01 to 2 (0.5 to typical)	0.01 to 1.4 (0.1 to 0.7 typical)	0.03 to 0.6	0.02 to 0.1
Bulk water content (wt %)	~ 8 to 14	~ 14 to 19	~ 18 to 25	725
Bulk wet density (kg m^{-3})	2 000–2 600	1 900–2 150	1 800–1 950	< 1 800
Surface flow rates (mm s^{-1})	1–5	2–50	150–1250	10–2 000
Typical length of flow (m)	10 to 300†	10 to 300†	100 to 400†	50 to 400†
Surface shear strength (MPa)	0.04–0.15	0.06 or less	Not measurable	Not measurable
Approx. bulk mean grain size (mm)	2 to 0.3	0.4 to 0.1	0.15 to 0.06	≤0.06
Flow character (laminar)	Shear in thin basal zone with override at head.	Rafted plug with shear in lower and marginal zones.	Discontinuous plug to shear through-out.	Differential shear throughout.
Grain support and transport	Gross strength.	Gross strength in plug; traction, local liquefaction and fluidization, grain dispersive pressures and reduced matrix strength in shear zone.	Reduced strength, traction, grain dispersive pres-sures; possibly liquefaction-fluidization, tran-sient turbidity.	Liquefaction; some traction; buoyancy (?).

† Maximum length of flow reflects boundary conditions of terminus region.

Table 9.3 Characteristics of sediment flow deposits in the terminus region of Matanuska Glacier, Alaska (after Lawson, 1979)

Lawson flow type	Bulk texture type 1) mean (ϕ) 2) std dev (ϕ)	Internal organization		
		General	Structure	Pebble fabric
I	Gravel-sand-silt, sandy silt (1) −1 to 2 (2) 3 to 4.5	Clasts dispersed in fine-grained matrix.	Massive.	Absent to very weak; vertical clasts. S_1† $\cong$ 0.49–0.55
II	Gravel-sand-silt, sandy silt, silty sand (1) 2 to 3 (2) 3 to 4	Plug zone; clasts dispersed in fine-grained matrix.	Massive; intra-formational blocks.	Absent to very weak; vertical clasts.
		Shear zone; gravel zone at base, upper part may show decreased silt-clay and gravel content; overall, clasts in fine-grained matrix.	Massive; deposit may appear layered where shear and plug zones distinct in texture.	Absent to weak; bimodal or multimodal; vertical clasts. $S_1 \cong$ 0.50–0.65
III	Gravelly sand to sandy silt (1) −2.5 to 2.5 (2) 3.5 to 2	Matrix to clast dominated; lack of fine-grained matrix possible; basal gravels.	Massive; intraformational blocks occasionally.	Moderate, multimodal to bimodal parallel and transverse to flow. $S_1 \cong$ 0.60–0.70
IV	Sand, silty sand, sandy silt (1) ⩾3.5 (2) ⩽2.5	Matrix except at base where granules possible.	Massive to graded (distribution, coarse-tail).	Absent.

Lawson flow type	Surface forms	Contacts and basal surface features	Pene-contemporaneous deformation	Geometry* and maximum observed dimensions (length × width, thickness, m)
I	Generally planar; also arcuate ridges, secondary rills and desiccation cracks.	Nonerosional, conformable contacts; contacts sharp; load structures.	Possible subflow and marginal deformation during and after deposition.	Lobe: 50 × 20, 2.5
II	Arcuate ridges; flow lineations, marginal folds, mud volcanoes, braided and distributary rills on surface.	Nonerosional, conformable contacts; contacts indistinct to sharp; load structures.	Possible subflow and marginal deformation during and after deposition.	Lobe: 30 × 20, 1.5; sheet of coalesced deposits.
III	Irregular to planar; singular rill development; mud volcanoes.	Nonerosional, conformable contacts; contacts indistinct to sharp.	Generally absent; possible subflow deformation on liquefied sediments.	Thin lobe; 20 × 10, 0.5; fan wedge; 30 × 65, 3.5; rarely, sheet of coalesced deposits.
IV	Smooth, planar; mud volcanoes possible.	Contacts conformable indistinct.	Absent.	Thin sheet; 20 × 30, 0.3; Fills surface lows of irregular size and shape.

* Length and width refer to dimensions parallel and transverse to direction of movement prior to deposition.
† S_1 is a normalized eigenvalue which gives the strength of the cluster of long axes about the mean axis (see Mark, 1973).

the sorting process. Bagnold found that the relative velocity of particles within the flow scales with their size and vertical velocity gradient:

$$U_p' = r_c (dU/dz) \qquad (9.14)$$

where r_c = clast radius
dU/dz = velocity gradient in the flow.

As the debris flows under shear, particles will migrate through the matrix to those zones of least shear rate. These are found at the upper surface of the flow and at its margins. This phenomenon can be expressed in terms of the ratio of the force exerted on a particle (F_p) by, and normal to, the boundary of the flow, and the resisting force (F_r) of the particle to movement:

$$F_p/F_r = k\,r_c (dU/dz) \cos \alpha_r \qquad (9.15)$$

where k = constant
α_r = angle made between two superimposed clasts and the direction of motion.

Type I flows give rise to well-defined depositional lobes (Figures 9.10 and 9.11) possessing many properties of the source region such as heterogeneous texture with larger clasts dispersed within a finer matrix, and with occasional distinctive lenses of more uniform sediments. The head or frontal part of the flow also retains its character with the larger individual blocks. Frequently, the action of surface meltwater and settlement may produce a fine-grained uppermost layer.

Figure 9.11A   Lawson Type I Flow. Depositional area is shown for a flow on Matanuska Glacier, Alaska. The source of the flow is in the background and there is minor channelling by sediments particularly to the right. Small meltwater channels have developed on the surface of the flow (to the left). Some reworking has taken place to the flow front where there is minor ridging. The scale bar is 3 m in length (courtesy D. Lawson).

Lawson Type II flows

The water content of Type II flows is higher at about 15–20% and acts to reduce sediment cohesion. Thickness is consequently less uniform; density and debris concentration somewhat lower. Flow velocities are in the order of 2–30 mm s^{-1} on slopes of 1–7°. Lawson reports that flow may be pulsatory, reflecting an incapacity in the supply of new sediment from ablating dirty ice to keep pace with discharge by the flow. 'Plug' type behaviour is common, in which the zone of strong shear, characteristic of the basal region in Type I flows, now extends into the margins leaving a central core or plug of unsheared material. Some degree of channel confinement occurs and run-out distance may be several hundreds of metres (Figure 9.11B). The flow of meltwater on the surface of the flow is usually more pronounced than in Type I. Some coarse sediment fractions may be rolled or dragged along by the force of the flow rather than being transported in suspension. In common with Type I flows a wide range of sediment sizes is apparent.

Deposition occurs in pronounced lobes as a result of greater fluidity and channellization. Due to pulsating flow, lobes are often stacked one on top of another and with lateral overlap (Figure 9.10). There may be considerable variation between the properties of active and previously deposited flows. The latter frequently comprise only the central plug or a heterogeneous and rather structureless mass. These distinctions usually reflect the changed water content in flows

Figure 9.11B Lawson Type II Flow (Matanuska Glacier, Alaska). Distinctive channelling is apparent (cf Figure 9.10) with flow from the source area in the background. The flow is of order 30 cm thick. Sediments are being transported in suspension and traction (the latter for the larger clasts). A thin sheet of meltwater over the surface of the flow is also shown. Pronounced mixing takes place below the step. Material from channel sides is being incorporated. Scale bar is 1 m in length: (courtesy D. Lawson).

after deposition. The more fluid the flow, the less homogeneous the resulting deposit may be due to expulsion of pore water and differential settlement: even graded sequences may occur. Lawson suggests that fluidization usually results in a textural coarsening while liquifaction (increasing pore pressure due to repeated loading and void ratio reduction) and shear produces sediment fining. The snout region may again consist of larger blocks of debris in a relatively structureless matrix – similar to Type I deposits. Surface meltwater and ponding gives rise to sequences of fine-grained sediments.

Lawson Type III flows

Water content of 18–25% by weight further reduces shear strength of debris and Type III flows are characterized by much of the flow being in shear with a weakly developed plug only present at lower saturations. Velocities lie between 150–1 250 mm s^{-1} and flow is nearly always channellized. Flows are thin due to higher fluidity and large clasts may penetrate through the upper surface. Lawson considers that Type III flows are highly erosive and degrade their channels. Due to a pronounced velocity gradient within the flow, size-sorting according to Equation (9.14) will take place. Textural features are related, in the first instance, to fining from the source region to the front of the flow. During deposition pulsatory stacking of successive flows may occur in a manner similar to Type II flows, but greater fluidity may generate overlapping fans (Figures 9.10 and 9.12). Coarse material will be deposited first at the head of a fan with finer material washed to distal regions. Some upslope imbrication may be witnessed. The stacking of flows gives rise to stratified units often possessing intermediate fine-grained meltwater sediments.

Lawson Type IV flows

These are the most fluid and mobile of glaci-sediment flows, possessing a water content in excess of 25% by weight. They are true 'slurries', with low shear strength, small thickness (20–100 mm) and high velocities (10–2 000 mm s^{-1} dependent upon slope angle) often maintained

Figure 9.12   Lawson Type III flow on coarse firn of Mitdalsbreen, Norway.

over long periods of several hours duration. Such flows are organized into distinctive, narrow channels and frequently fine-grained materials (fine sands and silts) predominate with small clasts. Deposition tends to occur when slopes reduce and such flows move into topographic lows forming a thin, continuous layer. Successive flows may build up sequences tens of metres thick of thin laminae. Texturally, such deposits range from massive to graded.

9.7.3 Sediment-flow fabrics

The sediment characteristics inherited by melt-out deposits from glacier ice are drastically altered or obliterated by flow processes. As the water content increases flow is more channellized, of higher velocity and clast long-axes become increasingly reoriented, especially where fan deposition takes place.

In the main part of the flow (assuming no 'plug' is present) a_l-axes align in the direction of flow and a_l–b_l planes are parallel to the flow surface or possess a dip up- or down-flow related to the bed topography (Boulton, 1971). In the head region (of all flows) a_l-axes may align transverse to the flow direction as longitudinal compressive strain becomes dominant, and a_l–b_l planes dip up-flow. Along the flow margins a_l-axes are again parallel to flow direction with a_l–b_l planes dipping in towards the body or axis of flow. Figure 9.13 illustrates contoured, equal-area fabric nets for deposited Lawson Flow Types I–IV in relation to the direction of motion before deposition.

9.7.4 Flow till and the sedimentation continuum

The mobilization of glacial sediments by meltwater, the flow of slurries and their deposition in a variety of environmental circumstances leads to a gradation of processes and resulting sediment facies. One end member (providing an example of a clear diagnostic group) is the true terrestrial till. The increasing importance of channellized, highly fluidized flow leads naturally into glaci-fluvial processes. Often Lawson Type IV flows discharge into super- or proglacial streams and frequently the distinction between highly water-saturated sediment flows and streams with very high debris concentrations becomes increasingly artificial.

Where a heavily ablating ice front terminates in a lake, a fjord or the open sea resedimentation by the flow processes discussed earlier in this chapter acquire characteristics of the newly dominant medium – the lake, the fjord or the ocean. Wave action, density currents, circulation patterns and many other factors introduce new and increasingly important aqueous attributes (Lowe, 1976a, b). Although transitional processes and facies will exist, succeeding chapters attempt to

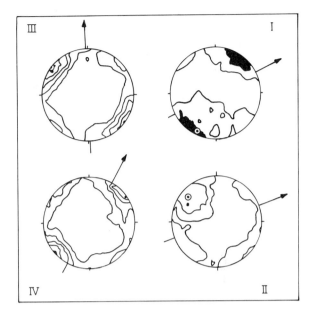

Figure 9.13 Contoured (2σ intervals) Schmidt equal area projections indicating the distinctive fabrics developed in Lawson flow types I–IV. Arrow indicates flow direction. From Matanuska glacier, Alaska (after Lawson, 1979).

deal with the key aspects of glacial sediments within the river, lake and sea environments.

9.8 Lithofacies models

Eyles *et al.* (1983) and Miall (1983) have erected a lithofacies sequence for use with glacial diamicts, following the pioneering work by Miall (1977, 1978) in defining lithofacies types upon lithological criteria for fluvial deposits. Their code is shown in Table 9.4 and is based primarily upon structural, lithological and bed-contact features. Although Eyles *et al.* (1983) support their scheme with a discussion of the likely genetic processes for temperate glaciers, cold ice and glaciers with complex basal thermal regimes and give various examples, their work will require future testing, development and refinement. Nevertheless their approach provides a refreshingly new direction for the advancement of knowledge, geared to process, of glacial sediments.

Table 9.4 Lithofacies code and accompanying symbols for diamicts based upon the scheme of Eyles *et al.* (1983).
Hyphens show possible combinations – second letter (m and c) indicates matrix or clast support (degree undefined). Third letter gives internal structure (m – massive, s – stratified, g – graded). Fourth letter (in parenthesis) suggests possible useful environmental characteristics – optional due to subjective nature (e.g. s – evidence of shear, r – resedimentation, c – current reworking).
The symbols are suggested as useful for construction of sedimentary logs.

Diamict, D

Dm	matrix supported
Dc	clast supported
D-m	massive
D-s	stratified
D-g	graded

Genetic Interpretation, ( )

D—(r)	resedimented
D—(c)	current reworked
D—(s)	sheared

Sands, S

Sr	rippled
St	trough cross-bedded
Sh	horizontal lamination
Sm	massive
Sg	graded
Sd	soft sediment deformation

Fine-grained (mud), F

Fl	laminated
Fm	massive
F-d	with dropstones

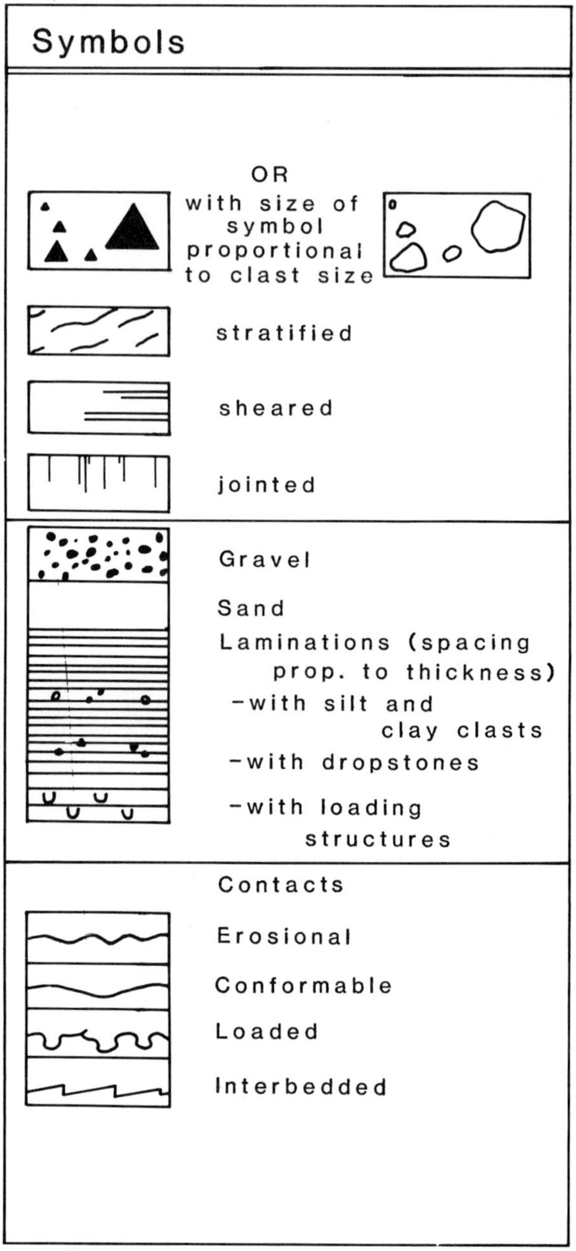

Symbols

OR with size of symbol proportional to clast size

stratified

sheared

jointed

Gravel

Sand

Laminations (spacing prop. to thickness)
–with silt and clay clasts
–with dropstones
–with loading structures

Contacts

Erosional

Conformable

Loaded

Interbedded

10 Glaci-fluvial processes and sedimentation

10.1 Introduction

In Chapter 2 the general disposition of water in glaciers and its subglacial flow regime were outlined. The erosive role of meltwater was discussed in Chapter 5. We now consider the manner in which sedimentary particles, gathered by meltwater and transported either solely within or discharged from an ice mass, are eventually deposited. It is often useful to distinguish between glaci-fluvial sedimentation in proglacial channels and that in close proximity to the glacier either at the glacier margin or within and at the glacier bed. Glaci-fluvial deposition is later discussed under the two simple headings subglacial and proglacial.

10.2 Sediment in glacial meltwater

Abrasion and ploughing-out of bed material by clasts held in basal ice, crushing of bedrock and attrition of clasts by strong particle–particle interactions produce a wide range of sediments at the glacier sole (Chapter 7).

These erosion products may be entrained by sub-glacial water and transported as suspended matter or bedload accompanying material removed directly by meltwater. Sediment stored in the lowermost layers of the glacier may also be melted out from the walls of meltwater channels, especially during tunnel enlargement under increasing discharge conditions in springtime. While much coarse sediment can be moved by the high flows in larger arterial subglacial channels, considerable quantities of finer debris may be carried in thin 'Weertman films' from over a much more extensive area of the glacier bed (Hallet, 1979a). The hydraulic factors relevant to entrainment in major glacial channels are little different from open-channel conditions, bearing in mind very high velocities (induced by strong pressure gradients and reduced channel friction), marked discharge variations (both seasonal and diurnal) and increased water viscosity due to low temperatures. Such factors are responsible for pronounced variations in the concentration and discharge of sediments in meltwater.

Increased sediment concentration due to greater viscosity occurs primarily in the upper levels of flow rather than close to the bed, and affects finer fractions (< 0.3 mm) more than coarser material (> 0.25 mm) or finer than 0.06 mm. Figure 10.1 shows a pronounced inverse relationship for sediment concentration and sediment discharge with water temperature. The scatter of values in Figure 10.1 giving rise to the envelope is probably due to variations in discharge. A second effect that results from diminishing water temperatures is a change in bed conditions of **loose-boundary channels**. Bedform characteristics observed at two different temperatures, 28°C and 4°C, but with approximately the same discharge exhibit a change at lower temperatures to smaller amplitude, longer wavelength bedforms corresponding to a decrease in bed friction and an increase in velocity. It seems likely that if the bed is modified by erosive processes with a fall in temperature, bed load discharge must also be affected.

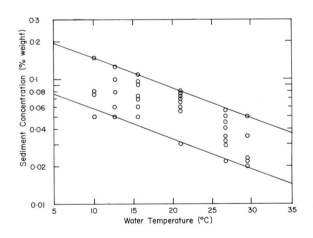

Figure 10.1 Inverse relationship between sediment concentration and water temperature demonstrating the effects of water viscosity for only a small range in water discharge which is responsible for the envelope of values. Data for the Lower Colorado River, USA (re-plotted from Lane *et al.*, 1949).

10.3 Transport and discharge

A common phenomenon encountered in glacial streams is that a large proportion of the annual sediment load is discharged during only a few days of very high, summer flow. Metcalf (1979) for instance, reports that the sediment transported during 5 minutes in midsummer (June) was equal to the total sediment yield for the month of January. The measurement of sediment budgets, therefore, has to take place on as nearly a continuous basis as possible, especially during the early summer, in order to sample peak discharge conditions. The strategy for sampling suspended sediments is discussed by Gurnell (1982) who suggests that even hourly measurements may be insufficient to record rapid flushes of sediment.

Almost all measurements related to the transport and discharge of sediments in glacial meltwater have been made at the portal or at some distance downstream from the glacier. Few detailed studies have been made within subglacial or englacial channels. Borland (1961) has shown that sediment yield falls rapidly with distance from the glacier so that measurements downstream may not be representative of subglacial conditions. Gurnell (1982) indicates that the location of a sampling site is important. Increasing distance from the glacier front leads to masking of the suspended sediment 'signal' as non-glacier sources of sediment (such as bank collapse) become influential. Since, however, one of the primary purposes of measuring sediment flux is to estimate material transfer out of the glacierized basin this may not be a serious disadvantage.

It is usual to separate specific yield or load of rivers into that suspended within the water and that in traction at the channel bed.

10.3.1 Suspended sediment

This component is relatively easy to measure in highly turbulent meltwater streams where sediments are more or less uniformly distributed (Østrem, 1975). Significant lateral variations in suspended sediment concentration may occur, however, over short periods (e.g. 30 secs) according

to Metcalf (1979). Sediment concentration may be deduced by water bottle sampling or semi-continuous vacuum liquid sampling (Collins, 1979a). The sediment flux is obtained by multiplying average sediment concentration by the water discharge obtained from a gauging station (Petts and Foster, 1985).

Østrem has shown that for some Norwegian glaciers, at least, there is no simple relationship between short-term suspended sediment load and water discharge. Some very general relationships are, however, clear (Figure 10.2). During the winter, when discharge is negligible, meltwater is usually low in sediments with a concentration of a few mg l^{-1}, and sometimes even completely clear of suspensoids. In the spring as melt increases both water and sediment discharge rise rapidly to several g l^{-1}. At this time, suspended sediment concentration is usually much higher than for the same discharge later in the season, due to the early flushing-out of easily transported materials produced by abrasion and crushing. By late summer much of the water-accessible sediment stored at the bed has been removed.

On a daily basis the quantity of suspended sediment shows pronounced variation. Figure 10.3A illustrates the change in suspended sediment over a 24-hour period in July 1966 at Storsteinsfjelljökull, Norway. Water discharge rose from 3 to 7.8 m^3 s^{-1} while sediment flux went up from 0.4 to 5.8 kg s^{-1}. It is important to note that the suspended sediment discharge peaks *prior* to the maximum water discharge. This phenomenon is common to most meltwater streams issuing from glacier termini and may result from the rapid suspension

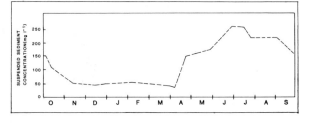

Figure 10.2 Annual discharge of suspended load in meltwater discharged from the Glacier d'Argentière (1967–68) (from Vivian and Zumstein, 1973).

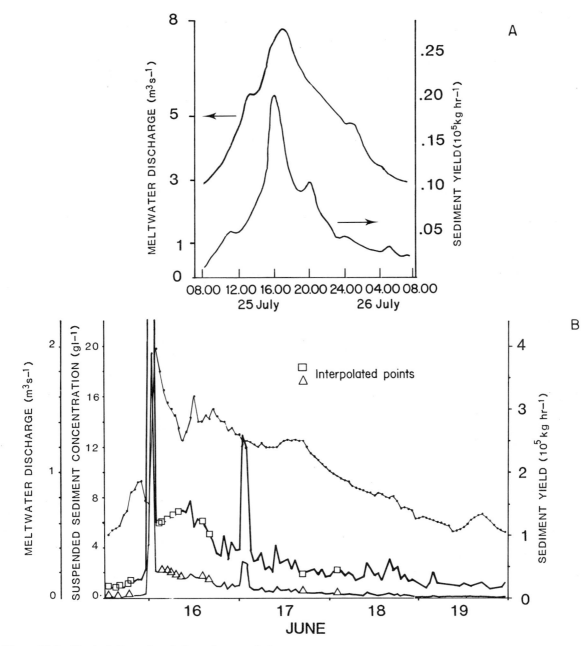

Figure 10.3 Typical diurnal variation of suspended sediment and water discharge. **A** from Storsteinsfjelljökull, Norway (from Østrem, 1975). Upper curve water discharge; lower curve sediment yield. Note the sediment flux peaks in advance of the water discharge; **B** from Tsidjiore Nuove glacier, Switzerland (from Beecroft, 1983) Upper curve water discharge; middle curve sediment concentration; lower curve sediment yield.

of the loose sediment at the glacier bed during the early part of the rising stage. Collins (1979a), however, reports daily sediment concentration maxima both before and after the water discharge peak for the Gornera in Switzerland. He also notes that, in general, highest water discharges are accompanied by sediment concentration minima. Such behaviour may again be related to sediment supply at the glacier bed.

Figure 10.3B illustrates suspended sediment and water discharge during a 4-day flood from Glacier de Tsidjiore Nuove, Switzerland (Beecroft, 1983). The sediment yield dropped rapidly from a peak of $70.7\,\mathrm{g\,l^{-1}}$ although it did not reach a level comparable with that before the outburst for several days. The water discharge hydrograph, however, peaks just after the suspended sediment maximum and shows a more gradual falling limb. This pattern further suggests the initial rapid removal of sediments from the glacier bed followed by exhaustion of the debris supply.

Major flood events appear to dominate suspended sediment transport. On Baffin Island, Østrem et al. (1967), for instance, report that about 60% of the total 1965 summer sediment load discharged from Decade Glacier was removed during a single 24-hour period. At Erdalsbreen in southwest Norway 50% of the sediment load was transported during 4–5 single days (Østrem, 1975). Beecroft (1983) noted that during a 4-day outburst from the Glacier de Tsidjiore Nouve in Switzerland in June 1981 (Figure 10.3B) the total suspended sediment yield (calculated as $\sim 1.7 \times 10^6$ kg) was equivalent to the total discharge of suspended sediments for the months of July and August 1978 – taken as typical of peak summer conditions.

10.3.2 Bedload

Material moved along the bed in traction or by saltation is more difficult to sample and measure than suspended sediment. Several attempts to estimate bedload transport in glaci-fluvial channels have been made on the basis of empirical formulae (e.g. Church, 1972). Some of the most detailed and useful studies, however, are those by Østrem and

his group (1975 and in Aarseth et al., 1980, pp. 27–33) They used a variety of techniques to directly measure bedload including the use of a bedload sampler. During the summer of 1969 at Nigardsbreen, Norway a 50 m steel fence was erected across the outlet stream of the glacier (Figure 10.4). The mesh size was such that it trapped all material > 20 mm. The accumulation of coarse bed deposits was measured by vertical probing twice a day at 176 points across the net. After three weeks, a major flood carrying large ice blocks breached the net and measurements were terminated. Between 24 May and 19 June 400 tonnes of material were trapped. 1 200 tonnes of suspended matter were discharged over the same period, which gave a ratio of suspended to bedload of 3 : 1. Østrem records that material in the size range 1–20 mm was not directly sampled (i.e. particles not suspended, possibly in saltation, and not trapped by the net) so the ratio 3 : 1 is a maximum.

An alternative, though less direct, method of estimating bedload was also conducted at Nigardsbreen. The meltwater stream issuing from the glacier enters a proglacial lake, Nigardsvatn, located about 1 km from the snout (ice front position of 1981) and forms a delta. Bedload is deposited here and annual surveys of the delta since 1969 indicate an average accumulation of 11 200 tonnes (range from 3 000–19 500 tonnes). This figure may include some suspended material entrapped amongst bed deposits in the outer part of the delta. Based upon these figures the suspended load may constitute 50%, or possibly 40%, of the total sediment flux. Church and Gilbert (1975) report the suspended component as comprising 80–90%. Such a range of values again reflects the variability of the glacial meltwater regime.

10.4 Sedimentation in glacier meltwater channels

Sediments transported by meltwater are deposited principally under conditions of diminishing or recessional flow. Deposition occurs in channels at the surface, within or at the bed of a glacier. The resulting sedimentary accumulations range from

Figure 10.4 Bedload sampling net constructed across the Nigardsbreen meltstream in 1969. Mesh trapped material coarser than 20 mm. Accumulation of debris was calculated by twice-daily soundings and levelling at 176 points along the net (courtesy G. Østrem).

thin layers or lags of coarse particles through discrete channel fills of heterogeneous or stratified material to extensive linear sequences. Because the environment of deposition is highly unstable, due to ice deformation and melting, it is rare that sediments and their associated primary structures are fully preserved for any length of time: they become quickly reworked by ice and water. Only in terminal zones of glaciers are such deposits regularly encountered, that is where basal sediments may be quickly uncovered by ice retreat or super- and englacial material is let down onto bedrock by vigorous ablation.

10.4.1 Deposition in tunnels

When sub- or intraglacial conduits are full of water no free upper surface may exist and flow approximates that found in pipes. Aspects of fluid flow in pipes have been given in Chapter 2 and may also be found in Graf (1971). In other instances flow is analogous to that in open channels. Sediments are transported, sorted and deposited but there is a constant interplay between erosion and deposition in glacial conduits, governed by flow regime, sediment concentration and particle size. The mean flow velocity ($\overline{U}_w$) in conduits may be expressed in terms of the empirical **Manning** relationship:

$$\overline{U}_w = (R''^{2/3} i^{1/2})/n' \tag{10.1}$$

where R'' = hydraulic radius
      i = hydraulic gradient
     n' = Manning roughness coefficent.

Typical values for n' (given in units of $m^{-1/3}s$) in natural open channels composed of cohesive sediments vary from about 0.025 in small, smooth-banked, straight stream sections of low slope, to 0.160 in channels with considerable roughness. It is often difficult, however, to determine n' in anything more than an empirical manner. Glacial meltwater channels cut in ice such as those found superglacially will, in general, exhibit only minor resistance (except in tight bends) and it is expected that n' will be low – maybe as low as $0.01\ m^{-1/3}\ s$. Such values have the effect of increasing flow velocity and turbulence and thus the sediment transport rate. Low values for n' are clearly not favourable for net deposition of sedimentary materials in a channel: particles at rest in the bed of an ice channel are more likely to be re-entrained due to low frictional coupling between clast and bed. Nevertheless, deposition does occur in ice channels in areas where flow is reduced by reduction in surface slope and thus hydraulic gradient. Glacial channels are often highly ephemeral with frequent, violent changes in discharge. Such behaviour leads to temporary sediment accumulations in active channels.

At the glacier bed sedimentation in open channels is subject to higher flow resistance as rock or till may constitute the bed with correspondingly greater roughness. n' may approach $0.1 \, \mathrm{m}^{-1/3}\mathrm{s}$ or more. Clarke and Matthews (1981), for instance, found that a Manning n' coefficient of $0.12 \, \mathrm{m}^{-1/3}\mathrm{s}$ gave an acceptable fit to discharge data for Summit Lake, Canada which drained through subglacial conduits. In a complimentary study, Clarke (1982) used a value of n' = $0.105 \, \mathrm{m}^{-1/3}\mathrm{s}$ in a simulation of discharge from Hazard Lake, Canada.

Where discharge is high and the surface gradient low, sedimentation may proceed at the bed. The balance between bed erosion and deposition may be described by a parameter ($\varXi$), following Allen (1971):

$$\varXi = \frac{dG^*}{dx} + \frac{1}{\overline{U}_w} \cdot \frac{dG^*}{dt} \qquad (10.2)$$

where G* = sediment transport rate
    = $[(\rho_c - \rho_w)/\rho_c] \, m' \, g \, \overline{U}_w$
    m' = dry mass of sediments
    ρ_c = density of sediments
    ρ_w = density of water
    x = distance in direction of flow
    t = time.

For deposition to occur, either or both of the RHS terms in Equation (10.2) have to be negative (i.e. dG*/dx < 0 and dG*/dt < 0). This can be achieved if the flow velocity falls (either at one place in a tunnel through time, or at any instant in time with direction of flow) thus reducing the value taken by G*. If dG*/dt > 0 or dG*/dt > 0 erosion will ensue. If both terms are equal to zero there is uniform sediment transport.

The bedforms and structures resulting from sedimentation in glacial tunnels differ little from those generated by open-channel flow. It is not intended to discuss in any detail the production of fluvial bedforms – adequate summaries are given in several texts (Reineck and Singh, 1980; Allen, 1982; Leeder, 1982; Collinson and Thompson, 1982).

The sedimentary units produced by tunnel deposition are usually linear, single, steep-sided ridges (Figure 10.5). They are composed of stratified sand and gravel (Figure 10.6A), which form tabular or longitudinal (sheet-like) sequences possessing parallel or cross-bedding with secondary bedforms such as ripples, dunes and graded bedding (Figure 10.6B). In some cases finer fractions of sand, silt or clay are noticeably absent. Current directions may be pronounced with little variability. Good descriptions of typical deposits are given by Banerjee and McDonald (1975), and Saunderson (1975, 1977).

Instructive experiments, specifically designed to investigate the movement and deposition of sediments in glacial pipes flowing full have been made by McDonald and Vincent (1972). These workers found that for flow up a slope, at a given discharge, increasing concentration of sediments generates a successive bedform sequence from **ripples** through dunes, dune planes, plane beds to dune planes. Beyond this last bedform entrainment and full suspension of particles ensues. No **anti-dunes** form, in constrast to the open-channel bedform sequence. The conditions required for anti-dune construction (i.e. **standing surface waves**) are suppressed by the ice roof of tunnels. Deposition in the McDonald and Vincent study was characterized by production of cut-and-fill successions forming large-scale **cross-stratification**, or sometimes parallel laminations with shallow downstream dips (Figure 10.7). A pronounced decrease in particle size was observed in the **foresets** in the downslope (downstream) direction associated with settling through the zone of flow separation. Slumping and avalanching on the foreslope were also apparent.

Figure 10.5   A sinuous ridge of glaci-fluvial sediments (esker) on the surface of Paulabreen, Svalbard resulting from division of a glacier meltwater channel or tunnel during recent ice retreat and progressive exposure by differential ablation. Height of ridge given by figure (circled).

Figure 10.6   Glaci-fluvial sediments. **A** Section through a sequence of sediments deposited in a tunnel, Paulabreen, Svalbard (detail from esker shown in Figure 10.5). Note pronounced vertical and lateral variations in grain size from cobbles to finer fractions which contain significant clay-size material (unit at knife). **B** An example of ripple-drift cross-lamination in glaci-fluvial sediments from Burke's pit between Huntley and Stittsville, near Ottawa, Canada. An increase in angle of climb can be observed combined with a decrease in grain size (from Rust and Romanelli, 1975).

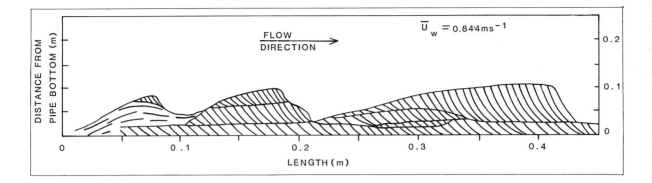

Figure 10.7 'Large-scale' bedforms (dunes) and internal stratigraphy resulting from experiments in pipes (after McDonald and Vincent, 1972).

Because the hydraulic mean depth was maintained at or near a maximum during sedimentation experiments, McDonald and Vincent found that excess energy was expended by thermal erosion of the tunnel roof (see Equation 2.9), rather than the removal of sediments. Shreve (1972) has suggested that secondary currents may be set up within a tunnel flowing upwards along the flanks of a growing sedimentary deposit as well as outward and upwards along the walls of the tunnel. In this way the pseudo-anticlinal bedding, revealed in many tunnel deposits, is related directly to water flow pattern and not simply to post-depositional disturbance.

Saunderson (1982) reports the earlier studies of Newitt *et al.* (1955) which suggest four principal transport regimes during pipe sedimentation. (i) Stationary beds occur at low velocities. Bedload is present and some particles move by saltation: the major bedforms are ripples. (ii) As velocity rises the bedload may begin to move or slide, particularly when composed of coarser fractions. This condition is characterized by the bed moving as a single unit along the channel bottom. Depending upon flow regime and sediment concentration the bed may be either continuously sliding or exhibiting jerky motion. (iii) Heterogeneous suspension of all particles takes place at higher velocities again, although a size sorting is present in which coarse material

moves closer to the channel bottom. (iv) Full homogeneous suspension with no vertical size gradation characterizes the highest velocities. Saunderson (1977) contends that sliding beds may be a distinctive sedimentological characteristic of subglacial tunnel deposition – a notion that has been disputed by Boulton and Eyles (1979).

10.4.2 Palaeohydraulic inference

The results of pipe and flume studies may be used to interpret water flow conditions in ancient glaci-fluvial deposits. Experimental relationships established between sedimentary characteristics (bedform dimensions, geometry, particle size, etc.) and flow regime are applied to measured field sections. Saunderson and Jopling (1980) and Saunderson (1982) have exploited this technique by using bedform stability diagrams to infer water velocity, depth, surface slope and a number of other hydraulic parameters for ancient **esker** deposits in southeastern Canada. In the cases they examined a free water surface in the palaeo-channel was assumed to be present. For the Brampton Esker, Ontario small, tabular units were 0.15 m average thickness (probably dunes migrating over the topsets of a micro-delta). Median grain size was 0.2 mm. Based upon the work of Sundborg (1967) and Ljunggren and Sundborg (1968) for entrainment and deposition these values give a threshold surface water velocity of ~ 0.35 m s^{-1}. Saunderson and Jopling (1980) corrected this figure upwards to 0.65 m s^{-1} to allow for full suspension of sand particles (a condition which they believe obtained) and relationship between surface and mean water velocity. Using empirically established associations between water velocity and water depth and particle size Saunderson and Jopling were able to estimate the flow depth as ~0.3 m. With mean flow velocity, water depth, and mean grain diameter it is possible to go further and evaluate **Froude** and Reynolds numbers, the **Chezy formula, Darcy-Weisbach equation** and also the bedload sediment discharge.

In englacial or subglacial tunnels flowing full, where hydraulic behaviour approximates pipeflow the experimental results of McDonald and Vincent (1972) may be used to infer palaeohydraulic conditions. McDonald and Vincent established a relationship between the height of dunes (l'), the hydraulic mean depth (J_d) and a water column depth in a circular pipe

of radius R_t. Typically the ratio (J_d/l') was found to lie between 1 and 3 which again allows dune height to be used as an estimate of the water depth in former glacial meltwater channels. J_d peaks when sediment thickness on the bed fills the pipe to 0.38 R_t depth and 0.78 R_t width. For tunnels of non-circular cross-section the wetted perimeter is twice the channel width and the actual water depth tends to $2J_d$ so that the water depth, insensitive to tunnel shape, is $2J_d - 3J_d$ and thus between $2 l'$ and $10 l'$ (where l' is the dune height). McDonald and Vincent (1972) applied these relationships to the formation of the Windsor Esker in Quebec. Sediments comprise large-scale, tabular, cross-stratified units of sand and gravel with an average thickness of 0.4 m. Such dune heights suggest a water depth within the meltwater conduit of 1–4 m at the time of formation.

10.4.3 Ripple-drift cross-lamination

Allen (1971) has made a detailed study of some of the smaller-scale structures exhibited by sediments deposited in both open channels and pipes, especially **ripple-drift cross-lamination**

Figure 10.8 Migrating ripples and cross-laminations generated by sediment deposition in a steady fluid flow (shown by alternating arrowed lines). The figure also defines several quantities and terms. ϕ_r = climb angle of ripples S_{ds} = suspended sediment deposition rate; U_w = fluid velocity, see text for further explanation; λ_r = ripple wavelength; l' = ripple height. (Insert defines horizontal, vertical components and resultant.)

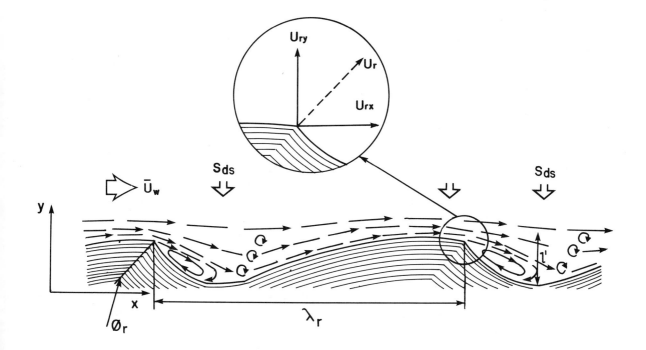

(Figure 10.6B). – a bedform feature not described from the McDonald and Vincent experiments yet frequently observed in glaci-fluvial and glaci-lacustrine sequences. Sediment deposited onto a surface composed of migrating ripples gives rise to a distinctive sequence of cross-laminations in which ripples are seen to climb over those downstream (Figure 10.8). Allen (1971) demonstrates that such laminations can be categorized by the climb angle of the ripples (ϕ_r) which is defined by:

$$\phi_r = \tan^{-1}\left(\frac{U_{ry}}{U_{rx}} + U_{rx}'\right) \qquad (10.3)$$

U_{ry} and U_{rx} = vertical and horizontal velocity components of the rippled bed respectively (Figure 10.8)

U_{rx}' = horizontal velocity component generated by local non-uniform or transient flow.

Thus the ripples climb in direct relationship to the deposition rate of sediment, orthogonal to the mean bed but inversely with velocity of the ripples. Allen (1971) defines both U_{rx} and U_{ry} in terms of stream flow, rate of bedload transport (G_b) and ripple geometry:

$$\phi_r = \tan^{-1}\left(\frac{l'\,\overline{S_{ds}}}{2G_b + \lambda_r\,\overline{S_{ds}}}\right) \qquad (10.4)$$

where l' = height of ripples
λ_r = ripple wavelength
$\overline{S_{ds}}$ = mean sediment deposition rate (see also Equation (10.2), Ξ term).

Due, however, to the changing flow pattern there is a difference in the ability of the current to suspend sediments over ripple crests and troughs. Over the upstream face flow is compressive such that deposition of some sediment is inhibited. On the downstream face flow is extensive and deposition is greater. The difference between upstream and downstream deposition rate (S_{du} and S_{dd}) can be expressed as:

$$S_{du} - S_{dd} = (\lambda_r/L_f)(\overline{S_{ds}} - S_{rd}) \qquad (10.5)$$

where L_f = length of the leeward face
S_{rd} = local deposition rate.

S_{rd} is thus a rather important factor in the development of climbing ripple cross-laminations.

The results of Allen's theoretical and laboratory work also suggests that there is a strong coupling between the angle of climb of ripples and sediment size sorting in the laminations. The coarsest material is found on the leeward side of a ripple (where the foresets develop) while the finest is usually located on the upstream face. The sorting is primarily imposed by those particles entrained into and in transit within the thin bedload layer just above and in contact with the ripple surface.

A relationship between sorting (S_e) and angle of climb (ϕ_r) for experimental cross-laminated deposits and samples from the Uppsala Esker at Löfstalöt, Sweden, gave the form: $S_e \simeq \tan^{1/2}\phi_r$.

A further interesting result from Allen's study relates to the rate of deposition (S_{ds}) in the Uppsala Esker and deductions therefrom on flow characteristics. By analogy with laboratory experiments pronounced variations in the angle of climb of ripples indicate that the meltwater flow regime forming the Uppsala Esker must have been unsteady and non-uniform – quite in keeping with what is known of glacial streamflow. It was possible to calculate the time periods involved in changes in flow of water in the glacial tunnel that formed the Uppsala Esker at Löfstalöt from the vertical variation of the sediment deposition rate S_{ds} in the cross-lamination by:

$$h_s = (1/\rho_c) \int_{t1}^{t2} S_{ds}(t)\,dt \qquad (10.6)$$

where h_s = vertical thickness of sediment layer.

Allen finds that the deposition periods are between hours to tens of hours – corresponding closely with the typical patterns of discharge fluctuations measured in meltwater channels.

10.4.4 Post-deposition modification

Sediments deposited from meltwater flowing in glacial channels of one sort or another accumulate in a highly unstable environment. Deposits may be rapidly modified by various external factors resulting in a spectrum of change from minor alteration of bedforms (such as faulting and folding) through loss of primary structures (large-scale slumping and mobilization of sediments) to complete obliteration by erosion.

For sediments deposited in glacial tunnels disturbance can arise from the creep-closure of the channel during winter months when water flow is at a minimum (Equation 2.8). In addition the motion of basal ice, especially by sliding over the bed, may lead to the deformation of sedimentary deposits. Most sedimentation occurs during recessional flow in the late summer. Deposits on the tunnel floor may be re-entrained (i.e. eroded and reworked) the following spring and summer once passages and conduits are reoccupied and reamed-out as water discharge rises.

These factors suggest that good preservation or survival of sediment bodies is likely *only* when the accumulation is small in size and located in the thinner marginal parts of ice masses where forward motion is small or the ice stagnant, tunnel closure rates are low, the ice front is retreating allowing exposure of sediments and where there is a high probability of deflection or change to meltwater channels. In terminal areas it is also possible that during the winter period sediment accumulations of high water content may become frozen with the penetration of cold air through tunnels and passageways. Where ice is thin (say 20 m or less) downward conduction of the cold winter wave may also assist freezing of sediments. This is only a secondary factor, however, and due to the lag in conduction through the overlying ice, maximum cold conditions in a shallow subglacial tunnel may be significantly out of phase with ice surface temperatures.

Superglacial or near surface englacial open-channel sediments may be subject to differential ablation – deposits protecting the ice beneath in the manner described by Østrem (1969), Loomis (1970) and Drewry (1972). Ablation leads to both the raising of the sediments above the surface (Figure 10.5) and disturbance of primary depositional structures by meltwater flow, and slumping on the flanks with differential movement

resulting in faults. Such activity is confirmed from numerous field studies. Jewtuchowicz (1965), for example, has described quite significant alteration to the attitude of bedding in sediments exposed in the forefield of Gåsbreen in Svalbard where melting of ice beneath an esker-like sediment body has been inferred. It is worth noting, however, that Price (1973), has argued for the let-down of sediment in former stream channels onto subglacial bedrock without significant disturbance during ice wastage. Unless an esker is of considerable dimensions in order to maintain structural integrity, or in an area of thin, slowly ablating ice, it would appear likely that little original structure will be preserved upon let-down. Resedimentation and flow may also occur to glaci-fluvial deposits due to ice ablation in a manner similar to till bodies as described in detail in Chapter 9.

10.5 Sedimentation in proglacial meltwater channels

Water and sediment gathered by the glacial drainage system discharge proglacially where net deposition takes place in channels by overbank flow and with the development and coalescence of fan-shaped sequences. In recent years detailed studies of proglacial, glaci-fluvial sedimentary processes have been undertaken commencing with the classic works of Hjulström (1954–57) and Fahnstock (1963). Excellent case studies are given by Church (1972), Boothroyd and Ashley (1975), Boothroyd and Nummedal (1978), Rust (1972, 1975), Rice (1982) and Fraser (1982). Much of the sedimentation occurring in pro-glacial areas is similar to open-channel deposition in non-glacial environments so that only key 'glacial' characteristics are described. Most proglacial streams exhibit pronounced changes in regime – glacier-controlled seasonal flow with high periodic discharge of water and sediment. These result in a somewhat unstable environment where there is relatively rapid evolution of channels and in which braiding is characteristic.

In areas where the glacier **forefield** comprises the low reaches of a valley, glaci-fluvial sediments will provide substantial alluvial infill giving rise to

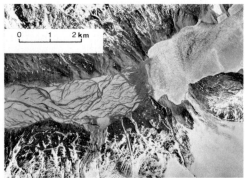

Figure 10.9A   Alluvial valley train infilling the Tasman Valley, Southern Alps, New Zealand. A primary west-bank meltwater distributary from the Tasman Glacier is shown. **B** Aerial photo of the extensive proglacial braided outwash delta and sandur at the head of Itirbilung Fjord, Baffin Island, NWT, Canada. The fjord lies to the right and is ice-covered. Note the small glaciers and terminal moraine complexes entering both the fjord and the valley. (Photo 11 June 1958 courtesy Canadian Dept. Energy Mines and Resources and J. Syvitski.)

valley trains or braidplains (Figure 10.9A). Where more extensive lowland areas receive outwash, either individual or interlapping alluvial fans **prograde** to form **sandar** and large-scale **braidplains** (Figure 10.9B).

10.5.1 Channel characteristics

Dominant features of the proglacial depositional zone are low gradient fans. Boothroyd (1976) indicates that gradients in the proximal zone are of the order 0.006–0.017 decreasing to 0.002–0.003 at the distal margin. Fans are prograded by ephemeral **bars** and channels, which constitute braided streams. Individual channels vary in size from a few metres to hundreds of metres in width with corresponding sizes to the bars and alluvial islands around which streams bifurcate. The depth of channels is typically a few metres.

The braided character of proglacial channels arises principally from the steep gradients of glacial valleys and glacier forefields upon which sediments are deposited, high sediment fluxes and marked changes in water discharge. For several proglacial outwash areas relations between discharge channel morphology (i.e. width, depth) and mean flow velocity can be established. While some similarities are apparent between glacial channels and temperate fluvial systems the width–depth ratio versus discharge is significantly different, characterized by an extremely low value of 20, and indicative of the presence of shallow braided streams. There appears to be little change in this ratio with increasing discharge.

There is often a clear relationship between size of sediment and channel slope and distance downstream from the glacier terminus – reflecting processes of clast attrition in highly turbulent channels (Figure 10.10). Rice (1982) has used the downstream decrease in particle size of the Sunwapta River braidplain to account for several aspects of the observed **hydraulic geometry**. Boothroyd (1976) suggests that segments of individual braidplain fans may be grouped by average clast size: sand-sized material (< 2 mm) in the distal region, sediments between 100 and 2 mm in the middle region and > 100 mm in the proximal zone.

10.5.2 Bedforms and internal structures

The sediments resulting from braided channel deposition range from boulders and cobbles through gravels to sands. Miall (1977, 1978, 1984) has given a useful summary of the primary sedimentary structures occuring in modern braided stream deposits with a brief interpretation of their

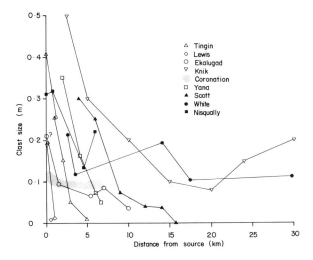

Figure 10.10 Relationship between clast size and distance from glacier terminus (stream source) for variety of outwash streams. Although there is a clear trend for size to diminish with distance some reversals are evident. These may relate to a new input of clastic material to the principal channel as described by Fahnstock (1963). Data from several sources.

Table 10.1 Lithofacies code and related structures for braided stream deposits (Miall, 1977 and Eyles *et al.*, 1983)

Facies identifier	Lithofacies	Sedimentary structures	Interpretation
Gm	gravel, massive or crudely bedded, minor sand, silt or clay lenses	ripple marks, crossbeds in sand units, gravel imbri-cation	longitudinal bars, channel-lag deposits
Gt	gravel, stratified	broad, shallow trough cross-beds imbrication	minor channel fills
Gp	gravel, stratified	planar cross-beds	linguoid bars or deltaic growths from older bar remnants
St	sand, medium to very coarse, may be pebbly	solitary or grouped crossbeds	dunes (lower flow regime)
Sp	sand, medium to very coarse, may be pebbly	solitary or grouped planar cross-beds	linguoid bars, sand waves (upper and lower flow regime)
Sr	sand, very fine to coarse	ripple marks of all types, including climbing ripples	ripples (lower flow regime)
Sh	sand, very fine to very coarse, may be pebbly	horizontal lamination parting or streaming lineation	planar bed flow (lower and upper flow regime)
Ss	sand, fine to coarse, may be pebbly	broad, shallow scours	minor channels or scour hollows
Fl	sand (very fine), silt, mud, inter-bedded	ripple marks, undulatory bedding, bio-turbation, plant rootlets, caliche	deposits of waning floods. overbank deposits
Fm	mud, silt	rootlets, desic-cation cracks	drape deposits formed in pools of standing water

genetic environment (Table 10.1). Five main processes are considered responsible for these: longitudinal bar formation, bedform generation and migration, channel scour and fill, low-water accretion and overbank sedimentation. Collinson (1978) suggests gravels will display **imbrication**, upward fining and horizontal bedding although some cross-stratification may be present. The sand units comprise **tabular sets**, cross-bedded sometimes with ripple-drift cross-laminations. Infilled channel sequences may exhibit clays and silts in sharp contact with gravels, all resting upon erosion surfaces. Since glacial streams exhibit marked discharge variations periods of low flow will give rise to **veneers** and **drapes** of mud and silt. As previously demonstrated there is often a clear relationship between size of sediment and channel slope and distance downstream from the glacier terminus.

Sediments of braided glacial streams are formed principally into a rather complex distribution of incomplete bars and channel fill. Although many types of bars have been described (e.g.

longitudinal, lobate, side, lateral, diagonal, point, scroll, spool, chute, etc.) Miall (1978) points out that only those giving rise to distinctive lithological assemblages or structures should be defined. Three principal types of channel bar are thus recognized (Figure 10.11), although it should be emphasized that the nomenclature for some of these features is not clearly defined and by no means widely accepted. To add confusion the term 'bar' is not necessarily ideal! For features which have spacings of 5–100 m, small height/spacing ratios and have characteristic flow velocities in the order of >0.3–0.4 and <0.7–$0.8\,\mathrm{m\,s^{-1}}$, Harris *et al.* (1975) have proposed the term **sandwave**. Such bedforms lie between the dynamic ranges (defined by channel depth, water velocity and sediment size) of ripples and dunes.

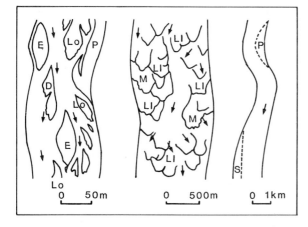

Figure 10.11   Principal types of bar or sandwave. The sandwave types are P, point; S, side; LI, lingoid; M, modified lingoid; Lo, longitudinal; D, longitudinal with diagonal flow; E, eroded sandwave remnant. Arrows indicate water flow direction (after Miall 1978).

Longitudinal sandwaves or bars
These features are sometimes also referred to as 'diamond bars' because of their distinct lozenge shape. They may reach a size of several hundred metres length with a vertical relief of over 1 m. Constituents are principally sand and gravel which are often massively bedded with tangential to planar cross-lamination of sand and exhibiting

reactivation surfaces. They commonly occur in mid-channel, are elongate downstream (that is, parallel to the flow direction) and migrate by the development of **slipfaces** along the bar edges although this is rare in gravel-dominated forms (Figure 10.12). In addition the top surface of such bars may be marked by secondary bedforms such as ripples and transverse ribs of coarser material. Rust (1978a) has suggested that sandwaves are stable bedforms even at flood stages. 'Diagonal' bars are longitudinal sandwaves which form obliquely to the principal channel trend and comprise an intermediate, geometrical category between longitudinal and transverse bedforms although their internal structure is similar to longitudinal sandwaves. They are often encountered where channel braiding is more pronounced and there is flow across the bar.

Transverse sandwaves or bars
Such bedforms are also known as 'cross-channel' bars and where they possess cusp or spoon-like shapes the term 'linguoid' sandwave is frequently applied (Figure 10.13). Bar length is up to a few hundred metres with a vertical relief typically of 0.5–1.0 m. They are found in coarse pebbly units of low sinuosity although more common in sandy alluvial channel networks. Walker and Cant (1979) suggest that transverse sandwaves possess a nucleus which is emergent at low-water stages and grows by the downstream prolongation of lateral 'horns'; upstream expansion by accretion of smaller migrating sand dunes; their upper surfaces tending to dip gently upstream. Bluck (1974) found that

Figure 10.12   Longitudinal sandwaves: morphology, water flow and structure (from Boothroyd and Ashley, 1975).
A aerial photograph of a number of longitudinal sandwaves or bars on the Upper Yana River, Alaska. Water flow is from left to right.
B diagrammatic interpretation of **A**, showing sandwave boundaries and water flowlines.
C internal structure illustrating characteristics of the slip-face and of planar cross-bedding. Large cusp-shaped ripples are present on the surface of the sandwave. A silt drape may be present overlying reactivation surfaces. There is only indistinct ripple cross-bedding.

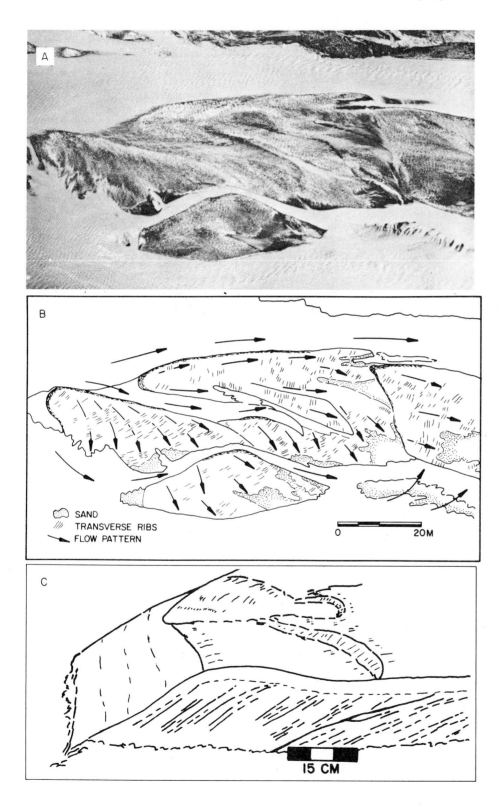

SAND
TRANSVERSE RIBS
FLOW PATTERN

0   20M

15 CM

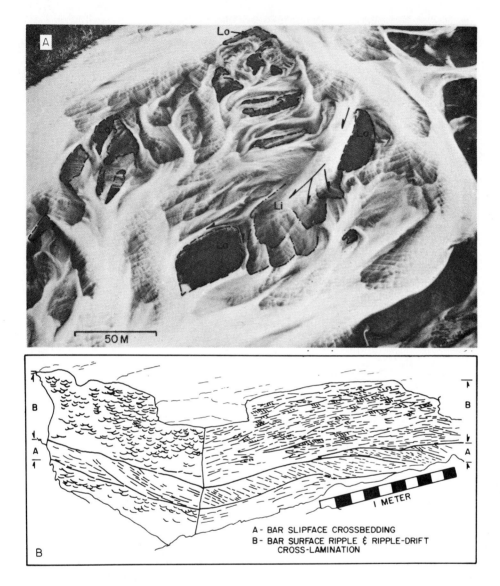

Figure 10.13   Transverse (lingoid) sandwave:
morphology, water flow and structure (from
Boothroyd and Ashley, 1975).
A Aerial photograph of transverse sandwaves (some
longitudinal sandwaves are also present) on the Scott
River, Alaska. Waterflow is from top to bottom of the
illustration.
B Internal structure of a transverse bar showing
characteristics of slip-face, large-scale cross-beds, and
ripple-drift cross-lamination.

transverse sandwaves which contain a significant fraction of sand-sized material exhibit laterally extensive progressive slip-face migration. This movement leads to the development of tangential to planar cross-bedding and pronounced ripple-drift cross-lamination. Transverse bars frequently occur in groups with out-of-phase relationships one to another (Miall, 1977) so that bars advance into the space between two preceding bars (Collinson, 1970).

Point, side and lateral sandwaves or bars
Such features originate in zones of low energy. They possess highly variable dimensions, from a few centimetres to kilometres in length (the latter for some side bars). The internal structure of such bars is complex – resulting from their history of coalescence of smaller bed units (e.g. dunes). Miall (1978) indicates that planar-tabular cross-bedding is common but may merge with trough cross-bedding resulting from scours or dune forms. Ripple marks and coarse-grained **lag** deposits may be present as well as drapes and fills of finer-grained sediments.

Fabric
Several other features of braided, bar and channel proglacial sedimentation are evident from the studies available. Strong orientation of large clasts takes place. According to Boothroyd and Ashley (1975), a predominant trend is evident for the long axes of oblate particles to be transverse to the channel (and therefore flow) direction. Rust (1972, 1975), however finds that the orientation of clasts shows a high degree of variability but with a fabric maximum close (i.e. parallel) to the flow direction as given by major channel features. The orientation of other structures (such as ribs and gravel stripes) appear to vary with discharge (Bluck, 1974). As most proglacial channels exhibit very high flow regimes, imbrication is commonly observed as a stable fabric characteristic. Rust (1975) found an average upvalley dip to coarse gravel clasts of $25° – 35°$ angle.

10.6 Facies associations and channel evolution

Several authors have attempted to summarize the complex sedimentation occuring in active braided glacial channels and develop generalized facies associations to describe the varying influences of channel depth, water discharge, sediment size and concentration (especially bedload). Boothroyd and Nummedal (1978) introduced the concept of 'core facies' and accompanying 'lateral and distal facies' which make up the complete facies assemblage (Figure 10.14), and based upon fans in southern Alaska and southern Iceland. Recognition is given in their scheme to a progressive downslope facies change from proximal conditions characterized by coarse gravel and few channels, to distal conditions with increased braiding and dominated by sand fractions. The concepts are tied into the lithofacies scheme of Miall (1977, 1978) given previously in Table 10.1.

10.6.1 Core facies

In the proximal and mid-slope fan region longitudinal bars and gravel sheets dominate with cross-stratification (shown in Figure 10.14). During recessional flow sandy foresets may be developed around the margins of bars in the mid-fan area. Bedforms, including dunes with trough–trough wavelengths of 0.6–6.0 m develop under these condition on the tops of some bars. In the distal zone transverse bars become dominant being dissected at low-flow stages with lobate slip-face development.

10.6.2 Lateral–distal facies

The core facies possess lateral and distal facies which are characterized by diagnostic sedimentary structures (Figure 10.14). In some regions vegetation may encroach and colonize the inter-stream areas. This occurs in Alaska as described by Boothroyd and Ashley (1975) and Boothroyd and Nummedal (1978). Extensive marsh and swamp trap fine-grained sediments during overbank flow stages. These in turn promote a change from

164

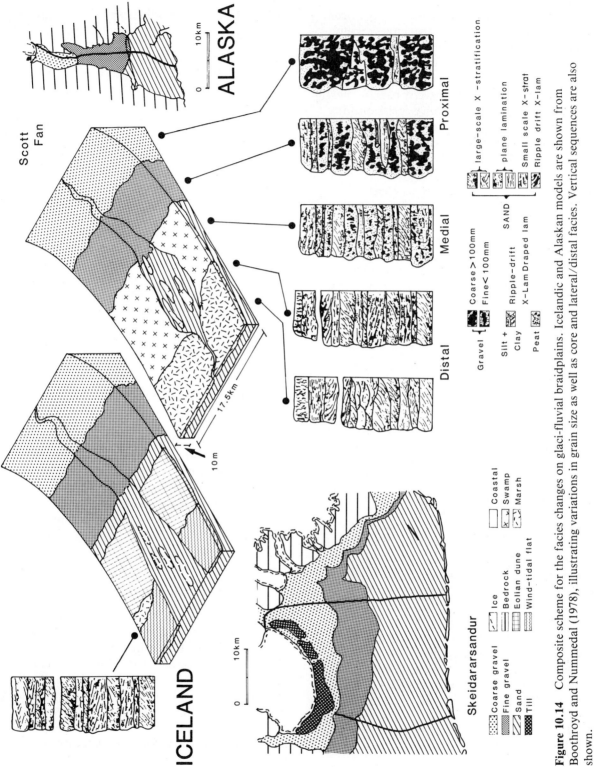

Figure 10.14 Composite scheme for the facies changes on glaci-fluvial braidplains. Icelandic and Alaskan models are shown from Boothroyd and Nummedal (1978), illustrating variations in grain size as well as core and lateral/distal facies. Vertical sequences are also shown.

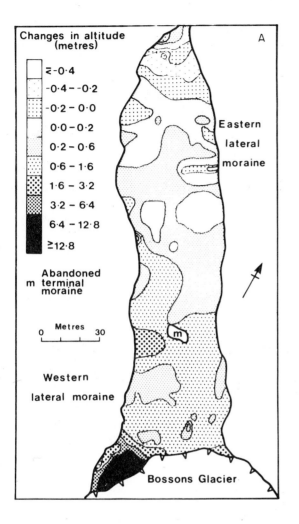

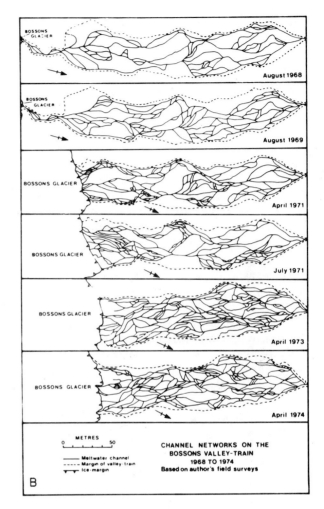

Figure 10.15   Changes to the braidplain of the Glacier des Bossons (from Maizels, 1979).
A Surface elevation changes between April 1973 and April 1974.
B Changes to the braided channel network between August 1968 and April 1974 following an advance of the glacier.

braided to **meandering** stream habit, with development of lateral and point bars. A layer of peat may be found on the tops of more stable inter-fan areas.

In regions where the environment is too hostile for extensive vegetation growth (such as southern Iceland, parts of the Canadian Arctic and Greenland) the inter-stream zone may be characterized by aeolian dunes or tidal flats. The former are usual of the upper, distal area; the latter at the lower, proximal fringe (Boothroyd and Nummedal, 1978). Intermittent shallow flooding of these zones may give rise to some cross-stratification. The tidal flats possess straight-crested and small lobate-shaped bars. Planar and cross-stratification are considered to be present with horizontal laminations and a fine-grained **flood-stage** cap.

10.6.3 Channel evolution

Major changes take place in channel pattern and surface elevation produced by strongly contrasting periods of erosion and deposition in response to water and sediment discharge and glacier activity. Naturally there is long-term evolution of the channel systems as deglaciation, for instance, takes place affecting the magnitude of water and sediment supply (Maizels, 1983). During such changes steep braided, low-sinuosity channels with a high width–depth ratio and carrying a large bedload are translated into deeper, more sinuous low-gradient systems with significant suspended loads.

In the short term there is a constantly shifting pattern of braiding leading to rapid evolution of most braidplains and sandur. Fahnstock (1963) made a detailed study of numerous sections of the White River, Washington which were resurveyed several times over a period of about two years. Overall change recorded in the river basin over a reach of 2 km was + 0.36 m, aggradation being noted at all but one of the of 12 cross-sections. This was equivalent to 48 000 m³. Most change occurred at a period of greatest discharge in late summer (June–July).

Maizels (1979) monitored changes in the braidplain of the Glacier des Bossons, following a rapid ice advance between late 1969 and the spring of 1971. Proglacial aggradation resulted in a net addition of some 7 500 m³ over a six-year period (mean increase of 1 300 m³ a⁻¹) although the annual variability was high with a net loss in some years. Most change was recorded in the proximal zone as shown in Figure 10.15, and on the west side where greater quantities of sediment were available and a steep, little-vegetated rock wall was present. The increased discharge of sediments from advancing Glacier des Bossons also resulted in an increase in the density of braided channels in the glacier forefield (Figure 10.15). According to Maizels the number of channel segments rose over the five years to April 1974 from 344 to 500, and the total length of channel from ~1 890 to 2 800 m. At the same time the number of bars increased from 58 to 160, although their size was reduced. Their overall shape was, however, maintained within 15%. There was a corresponding reduction in channel width from 2.0 to 1.2 m.

11   Glaci-lacustrine processes and sedimentation

11.1   Introduction

Lakes are a common feature of glacierized terrain and result essentially from the derangement and ponding of streams and rivers by ice and glacial deposits (Figure 11.1). Lakes form under a wide range of conditions and have been well described in other works (see Embleton and King, 1975; Maag, 1969; Liestøl, 1956). They are frequently found in subglacial and supraglacial locations (see also Chapter 2). Many are ice marginal (at the sides or close to the termini of glaciers); others may form in glacial deposits revealed by the melt-out of stagnant ice (creating **kettle ponds**) or by the blocking action of moraines and glaciers themselves. Many glacial lakes occupy basins eroded or over-deepened by glacial action. On the larger scale isostatic uplift following ice retreat can cause reversals and blocking of drainage resulting in the formation of large glacial lakes.

In general there is a complex spectrum of sedimentation in such lakes from marginal fluvial outwash through deltaic to deepwater lacustrine deposition.

11.2   Physical processes in glacial lakes

An understanding of the physical and chemical limnology of glacial lakes is important since they control sedimentation and the characteristics of the lake deposits.

11.2.1   Density

In large measure the chemical and physical behaviour of glacial lakes and processes of sedimentation are governed by differences in and changes to the density of lake waters. The density of pure water is $1\,000\,\text{kg m}^{-3}$ at a temperature of $+3.98°C$ (not the freezing point). The change of density with temperature is shown in Figure 11.2. Although differences are small they are nevertheless highly significant. Density is also increased by the concentration of dissolved salts

Figure 11.1   A Nigardsvatnet, Norway: a small glacial lake created by the retreat of Nigardsbreen (foreground) across a bedrock hollow in Nigardsdalen since about the 1920s. On the right-hand side there is a delta forming in the lake. Photo: August 1970. **B** Lake Joyce, Pearse Valley, Antarctica. A cold, amictic lake with a perennial ice cover, dammed by a small glacier to the right.

which, in cold glacial lake waters, is of the order of 10 to $100\,\text{mg l}^{-1}$. Density variations induced by salinity are therefore usually small. Increasing salinity, however, decreases the maximum density of water at a rate of $0.2°C\,\text{g}^{-1}$ as shown in Figure 11.2. Since salinities are typically low this effect is again of small magnitude.

Two other minor factors which have an effect on the density of water are the quantity of dissolved gas (such as O, CO_2, H_2S) and hydrostatic pressure. This latter quantity may depress the temperature at which the maximum density is encountered according to:

$$tm = 3.98 \text{ Deg} - 2.27 \times 10^{-3} (P_h - 1) \text{ MPa} \tag{11.1}$$

where P_h = hydrostatic pressure.

A final and quite critical factor which affects the *bulk* density of water is solid matter in suspension. The effect of sediment in lake waters and in streams entering lakes may frequently override those of temperature and salinity in controlling density, and thus govern the behaviour of water flow and sedimentation patterns. We shall deal with this aspect of suspended sediments in a later section (11.2.6).

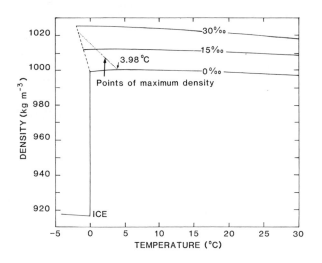

Figure 11.2   Density of water as a function of temperature at various salinities. Ice is also shown.

11.2.2 Thermal conditions

The heat absorbed and dissipated by water is fundamental to establishing the thermal structure, stratification, circulation pattern and ecological behaviour of glacial lakes.

A simple heat budget equation for a lake can be written as:

$$Q_T = Q_R + Q_h + Q_L + Q_g \tag{11.2}$$

where Q_T = change in heat stored in the lake
   Q_R = input from net solar radiation
   Q_h = exchange from **sensible heating**
   Q_L = specific **latent heat** exchange from evaporation or melting of surface ice
   Q_g = input from geothermal sources at lake bottom.

The input of heat will vary from season to season (although most glacial lakes experience only two major seasons – summer and winter) and between lakes, although differences are considerably smaller than in lakes from non-glacial environments. The contribution from solar radiation is usually the most important parameter.

11.2.3 Stratification

Thermally controlled density **stratification** is a common characteristic of most lakes although the degree to which glacial lakes stratify is usually considerably less and highly variable. In general, spring warming develops fairly uniform temperature conditions through the water body, close to 4°C, which enables easy mixing (or turnover) by wind energy. Summer heating establishes an upper layer of warm water which due to its lower density resists mixing. A temperature difference of only a few degrees can inhibit vertical circulation of the entire water column. At this time the lake water is separated into three primary regions (Figure 11.3). The upper layer, which may be only a few metres deep, is termed the **epilimnion**. It is composed of moderately turbulent water, circulating freely and, consequently, possesses a relatively uniform temperature. It is separated from the deeper cold and fairly undisturbed water of the **hypolimnion** by a marked gradient in temperature – the **thermocline**. The layer of water in which this thermal discontinuity occurs is the **metalimnion**. A density change also occurs across the metalimnion and is termed the **pycnocline**.

   In the autumn and winter surface waters cool,

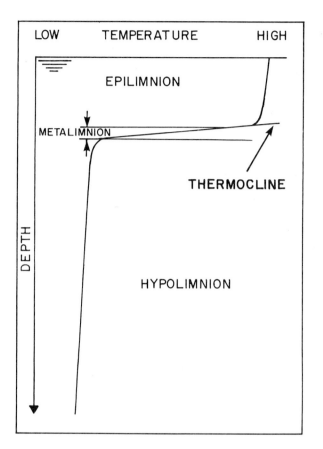

Figure 11.3   Characteristic thermal regions of lake waters.

become more dense and sink, eventually leading to complete turnover. In glacier-fed lakes the picture may be more complicated due to considerable quantities of low-temperature water contributed from ice melt. These ultra-cool surface layers inhibit stratification or, if they are input at depth, disrupt the hypolimnion. There are few studies of glacial lakes and it is difficult to assess how typical their profiles may be. Nevertheless, strong spatial and temporal contrasts in thermal structure are evident (Smith *et al.*, 1982).

In Malaspina Lake, Alaska, dammed among the southeastern margin of Malaspina Glacier, described by Gustavson (1975), summer temperatures close to the lake surface reach 5°C while at depth are typically <2° (and sometimes

down to 0.3°C). The maximum density is achieved in the upper 10 m giving reverse stratification. This pattern is quite different from that described by Campbell (1973) from the smaller South Cascade Lake, Washington located at the snout of South Cascade Glacier. In summer a pronounced epilimnion may be established while in the winter complex inverse stratification may be observed: colder (<4°C) but less dense water overlies warmer but denser water (at 4°C). Such conditions are not very stable and thermal resistance to mixing is weak. Small amounts of wind energy can easily disrupt the stratification.

The type of stratification found in lakes in which a hypolimnion forms has been classified by Hutchinson (1957) in accordance with the mixing circulation pattern. Glacial lakes of concern to this chapter exhibit **amictic**, cold **monomictic, dimictic** and occasionally transient **polymictic** characteristics [mictic = mixing].

Amictic lakes are found almost exclusively in Antarctica, northern Greenland and Arctic Islands of Canada. They possess a perennial ice cover through which solar insolation and conduction take place. There can be little direct wind coupling though wind and barometric pressure will strain and flex the ice lid and induce some oscillating motion in the water column. In exceptional circumstances such as at Lake Vanda in Wright Valley, southern Victoria Land, Antarctica the elevated geothermal heat flux raises bottom temperatures as high as 25°C and gives rise to remarkable inverse thermal stratification.

The water in *cold monomictic* lakes is usually at or below 4°C and possesses only a weakly developed thermocline and has *one* period of turnover in the summer. Such lakes occur primarily at high latitudes and high altitudes, often in proglacial areas and frequently in contact with glaciers. Surface temperatures in the summer are mainly <10°C but may rise to 10–15°C when the ice cover is temporarily melted for 2–3 months. Glacial lakes are subject to considerable variations in their thermal balance from year to year due, in most part, to the timing of break-up of the ice cover. Monomictic characteristics may thus be evident during one year while in the succeeding year a lake may exhibit a dimictic or even polymictic aspect.

Dimictic lakes are the most common type of lake. They turnover *twice* a year, in the spring and autumn, and are, in most cases, normally stratified in summer and inversely stratified in winter. South Cascade Lake in Washington exhibits such characteristics in some seasons.

Polymictic behaviour may characterize certain cold glacial lakes if conditions of warming, influx of water and wind coupling prevent stratification and induce continuous or semi-continuous turnover at temperatures close to 4°C.

11.2.4 Circulation and water movement

The movement of water in proglacial lakes is usually achieved by waves and currents, tides being of relatively little significance.

Waves are generated by wind-coupling to the lake surface, by the calving of large blocks of ice in lakes marginal to glaciers, and from barometric pressure variations. Waves are responsible for suspended sediment transport within the body of the lake and along lake shores, their effect being dependent upon predominant wave direction, shoreline orientation and fetch. Waves are responsible for mixing the upper lake layers (epilimnion), especially those generated by iceberg calving events. Where there is significant vertical penetration and disturbance of the water column by icebergs such activity works against temperature stratification. Glacial lakes adjacent to glaciers or ice caps may experience the effects of persistent **katabatic winds**. Along with iceberg calving (and low atmospheric pressures) such winds can be responsible for producing oscillating waves or **seiches** in the lake which affect the motion of the entire water body. Katabatic winds induce tilting of the lake surface and thermocline, piling-up water downwind. When the wind abates the water surface is tilted back but, due to the great momentum, oscillation occurs about a nodal point. Such seiche behaviour may occur at the surface or internally.

Geostrophic effects on seiche activity cannot be ignored. The **Coriolis force** (a function of latitude and current velocity) induces deflection of a moving body of water – to the right in the northern hemisphere and to the left in the southern hemisphere (zero at the equator). Hamblin and Carmack (1978) have investigated this effect in detail with a model applied to Kamloops Lake in British Columbia.

11.2.5 Water inflow and mixing

The input of water and sediments to the lake from melt streams is one of the most important factors to be considered in glaci-lacustrine sedimentation. The nature of sediment discharge from glaciers has been outlined in Chapters 2 and 10 and these conditions and quantities should be borne in mind in this section. In the first instance it is important to identify the location of the water inputs to the lake.

If the lake is some distance from the glacier terminus streams will bring water to the lake at a high (surface) level with respect to the lake basin. Where ice-damming occurs, however, glacial meltwater may be discharged at the glacier base or at some intermediate height through englacial conduits. Although difficult to identify, upwelling of groundwater in front of a glacier due to the reduction in cryostatic pressure (often called a 'fountain effect') may frequently contribute water to the lake bottom.

The entry of water from a glacial river or a glacial channel into a lake can be understood by analogy with the outlet of a pipe into a standing body of water. An early application of such theory was made by Bates (1953) and more recently by Wright and Coleman (1973), Fisher *et al.* (1969) and Wright (1977). Near the stream outlet (as shown in Figure 11.4) the boundary between the moving fluid and the relatively still lake water is well defined and there is a pronounced velocity gradient between the two. Turbulent eddies are created if the velocity shear overcomes any stability arising from density stratification. The ratio of these two factors defines the **Richardson number** (R_i), and thus

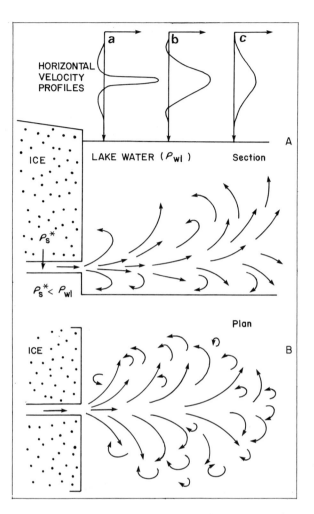

Figure 11.4   Pattern of water diffusion into a glacial lake from a subglacial tunnel shown in section (**A**) and plan (**B**). The density of lake water is ρ_{wl} and that of the influent stream ρ_s*.

Three horizontal velocity profiles are shown. These could be superimposed over the plan of water outflow in **B**. They show also the mean velocity of the water mass is reduced upon entering the lake but that an increased quantity of water is in motion so that momentum is preserved.

determines whether entrainment and diffusion of the incoming stream will take place in the lake water.

It is often useful as in the case studied here to define a Richardson number (see also Equation 12.1) related to an upper layer flowing relative to a lower layer of different velocity and density:

$$R_i = \frac{(\Delta\rho/\rho_{wl}) \, g \, h_{di}}{U_{w*}^{\,2}} \qquad (11.3)$$

where h_{di} = depth of the upper flow of influent stream water

   ρ_{wl} = density of lake water

   $\Delta\rho$ = density difference (between influent stream water and lake water)

   U_{w*} = velocity of influent stream water relative to lower layer.

The greater the density difference between inflowing stream waters and those of the lake, the larger the value of R_i and the more difficult any turbulent exchange and mixing between the two. The incoming stream water will, under these circumstances, maintain its integrity within the lake as a discrete density current. Such activity is important in the resulting pattern of sedimentation from density currents. The Richardson number is also important for hydrographic processes in estuaries and is further discussed in Chapter 12. If R_i is low the two water bodies will mix by entrainment and diffusion and create a layer of intermediate velocity. The mixed boundary becomes less pronounced as the velocity gradient, between incoming and lake water, diminishes away from the stream infall. Full homogenization of influent and lake water may take place in the absence of stratification as described for Lower Waterfowl Lake in Alberta by Smith *et al.* (1982).

From the velocity profiles shown in Figure 11.4 it is possible to see that speed of the incoming stream is reduced but that an increased quantity of water is in motion: the momentum of the incoming water mass is preserved. In two dimensions this is equivalent to the integral of the velocity–width profile [∫ U_w dw] being identical at a, b and c. No pressure changes are associated with the dissipating stream so no slope is imparted to the lake surface.

If momentum is preserved, kinetic energy of the incoming stream is, in contrast, dissipated and converted into low-grade heat by an amount $U_w^2/2g$. This reduction in kinetic energy leads to the settling of any suspended particles by reducing stream transporting power.

11.2.6 Over-, inter- and underflow

The relative density and hence buoyancy of stream water entering a glacial lake will determine into what position within the water mass it will initially flow. The density of water, as described in Subsection 9.2.1, is principally dependent upon temperature, salinity and suspended sediment content.

Many of the key physical processes of flow into lakes are similar to those occurring in estuaries where a substantial body of observational data has been collected and a good theoretical understanding of water movement developed. As these will also be considered in the next chapter only a general account of flow behaviour is given. Three principal modes can be recognized.

Overflow (hypopycnal)

When the density of the overflowing water from a channel of cross-sectional width (W_x) is less than that of the lake water ($\rho_s^* < \rho_{wl}$) it spreads out (diffuses) in a turbid plume on and somewhat beneath the surface. According to Bates (1953) the flow is analogous to a two-dimensional (plane) jet in which the outward diffusion distance is $10^3\,W_x$. Lateral spreading is also restricted and takes place at a rate $0.2\,W_x$ per width equivalent in the downflow direction (x). More rigorous numerical studies of sinking and floating plumes have been undertaken in recent years by several workers (e.g. Svensson, 1978; Raithby, 1976; Bork, 1978) which show the important complications introduced by vertical and horizontal turbulent eddies and heat losses.

Turbulent exchange may take place at the base of the outgoing prism or plume of water if the density contrast is not pronounced and dU_w/dz is large (i.e.

low R_i). Such phenomena are also dealt with in Chapter 12. In other cases, where the water layer densities are strongly contrasting, there is little turbulent diffusion and some entrainment will take place.

If low-density, buoyant water is input from a subglacial conduit at the base of a glacier it will rise to the surface of the lake and there spread out in a similar manner. The distance from the glacier front at which the overflow commences will depend upon the stream trajectory governed by velocity of the jet and the density contrast.

Hypopycnal flow is observed in glacial lakes but some workers (e.g. Smith, 1978) believe that such conditions are not as common nor as well developed as inter- or underflow. Typical conditions favouring overflow are periods of low suspended sediment discharge from streams (such as in the spring and autumn), where intermediate ponds have removed quantities of sediment and when stream waters, from non-glacierized catchments especially, have been significantly warmed relative to lake water. It should be remembered that density differences between inflowing and lake waters vary considerably and generalizations even about one lake during one season cannot necessarily be made. Awareness that flow behaviour and hence sedimentational characteristics can, and do change dramatically, will measurably assist in understanding glaci-lacustrine phenomena (see Smith, 1981).

To illustrate more precisely the nature of overflow conditions it is instructive to examine specific cases. Gustavson (1975) reports a stream discharging into Malaspina Lake, Alaska (Figure 11.5) whose suspended sediment concentration is low (due to stilling in smaller lakes upstream) and approximately equal to that of the lake. No surface turbid sediment plume was visible. Temperature measurements indicate that the stream was warmer than the lake thus producing hypopycnal flow – reflected in the velocity profile which shows almost stagnant water at depth in the lake and maximum movement towards the surface (Figure 11.5). Mathews (1956) also reports periodic overflow conditions in Garibaldi Lake, British Columbia, and Smith *et al.* (1982) occasional overflow behaviour from Hector and Bow Lakes in Alberta.

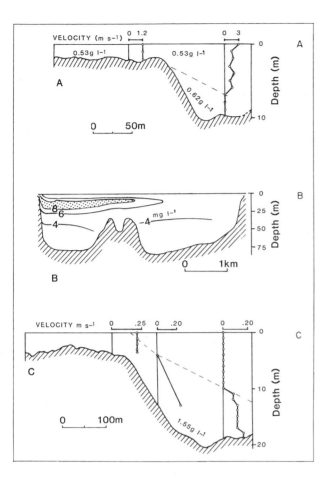

Figure 11.5 **A** Overflow conditions in glacial lakes. Section through Tarr Stream and into Malaspina Lake, Alaska. Vertical velocity profiles are shown at two locations. Suspended sediment concentrations in the stream and lake are given (from Gustavson, 1975). **B** Interflow conditions in glacial lakes. Longitudinal section of Hector Lake, Alberta. Shaded areas show zone of highest suspended sediment concentration and lowest transmissivity which lies on the south side of the lake (from Smith *et al.*, 1982). **C** Underflow conditions in glacial lakes. Section through Russell Stream and Malaspina Lake, Alaska showing vertical velocity profiles. Suspended sediment concentration in lake is shown (from Gustavson, 1975).

Interflow (homopycnal)

If influent water possesses a density similar to or greater than the epilimnion but less than that of the metalimnion or hypolimnion the water will enter as a plume at intermediate depth. According to Bates (1953) such conditions are illustrated more clearly by considering the inflow as a three-dimensional (axial) jet in which mixing takes place more readily than in plane jet flow. As a consequence energy is readily dissipated and the interflow plume will extend only to some $10^2 W_x$. Gilbert (1972) and Church and Gilbert (1975) consider interflow a very common phenomenon of deep glacial lakes, although the condition may not be sustained over long periods of time.

Factors which affect the development of continuous interflow are diurnal variations in discharge, suspended sediment content and temperature of glacial lake waters. Mathews (1956) reports the occurrence of interflows in Garibaldi Lake while velocity–depth measurements along the ice margin of Malaspina Lake demonstrate the presence of several interflows with speeds up to $10 \, \mathrm{m \, s^{-1}}$, interpreted by Gustavson (1975) as discharging from englacial or subglacial conduits. Figure 11.5B shows longitudinal cross-sections of Hector Lake, Alberta indicating the presence of a well defined, shallow interflow identified from measurements of conductivity and the concentration of suspended sediment.

Underflow (hyperpycnal)

Underflow is frequently claimed as the most common type of density flow in glacial lakes: incoming meltwaters possess densities greater than those of the lake. This results from both cold meltwater and high suspended sediment content and is most prevalent during summer months. Smith (1978), however, asserts that the number of studies of glacial lakes is small and such generalizations are, therefore, unsupported.

Where underflow can be demonstrated the cold influent water sinks down the sides of the lake basin and thence outward for a considerable distance over the lake floor. Church and Gilbert (1975) report underflows that extend 5–6 km from the source stream in Lillooet Lake, British Columbia.

The flow characteristics will again be analogous to a plane jet.

The descending water may behave as a turbidity current in which there is a turbulent suspension of entrained particles. Considerable treatment of the mechanisms and resulting deposits **(turbidites)** of turbidity currents is available in the sedimentological literature and interested readers are referred to the detailed work of Middleton (1966, 1967, 1970) and Middleton and Bouma (1973). Many features of these **gravity flows** are similar to the motion witnessed in snow avalanches.

Two types of turbidity currents may be defined, low-density and high-density, depending upon the concentration of suspended sediments within the flow (which may range from $100-350\,\mathrm{g\,l^{-1}}$). The division between the two may be placed at the density imparted by a suspension of sand-size quartz grains ($110\,\mathrm{kg\,m^{-3}}$). According to Friedman and Sanders (1978) lacustrine turbidity currents usually fall into the low-density category, although this division cannot be held as exclusive.

The frontal lobe or head of the descending gravity flow will move downslope with a mean velocity governed by the flow resistance at the bottom and between the two fluids:

$$U_h = k_g (\Delta \rho / \rho_{wl}\, g\, h_l)^{1/2} \qquad (11.4)$$

where k_g = velocity coefficient related to bed
            slope (0.7)
        g = gravitational acceleration
        $\Delta \rho$ = density difference between
            turbidity current and lake water
            (ρ_{wl})
        h_l = thickness of the frontal lobe.

Behind the head of the current is the main body of the flow which travels at a higher velocity with resulting mixing or transfer. The head region is characteristically one of net erosion for underlying strata. In a lake the velocity of the head may be maintained by steady flow down the slope and its magnitude is given by:

$$U_h = [8g/k' + k'')\, \Delta \rho\, h_l'\, \beta] \qquad (11.5)$$

where k', k'' = drag coefficents for flow over

the bed and at the fluid interface
h_l' = thickness of the zone of uniform
            flow
β = bed slope.

Velocity profiles were made by Gustavson (1975) in Malaspina Lake, Alaska along the outflow of Russell Stream and by Smith *et al.* (1982) in Peyto Lake, Alberta which identify and illustrate underflow conditions (Figure 11.5).

11.3 Chemical and biological processes in glacial lakes

Neither the chemical nor biological characteristics of glacial lakes have been investigated in a systematic manner and few detailed studies of individual lakes are available for illustration. Although the roles of lacustrine chemical and biological process are of importance in the overall pattern of sedimentation this section is not intended to provide a comprehensive review of them but merely to draw attention to salient aspects. Certain biological activity is discussed later in this chapter in relation to specific processes.

The chemical composition of glacial lakes depends primarily upon the input of chemical species from influent streams and rivers and their characteristics have already been described in Chapter 5. Direct precipitation from the atmosphere into lakes is also important as is shoreline erosion which stirs sediments and allows **littoral** mixing. Surface evaporation may be a dominant process in controlling the composition of certain lakes where there is little stream input. The highly saline lakes of certain ice-free areas of Antarctica result from this process and achieve salinities up to $500\,\mathrm{g\,l^{-1}}$. Minerals may also originate within the water column itself by precipitation, absorption or organic activity. In addition, exchanges take place with lake-floor sediments by diffusion and advection and thus control the flux of nutrients (e.g. nitrate and phosphate) and silica. Jones and Bowser (1978) discuss the principal transport and reaction paths for lake waters.

Most glacial lakes are biologically poor and may thus be described as **oligotrophic**. The frequent occurence of high salinities and the impermanence of glacial lakes make them severe environments for colonizing organisms which must withstand high **osmotic pressures** or dessication and temperatures to below –40°C (Heywood, 1972). In many lakes the growing season is severely curtailed by either the freezing of the water body or snow cover which reduces **irradiance** below that required for **photosynthesis**. In addition, at high latitudes light conditions are seasonally variable and unfavourable for photosynthesis during several months of the polar night. The influx of large quantities of sediments is also unfavourable for certain organisms (especially **benthic forms**) which may become rapidly silted where sedimentation rates are high. Nevertheless fine-grained bottom sediments may be organically rich.

Heywood (1972) comments that in spite of such adverse conditions lakes may support a significant biota and abundant communities of **plankton** have been reported for many hundreds of lakes in western Canada. Similar conclusions have been drawn in regard to Antarctic lakes although the latter are typically less rich and diverse.

Despite a modest biological presence in glacial lakes the direct effects of organisms in sedimentation are difficult to evaluate and probably highly variable. It is clear, however, that in comparison with temperate lakes there are few reports of significant accumulations of biogenic material in glacial lake sediment sequences, and biological disturbance (**bioturbation**).

11.4 Sedimentation

11.4.1 Sediment input

Sediment in suspension and in bed traction in glacial meltwater streams provides the primary source material for deposition in glacial lakes. As outlined, additional contributions may be derived

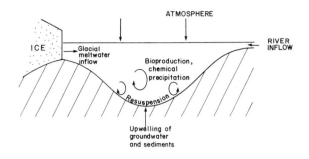

Figure 11.6 General sources of sediment in a glacial lake.

from atmospheric precipitation (including volcanic events), hydrochemical precipitation and biogenic activity, upwelling of material from groundwater flow and resuspension from bottom current activity. These sources are summarized in Figure 11.6.

For glacial lakes stream input and resuspension are considered the most important. Details of sediment characteristics and transport in proglacial streams were discussed in the previous chapter. It is again important to recognize, for the purposes of lacustrine sedimentation, that the glaci-fluvial environment is highly variable and this has led Smith (1978) to designate five inflow regimes: diurnal, sub-seasonal (of a few hours to days duration due to changes in local weather conditions), seasonal (related to snowmelt in the spring and glacier ice melt in the summer), annual and exceptional events such as jökulhlaups. Variations are related to complex hydrological conditions within the glacier of melt, storage and flow routeing (see Chapter 2). The distance from neighbouring glacier meltwater sources will also affect sedimentation in proglacial lakes by altering sediment load characteristics. This will also be apparent if a single stream feeds through a series of lakes where deposition will progressively scavenge suspended sediments (Smith, 1981) (Figure 11.7). The bulk of the sediment discharge in glacial streams occurs during the summer melt period when terrigenous clastics are contributed to the

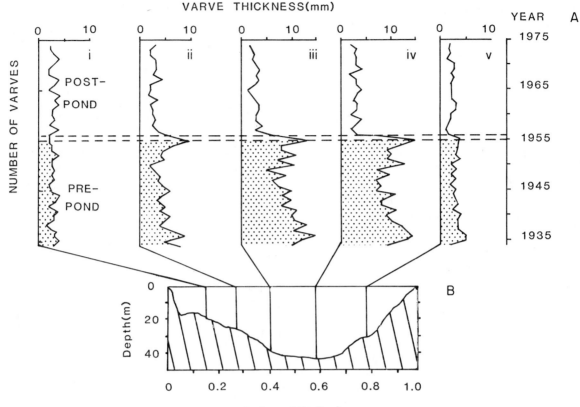

Figure 11.7 **A** Vertical distribution of the thickness of varves in five cores from Bow Lake, Alberta extending over a period of 40 years. In response to the formation of a pond the thickness of varves decreases abruptly after 1955 in cores iii and iv, less so in ii and virtually not at all in i and v. **B** Transverse section of Bow Lake showing the sampling stations (after Smith, 1981).

lake. The concentration of suspended sediments into glacial lakes is thus highly variable with values as low as a few mg l^{-1} and up to tens of g l^{-1} in extreme cases. The quantity of bedload, an equally important contribution, is very much more difficult to estimate but as suggested in Chapter 10 may be at least 50% of the total load carried by the stream.

Sediments in suspension alter the bulk density of inflowing stream waters, and thus depositional patterns. It is likely that suspended sediment content is the single most important factor in controlling the type of flow within the lake basin – frequently outweighing the effects of temperature. Only when sediment concentrations are very low such as during the winter season may temperature control become important.

11.4.2 Settling of sedimentary particles

Particles introduced into a water body such as a lake, fjord or estuary will settle out and give rise to distinctive sedimentary patterns on the bed. Although the principles of the settling process are relatively easy to comprehend the exact formulation of applicable relationships is highly complex and depends upon such factors as the size and shape of the particles, their density and concentration, the viscosity of the immersing fluid and its characteristic turbulence.

When the density of a particle (ρ_c) exceeds that of the surrounding fluid it settles through the water. In a still, unstratified fluid of density (ρ_{wl}) a constant terminal **fall-velocity** is reached when the net gravitational force on the particle (of radius r_c) due to its submerged weight $Y = [4/3\pi r_c^3 \, g \,(\rho_c - \rho_{wl})]$ is just balanced by the drag force. The drag term depends upon the Reynolds number. For very low Reynolds numbers, where fluid behaviour is laminar (i.e. $R_e < 500$), the drag can be calculated from **Stokes' Law** equation assuming a small (10^{-1} mm) spherical particle:

$$F_d = 6\pi \, \eta_w \, r_c \, U_e \qquad (11.6)$$

where η_w = fluid viscosity
            U_e = velocity of the sphere relative to the fluid.

At higher Reynolds numbers and those likely to be experienced in sedimentation the Newton equation is applicable:

$$F_d = \pi/8 \,(D_g \rho_{wl} \, 4 r_c^2 \, U_e^2) \qquad (11.7)$$

where D_g = drag coefficient.

Combining Equations (11.6) and (11.7) above, the fall velocity becomes

$$U_t = 4/3 \,(g/D_g) \, 2r_c \left(\frac{\rho_c - \rho_{wl}}{\rho_{wl}} \right) \qquad (11.8)$$

D_g is difficult to determine and will depend upon Reynolds number and particle shape. Figure 11.8A shows the drag coefficient as determined in relation to various theoretical and empirical estimates. By determining the submerged weight of the particle the approximate Reynolds number can be obtained. Figure 11.8B gives the fall velocity, U_t, as a function of particle size in water. It can be seen that water temperature has an effect through its influence on fluid viscosity.

Figure 11.8   Fall velocity of quartz spheres in water for various temperatures and Reynolds Numbers (after Rouse, 1972). The inset **A** shows drag coefficient of spheres as a function of Reynolds Number for various experimental determinations. D = Discs, S = Spheres.

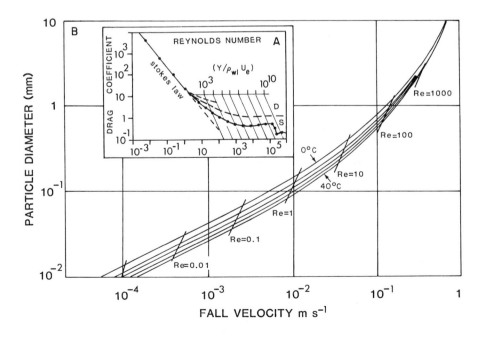

Only single particles have so far been considered. A high concentration of sediments will affect the terminal velocity as they tend to fall in a group with higher U_t.

The settling velocity of small particles is important since in the presence of a horizontal current as experienced in lakes it will determine the distance sediments are advected before deposition and thus the resulting pattern of lake-floor deposition. In glacial lakes a complicating factor will be water stratification. Changes in water density due to salinity and temperature alter the term ($\rho_c - \rho_{wl}$) and thus slow or speed the descent of particles. It is possible, therefore, (following the suggestion of Gilbert, 1982) to use entrainment and suspension functions developed from streamflow studies to provide order-of-magnitude estimates for the size of particles transported or the velocities required in carrying material some distance away from the infall of lakes (or fjords).

11.5 Facies characteristics

Lacustrine sedimentation will reflect the dominance of and fluctuations between underflow, interflow and overflow conditions. Nevertheless there will be a systematic reduction in **flow competence** away from the stream infall. Sediments will settle out progressively as the lake buffers the most distal regions from input variations. Such activity results in a very general

graded sequence characteristic of glaci-lacustrine deposits (Figure 11.9). Gravels, from braided streams, form topsets to outwash deltas in the proximal lake zone. They grade into steep foreset beds of sand and gravel with distinctive ripple-drift cross-laminations and draped laminations. Further out such sequences merge with weakly graded and alternating strata of pro-delta **bottomsets** which eventually blend, in more distal areas, with lake floor deposits of progressively weaker lamination.

11.5.1 Basin margin sedimentation

If incoming glacial stream waters are either less dense or as dense (hypopycnal or homopycnal) an arcuate delta will usually form at the lake margin, possessing strong similarities with marine deltas. The dispersive, reworking role of waves and tidal processes are, however, significantly reduced in glacial lakes. Bedload is rapidly deposited close to the river mouth as velocity diminishes, and does not feature beyond the point of density separation. In this zone topset

Figure 11.9 Typical sequences of sedimentation in glacial lakes. Note the gradation from glaci-fluvial deposits through deltaic units to the pro-delta area and thence into deeper lake bottom sediments. There is usually a progressive decrease in grain size from the stream to centre of the lake interrupted only by slumping or sliding on the delta slope.

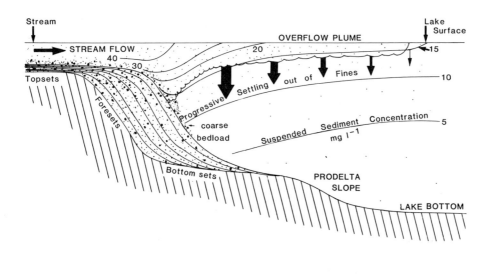

Figure 11.10 Ice-marginal deltaic sequence showing foreset beds in former glacial Lake Hitchcock, Connecticut Valley, Massachusetts, New England. Trash in foreground is accidental and non-glacial.

beds form an extension of the braided stream channel with gravel bar sequences dominated by crude planar bedding (Gustavson *et al.*, 1975). Coarse bedload will mix with finer sandy material generating foreset beds (Figure 11.10). If the coarse grade material is dominant higher slopes, close to the angle of rest (i.e. up to ~30°), may be sustained, while sands usually dip at less than 15° and are characterized by ripple-drift cross-lamination (Jopling and Walker, 1968). Further out in the distal zone of the delta slopes diminish to <5° and bottomset beds are constructed, usually as a graded sequence, which merge with true lake bottom sediments (Figure 11.11A).

In lakes where there is pronounced underflow the descent of turbidity currents down the basin sides may significantly inhibit deposition and in places cause active erosion. The bulk of suspended material and a large quantity of the bedload is transferred directly to deeper parts of the lake floor where a distinctive sediment suite is produced, characterized by ripple-drift and cross-lamination (Figure 11.11A and B). Jopling and Walker (1968) and Gustavson *et al.* (1975) describe units of ripple-drift cross-lamination and draped lamination which characterize the distal portions of the pro-delta slope. The former constitute sets of climbing laminae with either well or poorly developed stoss-side laminations

(Figure 11.12). Draped lamination comprises parallel laminae deposited from suspensions of sand, silt and clay which are draped over underlying bedforms (usually ripple-drift cross-lamination) (Figure 11.11B).

Houbolt and Jonker (1969) indicate that persistent underflows eventually develop subaqueous fans in front of the stream mouth with systems of channels, lateral **levées** and lobes. Gustavson *et al.* (1975) discuss the form of marginal glaci-lacustrine sediments which accord with this picture. They found that the major stream entrant to a glacial lake is likely to be braided. Each of the individual **distributaries** contributes a plume of sediment to the construction of a composite delta with individual lobes extending out across the lake floor from the delta front, thinning and widening with distance. Bottom topography and the overall configuration of the delta constrain the size and shape of the lobes. Several authors have also described the failure and slumping of basin margin sediments, including the fronts of prograding deltas (Shaw, 1977b; Shaw *et al.*, 1978; Gilbert, 1972). This phenomenon, which disturbs the more regularly graded sequences, may result from changing lake levels. Migration of the distributaries in response to glaci-hydrological conditions leads to the widespread progradation of the delta by overlap of adjacent and migrating lobes. The changing pattern of lobes in turn alters bed topography so there is a feedback effect to the process of sediment transport/deposition by underflows.

11.5.2 Lake floor sedimentation

Away from the lake margin, clastic sedimentation occurs by progressive settling-out of material carried in suspension. Smith (1978) in his study of Hector Lake, and Smith *et al.* (1982) for several other glacier-fed lakes in Alberta have clearly demonstrated the down-lake change in sedimentation rate and grain size (Figure 11.13).

Turbidity currents, generated by underflow or by episodic slumping, carry coarse material to the lake floor and develop a laminated (sometimes even micro-cross-laminated) deposit of sand

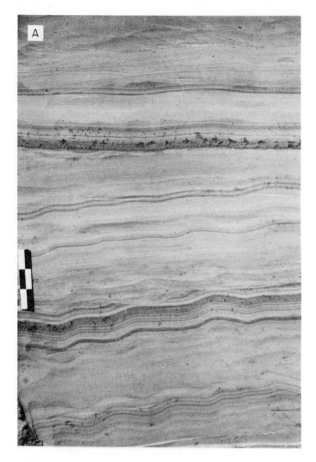

Figure 11.11 Glaci-lacustrine sediments.
A From shoreline of Malaspina Lake. The figure shows glaci-lacustrine pro-delta slope sediments which consist of fine sands and silts deposited as a sequence of ripple-drift overlain by normal and reverse-graded undulatory beds. The scale is 150 mm (from Gustavson, 1975).
B Sediments from Bennett's Brook delta, glacial Lake Hitchcock. Draped lamination and ripple-drift lamination form a sequence which commences with a unit of draped lamination shown at the level of the middle of the arrowhead. A second unit of draped lamination occurs at the top of the section. Two thin clay layers occur in the bottom draped lamination unit, but not in the top one. Water flow was from left to right. Scale is 300 mm (from Gustavson et al., 1975).

Figure 11.12 Draped laminations comprising parallel laminae overlying ripple-drift cross-lamination with a nearly vertical angle of climb. Most laminae are devoid of any slip-face development. Flow was from left to right. Scale is 300 mm (from Gustavson et al., 1975).

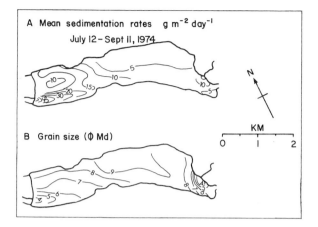

Figure 11.13   Pattern of sedimentation rate and grain size in Hector Lake, Alberta, Canada. Mean sedimentation rates determined by bottom sediment traps. Median grain diameter also shown (from Smith, 1978).

and silt. The thickness of such units is often $10-10^2$ mm with a marked decrease in the thickness of laminations away from the inflow zone (Figure 11.14). Data provided by Smith (1978) and Gilbert (1975) on **varve** thicknesses have been normalized and plotted in Figure 11.15. A clear negative exponential trend of thickness with distance is apparent.

Smith *et al.* (1982) emphasize the importance of lake bottom topography in determining the pattern of sedimentation where underflow activity is dominant. **Transmissivity** cross-sections of Peyto Lake, Alberta picked out the sediment transport paths commencing on the eastern side of the proximal basin adjacent to the river infall. The sediment plume then spread out and crossed the proximal basin to the western side, passing through a narrow channel into the distal basin and then returned to the east bank and the lake outfall. The persistent increase in the minimum value of transmissivity of lake water with distance from the river mouth was due to the progressive settling out of sediment to the basin floor.

When and where underflow activity is not evident (such as during winter months or due to fluctuations in discharge) simple vertical Stokes'-law settling of particles takes place

Figure 11.14   Sections of rhythmites from Malaspina Lake, Alaska. Three cycles of upward-fining sediments are shown, each representing one year's sedimentation and hence are 'varves'. There are numerous graded beds in each sequence (from Hartshorn and Ashley, 1972).

producing a weak grading of fine silt and clay. These units, only a few millimetres to centimetres thick, grade from silty-clay at the base to fine clay at the top and often terminate abruptly by a new underflow influx of coarse material (giving a sharp contact). In the most distal areas variations in sediment inflow may be sufficiently damped-out to give rise to homogeneous fine clay.

The processes described above give rise to a gradual facies change from the units characteristic of the pro-delta region to true lake-floor deposits. Cross-laminated and sometimes draped laminated units which merge into finer-

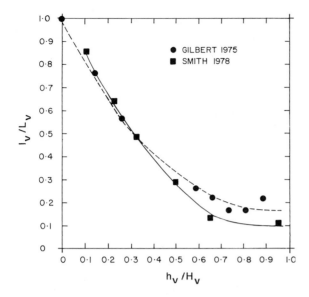

Figure 11.15 Normalized varve thickness (h_v = individual varve thickness, H_v = total thickness of varves) as a function of normalized distance from water infall (l_v = distance to sample point, L_v = total lake length). Based upon Canadian data from Gilbert, 1975 for Lake Lillooet, British Columbia; and Smith, 1978 for Hector Lake, Alberta.

grained laminated and weakly graded sequences of the pro-delta zone thus wedge into fine-grained sequences exhibiting **rhythmic** structure (Figure 11.14). The pattern sometimes shows pronounced rippling and sometimes cross-lamination indicative of deposition from underflows. Grading is evident due to the weakening, migration or reduction in suspended sediment concentration of the underflows. Smith (1978) has recognized a fourfold facies grouping of such laminated deposits based upon distance from the inflow: proximal, immediate proximal, immediate distal and distal.

An intriguing aspect of fine-grained sedimentation has been identified and discussed by Smith and Syvitski (1982) in an attempt to explain settling rates of fine silt and clay-sized material which are faster than those predicted by Stokes' law. According to data from Bow Lake, Alberta, for instance, suspended particles travelling with an interflow velocity of ~ 200 mm s^{-1}

would be discharged from the lake after 2.1 days. During this time particles < 4 μm (8ϕ) would settle vertically by less than 1.7 m and thus the bulk of suspended clay particles should be transported out of the lake. Yet the floor of Bow Lake is composed of considerable quantities of clay and such material is known to settle out in significant quantities from direct measurements in sediment traps. It is also clear that the rate of sedimentation of fine material is largely independent of total sedimentation rates and water input. Smith and Syvitski (1982) reject several hypotheses for explaining this paradox – overturning of water in the autumn to bring fine material caught in the epilimnion towards the lake bottom, and **flocculation** (see Subsection 12.4.1). The authors suggest biological intervention may provide an answer. Many cold-water glacial lakes possess significant plankton communities (Anderson, 1971, 1974). Certain free-swimming and **pelagic crustacea** such as **copepods** ingest suspended mineral particles. This material is later egested as *fecal pellets* which are ovoid in shape, range in size from 100–250 μm (350 μm maximum) with a dry weight of ~ 30 μg. The settling rates of such pellets in water at 5°C are between 0.9 and 4.6 mm s^{-1} – equivalent to a quartz sphere of diameter 43–96 μm. Thus pelletization may be responsible for increasing the downward flux of suspended fine-grained sediment to the lake floor – disrupting and confusing the 'normal' pattern of bottom sedimentation.

It is clear that at any location on the lake bottom the presence of the alternations of fine sediments from settling and coarser materials from underflows will be determined by: distance from the lake margin, the vigour and shifting habit of turbidity currents (in part controlled by subaerial, glacial and hydrological conditions), and lake water stratification. A useful summary of various facies in relation to water and suspended sediment is given by Stürm (1979) in Figure 11.16.

Complex events, therefore, on timescales varying from very much less than one year to several years may create rhythmic sequences, and suggest considerable caution is required in the

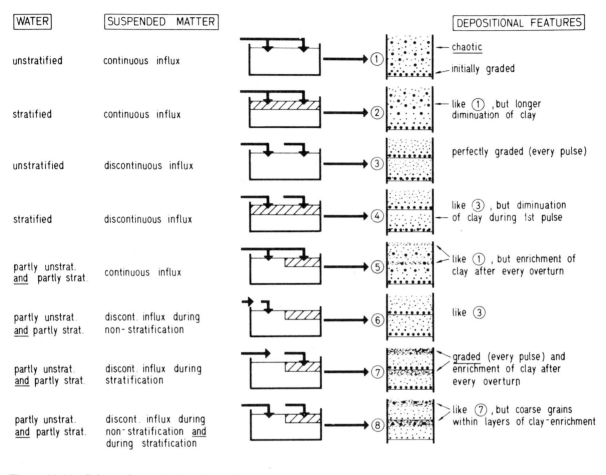

WATER	SUSPENDED MATTER		DEPOSITIONAL FEATURES
unstratified	continuous influx	①	— chaotic — initially graded
stratified	continuous influx	②	— like ① , but longer diminuation of clay
unstratified	discontinuous influx	③	perfectly graded (every pulse)
stratified	discontinuous influx	④	— like ③ , but diminuation of clay during 1st pulse
partly unstrat. and partly strat.	continuous influx	⑤	like ① , but enrichment of clay after every overturn
partly unstrat. and partly strat.	discont. influx during non-stratification	⑥	like ③
partly unstrat. and partly strat.	discont. influx during stratification	⑦	graded (every pulse) and enrichment of clay after every overturn
partly unstrat. and partly strat.	discont. influx during non-stratification and during stratification	⑧	like ⑦ , but coarse grains within layers of clay-enrichment

Figure 11.16 Schematic models for depositional features resulting from variable influx of suspended sediments and variable water stratification for an oligotrophic lake. Random dashes distinguish clay; light dots are for silt and heavy dots for sand. Example 7 represents an ideal varve (from Sturm, 1979).

interpretation of traditional varve couplets. In order to establish annual change Østrem (1975) has shown that it is necessary to use several methods in addition to the simple counting of couplets. These include grain-size variations, **X-ray diffraction** to point up differences in summer and winter laminae and any historical documentation on the development of a particular lake basin.

11.5.3 Littoral sedimentation

Shoreline processes in glacial lakes are not dissimilar to those from other environments. Wave activity, however, is usually inhibited for part of the year by the presence of an ice cover. Katabatic winds, however, may set up seiches and more modest wave activity which suppress longshore sediment transport. The effects of movement of the ice cover against the shore, are to produce small ice-push features. According to Pyokari (1981) considerable shoreline damage may result from ice-push action with over-thrusting, simple horizontal thrusting, intrusion activity (small boulder, fold and wedge ramparts) and the ploughing of littoral sediments. Ice-push ridges may reach heights of up to several metres. Two principal mechanisms are responsible for driving ice into the shoreline – thermal ice expansion (Alestalo and Haikio, 1979) and wind coupling. The former is inhibited if there is significant winter snow cover to insulate lake ice from low winter air temperatures. Wind drift of ice floes against the shore is more pronounced in large lakes.

Ice-push features may be subsequently modified by wave action. Several papers dealing with such process, principally in the marine environment, have been given in a volume edited by Dionne (1976) and are further discussed in Chapter 14. Some entrainment, by wave stirring and freezing, into lake ice can redistribute nearshore sediments and is dealt with in more detail in Chapter 14. Minor ridging in lake ice can produce keels which scour bottom sediments, activity which is similar to that produced by grounding of icebergs. The fluctuations of lake levels is also an important factor in the development of nearshore sedimentation patterns. Ice-dammed lakes may exhibit periodic drainage while others, stable for many years, may drain suddenly with important consequences for sedimentation. It should be mentioned, however, that the draining of such glacial lakes offers a unique opportunity to inspect and sample the various facies discussed above.

12 Glaci-estuarine processes and sedimentation

12.1 Introduction

Glaciers may terminate in the sea at the head of fjords and coastal inlets (Figure 12.1). In some circumstances where a glacier reaches the fjord mouth, as is frequently found around Antarctica, or where water depths are particularly great in the inner fjord reach (the case for many inlets around Greenland and some Canadian Arctic islands) the glacier comes afloat for a short distance. In other cases, however, glaciers are grounded on the sea floor in typically less than 200 m of water. These latter conditions are found in British Columbia, parts of southern Alaska, Svalbard and Chilean fjords. A major difference between these two groups of glaciers is the significance of subglacial melting in sea water. Sedimentation in such glacier-dominated **estuaries** is dependent upon the transport of debris, which is in turn controlled by the supply of detritus and fjord hydrography.

12.2 Sources of sediment

Sediment may be supplied to the marine environment from a variety of sources and by a

Figure 12.1 Tidewater glaciers terminating in fjords.
A Ferrar Glacier tongue, southern Victoria Land, Antarctica. The lower part of the glacier supports melt pools in the summer and is here fringed with fast ice. Taylor Glacier, terminating on land, in the adjacent valley (right) is shown with Lake Bonney in front of the snout.
B Snout of a small glacier entering Scoresby Sund, east Greenland. Note the heavy crevassing and the medial moraine.

number of mechanisms, the principal ones being illustrated in Figure 12.2. Streams enter the fjord carrying quantities of debris entrained during their routeing from basins inland of the flanks of the fjord. If an ice mass is present within the catchment it may constitute a more or less significant source of streamflow and sediments. As noted in Chapters 2 and 10, such glaciers will also play an important role in influencing the discharge of streams. Otherwise debris will be entrained during the passage of rivers through adjacent terrain where a considerable volume of sediment resulting from recent deglaciation is often present. The role of water discharge from streams and rivers on fjord hydrography will be discussed in Section 12.3.

Figure 12.2 Composite model of the several processes operating in the glaci-estuarine environment. Input of sediments from glaciers, wind, streams, rockfalls and sea ice are described in detail in the text.

Many fjords and glacial estuaries are steep-sided from recent glacial erosion, and many are found in mountainous coastal terrain. Sub-aerial rockfall and dirty snow-avalanching off valley sides are common. Such activity, which may be seasonal, either prograde submarine **talus** terraces into the fjord (Prior *et al.* 1981; Gilbert, 1982) or carries debris across winter sea ice to be further transported and later deposited during early summer break-up and melt.

Tidewater glaciers entering fjord waters are usually another source of sediment. Excepting numerous coastal localities of Antarctica, such ice masses will exhibit pronounced summer ablation with the release of sediments in the ice to adjacent waters. Calving is also a frequent occurrence and icebergs containing debris will float out where they too will release their sediment load to the ocean floor (Ovenshine, 1970). The special process of sedimentation from icebergs is dealt with in

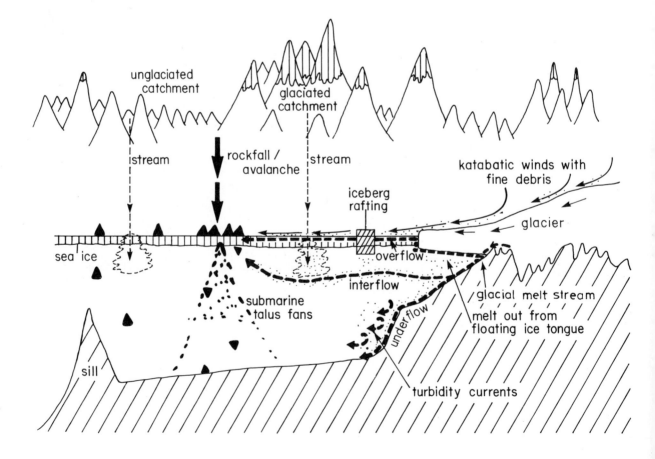

Chapter 14. Fine-grained material may be carried by strong, often katabatic winds onto winter sea ice or directly into fjord waters. Church (1972) and Gilbert (1982) suggest that some silt and fine sand in fjords may be derived from wind-borne debris from areas of exposed glaci-fluvial outwash.

12.3 Fjord hydrography and associated processes

Fjords represent one class of estuaries in which very deep water (up to 1 000 m) and a restriction near the mouth (a sill) are usually present. They are essentially semi-enclosed, coastal water bodies with free connection to the open sea. Fjord circulation is driven by the input of fresh water from the melting of glacier ice, melting of sea ice, direct precipitation and discharge of rivers into the fjord. Lake and Walker (1976) provide a useful case study of the

freshwater input to a typical fjord in Arctic Canada. Their results are summarized in Figure 12.3.

12.3.1 Water stratification and estuarine circulation

One of the principal factors affecting the temperature and salinity structure of fjords and distinguishing them from other estuaries is seasonal sea ice, but relatively few studies have been made of winter hydrographic conditions. In extreme cases fast ice may be retained within sheltered high-latitude fjords for many years (in Antarctica, northern Canada and Greenland). In other fjords, little or no sea ice may form at all (Chile, Norway, British Columbia, New Zealand). The effects of sea ice on fjord hydrography are quite pronounced. As sea ice forms in the autumn salt is rejected into the subjacent water, a process well described by Cox and Weeks (1974). The resulting high-salinity water is unstable and as it sinks induces gravitational **convective mixing** in the upper layer (to a depth of a few metres beneath the ice) a process which will tend to inhibit settling of suspended sediment. Most first-year sea ice has an

Figure 12.3 Freshwater inputs into a typical glacial estuary, d'Iberville Fjord, Ellesmere Island, NWT, Canada. The size of the various symbols is proportional to their relative contributions (after Lake and Walker, 1976).

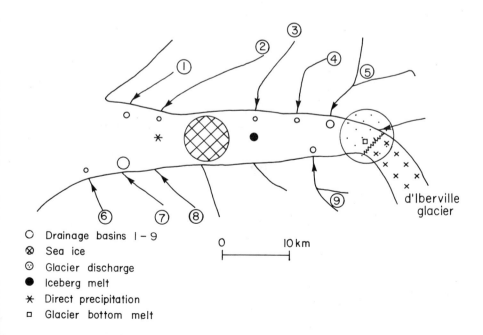

O Drainage basins 1 – 9
⊗ Sea ice
⊙ Glacier discharge
● Iceberg melt
✱ Direct precipitation
▫ Glacier bottom melt

average salinity of only a few parts per mil, decreasing with increasing thickness (Sinha and Nakawo, 1981). When the sea ice and possible snow cover warms and breaks up in the early summer the water released is of very low salinity.

Superimposed upon this stratification is the effect of rivers and melting tidewater glaciers. Significant freshwater input occurs only in summer and creates an **estuarine circulation** of varying importance in which an upper freshwater layer, moving down-fjord, is compensated at depth by a landward flow of saline water. Figure 12.4 depicts a longitudinal section of a fjord in which the cold water prism, extending down-inlet from a tidewater glacier, can be clearly identified. The thickness of the surface layer is usually in the order of a few metres, attenuating seawards as the fresh water mixes with the underlying saline waters.

Mixing of these different water masses takes place by two mechanisms: entrainment and

Figure 12.4 Longitudinal hydrographic section through a typical glacial estuary (based upon d'Iberville Fjord, Ellesmere Island, NWT). Temperatures are given at spot soundings and have been contoured; shading shows cold, freshwater prism from glacier input at the head of the fjord. Scattered salinity measurements are shown in ppm. Note the presence of a major threshold at the mouth (after Lake and Walker, 1976).

diffusion. Diffusion occurs between two density strata which exhibit turbulent flow. Although equal quantities of water are exchanged between the layers there is a net loss of salt from the lower saline horizon. Dyer (1973) and McDowall and O'Conner (1977) indicate that in a stratified fjord or estuary the density gradient between saline and fresh water tends to inhibit the exchange of momentum by turbulent flow. A critical velocity shear across the boundary of the two layers is, therefore, necessary in order to induce mixing: thus giving rise to the entrainment mechanism. The critical condition for entrainment is governed by the density stratification and can be considered as the balance between density stabilizing forces and the destabilizing velocity shear. As shown in Subsection 11.2.5 this condition is described by the Richardson Number R_i, written out here in a more general form than Equation (11.3):

$$R_i = (g/\rho_w)\,(d\rho_w/dz)/(dU_w/dz)^2 \qquad (12.1)$$

where   g   = acceleration of gravity
        ρ_w = water density
        (dU_w/dz) = critical velocity shear.

When $R_i > 0$ stratification is considered stable, and when $R_i < 0$, unstable. At $R_i = 0$ the water is neutral and unstratified. Flow is usually laminar below $R_i \simeq 0.25$ and turbulent above it.

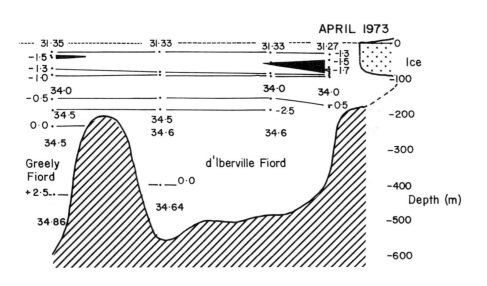

The influence of density stratification may be alternatively examined by comparing the outward flow velocity of a freshwater prism to the upstream propagation velocity of a travelling wave along the density interface (pycnocline):

$$F_i = U_{w*}/[(\triangle\rho/\rho_u)g\,d_i]^{1/2} \qquad (12.2)$$

where $\triangle\rho$ = density difference between water layers

ρ_u = density of the upper layer

d_i = depth of the upper layer

U_{w*} = velocity of the upper layer relative to lower layer.

F_i is the interfacial Froude number. When $F_i \rightarrow 1$ the upstream propagation of waves along the interface of the two layers is inhibited, they increase in amplitude and, as they break, give rise to pronounced vertical mixing by entraining small amounts of underlying saline water. The net result is a quasi-vertical flow of water, called the entrainment velocity, U_{en}:

$$U_{en} = V_n(U_w - 1.15\,U_{cw}) \qquad (12.3)$$

where U_{cw} = critical velocity of an upper layer for breaking of internal waves

V_n = coefficient proportional to $\triangle\rho/\rho_w$ and thus to F_i.

Entrainment results in an increased discharge, down the estuary, of water in the upper layer. In a fjord where the channel sides may be nearly parallel the increase in discharge is also reflected in a velocity increase (Figure 12.5). To compensate for water entrained and carried seawards a landward flow is established just below the pycnocline in the salt wedge. These two contraflow components comprise estuarine circulation.

Figure 12.5 illustrates a velocity profile for a typical fjord fed by glacial meltwater exhibiting a marked seasonal discharge. In the summer with high runoff ($10-10^2$ m^3 s^{-1}) the surface layer moves out rapidly and there is considerable entrainment. The compensating current is consequently well developed. In winter when discharge is low little entrainment occurs and the velocity profile shows only a weak counter-

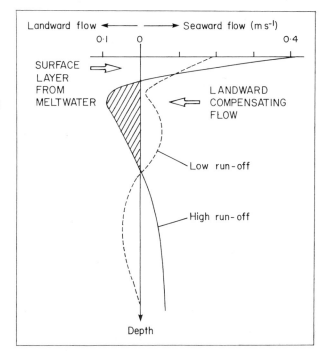

Figure 12.5 Vertical velocity section through the upper levels of a typical fjord to illustrate the principal characteristics of estuarine circulation in a glacial environment. Meltwater is added to a surface layer from the landward side (left). At depth this outward flux is compensated by a landward flow (at deeper levels when run-off is low).

current, due to tidal influence. Elverhoi et al. (1983) report a surface layer, 5–10 m thickness, of salinity 31–32‰ and temperature between 2° and 4°C for Kongsfjorden, Spitsbergen while Schei et al. (1979) noted an upper layer of salinity 5–20‰ near the head of Van Mijenfjord, Spitsbergen.

If the fjord threshold or sill is deep, circulation will not be significantly restricted, but where the sill is high replenishment by the deeper, saline water becomes difficult and only occasional. This situation may give rise to **euxinic** conditions where bottom waters in the fjord basin are isolated and renewal from the open sea limited (see Section 12.4.4). Bottom water will also be cold in summer at between −1 and +1°C.

12.3.2 Tidal effects

Tidal influences, which may be highly variable, drive **barotrophic currents** and circulation in fjords. During summer months runoff from stream catchments and contribution from glaciers may be dominant, while in the long winter period currents are forced principally by the tide. In Maktak, Coronation and North Pangnirtung Fjords on the east coast of Baffin Island, NWT, Gilbert (1982) reports only weak tidal action – surface currents of ~ 2.8 mm s^{-1}, and a tidal excursion of <0.12 km. Mathews and Quinlan (1975) report mean diurnal tide ranges of up to 5 m in Glacier Bay, Alaska which give rise to surface current velocities of 3 m s^{-1}.

Tidal movements will have an important effect in modulating the outflow of rivers and streams into an estuary or fjord and are possibly important for bottom sediment processes (Powell, 1983). Relling and Nordseth (1979) report a strong tide-induced variation in the velocity of a river discharging into Gaupne Fjord in western Norway. Variations were found to be in the order of 1 m s^{-1} with importance for the transport into and deposition of sediment within the fjord.

12.3.3 Coriolis effect

In controlling the pattern of sedimentation in fjords considerable influence is attached to the Coriolis force. Streams entering a fjord are commonly seen to be deflected (towards the right-hand shore in the northern hemisphere and towards the left shore in the southern hemisphere) – indicating the effect of the earth's rotation. This phenomenon is not restricted to fjords but is also well documented from large lakes. Tongues or wedges of coarser sediment are frequently seen to extend along the sides of fjords and lakes corresponding to the deflected plumes and there is often pronounced asymmetry in bottom topography as a consequence. Interested readers are referred to Hamblin and Carmack (1978) for the further investigation of the Coriolis effect for Kamloops Lake in British Columbia.

12.3.4 Sediment input

Streams entering fjords bring with them considerable quantities of sediments in suspension and as bedload. Although the estimation of bedload is difficult (as discussed in Chapter 10) coarse material plays a significant role in proximal glaci-estuarine deposition, particularly deltaic progradation. Unfortunately there are few studies of the details of bedload sedimentation from glacial streams in fjords. Suspended sediments, however, have received significantly more attention. Turbid water is often visible as murky plumes extending from the fronts of tidewater glaciers and from the entrance of streams into the fjord (Figure 12.6).

Figure 12.7 illustrates the variation in suspended sediment concentration with distance in Howe Fjord, British Columbia. From Chapter 10 it is clear that the concentration of suspended sediments entering the fjord will vary considerably both seasonally and on shorter time-

Figure 12.6 Suspended sediment plume issuing from the front of Idunbreen, an outlet glacier of Vestfonna in Nordaustlandet, Svalbard, width of ice front is 4.5 km. Note two superglacial lakes. (courtesy Norsk Polarinstitutt, Oslo).

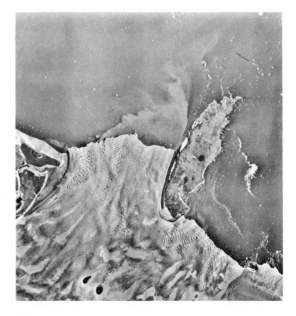

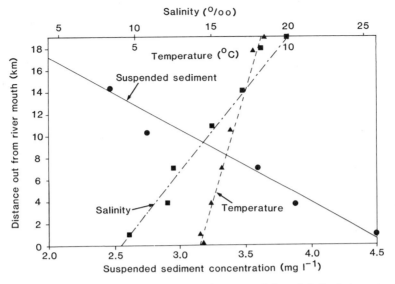

Figure 12.7 Reduction in suspensoid concentration in fjord waters as a function of distance from the source of stream input. Temperature and salinity variations are also shown (after Syvitski, 1980).

scales, related to discharge events (and ultimately to meteorological factors) – a situation confirmed by Syvitski (1980). Elverhoi *et al.* (1980) were able to estimate the likely sediment concentration (C_s) for meltwater streams issuing into Kongsfjorden from Kongsvegen by measurements of salinity and suspended matter at the glacier terminus:

$$C_s = C_s' \left(\frac{S_{alb}}{S_{alb} - S_{alv}} \right) \tag{12.4}$$

where C_s' = measured suspended sediment concentration at the glacier front
S_{alb} = salinity of deeper ocean water in fjord
S_{alv} = salinity of the upper, brackish water layer at the glacier front.

Equation (12.4) assumes that the suspended sediment concentration of meltwater relative to brackish water in the fjord is proportional to the dilution of sea water by glacial waters. A value of 2.5 g l^{-1} was found for C_s. More recently Elverhoi *et al.* (1983) report *direct* measurements of suspended matter in the overflow plume in Kongsfjorden. Adjacent to the glacier values

between 0.3 and 0.5 g l^{-1} are typical decreasing to 1–5 mg l^{-1} in the central and outer parts of the fjord.

In most cases the density of the inflowing fresh meltwater due to suspended sediment is insufficient to prevent the water forming an overflow. Even if water is debouched from a submarine tunnel at the base of the glacier it is likely to rise to the surface. The forward velocity of such freshwater plumes is of the order of a few tenths of m s^{-1} but diminishes down-fjord where sediment progressively settles out and mixing takes place (see Figure 12.7). Some streams issuing into a fjord with a high sediment load may plunge beneath surface waters and form interflows or even underflows. Such processes are described by Gilbert (1982) and were outlined in relation to lacustrine activity in Subsection 11.2.6. Underflows are more likely to occur where inflowing streams have entrained considerable quantities of material during their passage over fine-grained outwash deposits.

12.3.5 Flocculation

Important for sedimentation in fjords and estuaries is the process of aggregation of micron and sub-micron size particles (typically clay minerals such as montmorillonite, illite and kaolinite) and known as flocculation. The effect of coagulation is to increase the size and density of flocs (or floccules) so that they begin to settle out.

Flocculation results from a reduction in the naturally occurring repulsive forces that exist between two clay particles. This can be brought about by the introduction of or increase in concentration of an electrolyte (such as sea water), a rise in temperature or an increase in ion valence. A reduction in pH, anion adsorption, dielectric constant or the size of the hydrated ion will also tend to bring about coagulation. The net result is for the negatively charged *faces* of clay minerals (brought about by **isomorphous substitution**) to join with positively charged mineral *edges*.

In order for this type of an electrostatic bond to be created the clay minerals must come into close proximity (i.e. less than a few hundred Å). Several processes lead to the collision of particles and subsequent flocculation and the three most important are random Brownian motion, low velocity gradients in the water which may move certain particles more rapidly and cause collisions (at high shear rates, however, floccules may actually be disaggregated) and finally by the scavenging effect of larger settling particles (Lerman, 1979).

The probability of particle collision and subsequent flocculation is dependent upon particle size and concentration such that the rate of coagulation is:

$$F_c = f \{P [r_1, r_2, r_3. . . r_n] C_{s1}, C_{s2}, C_{s3}. . . C_{sn}\}$$
$$(12.5)$$

where r_n = particle radius
  C_{sn} = particle concentration
  $P (r_1, r_2, r_3 . . .r_n)$ is the probability function for collisions between particles of various radii.

Brownian motion is, in general, the least effective process and is temperature dependent with up to 50% fewer collisions in cold (and hence more viscous) water. Concentration of the electrolyte (reflected in sea water salinity) also governs the probability of flocculation. In general, when salinities reach the order of 3–4 ‰, illite and montmorillonite can be flocculated. Kaolinite is flocculated at much lower concentrations. Floccules may reach sizes of several tens of microns but because of their very high void ratios (as high as 2–4) initial bulk densities may be low

(1 100 kg m^{-3}). As water slowly fills the voids, density may rise to 1 500–1 600 kg m^{-3} and the particle may then sink through the water column.

Numerous workers have reported peaks in suspended sediment concentration at intermediate depths and in close proximity to changes in salinity, temperature and density (i.e. halo-, thermo- and pycnoclines). Relling and Nordseth (1979), for instance, demonstrate a threefold increase in suspensoid concentration above the halocline at a depth of ~ 5 m in Gaupne Fjord. Concentrations can be as high as several hundred mg l^{-1} with a commensurate decrease below the halocline. Such layers are usually termed turbidity maxima and result from processes which inhibit uniform settling. There is, as a result, a greater advection of fine material down-fjord and a change in the pattern of bottom sedimentation. Lerman (1979) has outlined various factors which control settling velocity and which may give rise to such concentration peaks. Density differences across the pycnocline are of relatively minor importance (except for very saline conditions) and not as important as temperature-controlled water viscosity variations across the thermocline. Trapping by small-scale convection cells can produce particle concentration, and small settling velocities. Another explanation relates directly to flocculation of clay particles. In settling from surface fresh or brackish water the large pore space in floccules is gradually filled with low-density fluid which, at first, inhibits settling into the denser, more saline water beneath. The fresh porewater is gradually replaced by higher-density sea water and floccules subsequently settle at a faster rate. The net result is to create an intermediate zone of higher concentration.

12.4 Pattern of glaci-estuarine sedimentation

The processes described above control the deposition of glacial sediments in fjords and glacial estuaries. There appear to be significant differences between estuaries occupied, in part, by tidewater glaciers and those where they are absent such as fjords headed by glaci-fluvial outwash and fed by meltwater streams. A variety of factors (hydrology, catchment area, sediment supply, etc.) govern rates of deposition which can vary substantially not only within a fjord or estuary but also between adjacent fjords. Sedimentation from debris-charged glacial water is complicated by other processes such as submarine slumping on steep slopes (sometimes generating small turbidity currents), ice rafting, the addition of wind-blown material, and bioturbation (Figure 12.2).

One most important factor is the time-varying character of glaci-estuarine sedimentation. It has already been shown that there are strong temporal changes in many of the key processes controlling deposition. Some exhibit a marked periodicity, such as tidal effects or stream flow, others are more random in nature like sudden ice front advances, ice calving events or wind-driven seiches. Short-frequency events, typified by summertime diurnal changes in discharge from glacial streams, are matched by longer-term fluctuations (for instance glacier retreat). Such variability cannot always be modelled satisfactorily and in discussing patterns of sedimentation only relatively simple generalizations can be made, or very specific case studies presented.

12.4.1 Fjords and estuaries with tidewater glaciers

Sedimentation in this environment has come under increasing scrutiny and recent reviews are provided by Molnia (1983) and Powell (1983, 1984) (Figure 12.8). Close to the front of tidewater glaciers sediments will tend to be coarse-grained gravels and sands with occasional muds although the pattern is complex (Molnia, 1983). Some material of this kind is produced by tidal reworking of till left by glacier retreat. Piles of coarse-grained clastics also result from the deposition of superglacial debris from retreating ice. Material is also melted directly from debris horizons in basal ice layers of ice cliffs. In addition hummocky push-moraine topography

may be evident due to small oscillations of the ice front. Kongsvegen in Svalbard advances 30–50 m during the winter and retreats (by calving) 50–100 m in the summer (Elverhoi et al., 1983). Meltwater streams issuing from the ice will contribute significant quantities of sand and clay-sized sediments, often mantling till, occasionally forming laminated sequences and sometimes pseudo-esker-like bodies (see mechanisms described in Chapter 10). Such sediments may

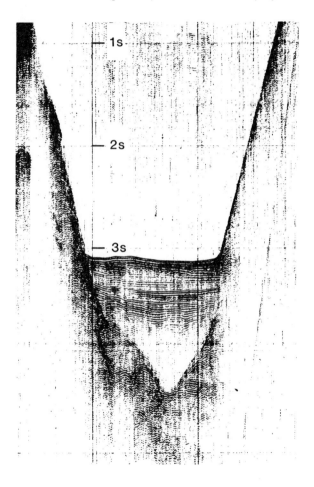

Figure 12.8 Seismic 'sparker' reflection profile across Muir Inlet, Glacier Bay, Alaska.
The profile illustrates near-horizontal seismic reflections which are interpreted as sediments deposited in the fjord following ice retreat. The sediments infilling the lower left of the trough have been interpreted as subglacial deposits or remnants from a former ice advance (illustration courtesy Ross Powell).

extend several hundred metres from the glacier terminus depending upon current velocities. The water content of such deposits may be high and **load structures** will often be present. Ice rafting may or may not be prominent. Elverhoi *et al.* (1980) report low IRD in Kongsfjorden due to both the rapid evacuation of calved icebergs and the high deposition rate of finer-grained material. An analogous situation has been reported by Molnia for the Glacier Bay area, Alaska.

Sands frequently alternate with layers of finer-grained materials and there is usually considerable spatial variability, related to both the number, position and lateral migration of outflow streams from the glacier face. Gilbert (1982) reports pronounced sediment laminations in front of Coronation Glacier, Baffin Island ranging in size from 10 mm to thin grain-sized **partings** attributable to changes in discharge from glacier outlets. Periodic events such as the calving of ice blocks from the glaciers snout may churn bottom sediments causing resuspension and their outward dispersal. Current velocities decrease

from the ice front and there is a progressive loss of coarser fractions by settling from over- or interflows. Sands grade into soft muddy sediments often with little lamination or structure. Such homogeneity may result from the deposition of flocculated clays. Scattered pebbles and coarse grains may be present in small quantities as a result of ice rafting from either icebergs or winter sea ice. Figure 12.9 shows the grain size characteristics of bottom sediments for Kongsfjorden, Svalbard which appear typical for glaci-estuarine bottom sediments. Powell (1981) finds that if an ice front is retreating very slowly submarine glacial streams will build fans and deltas and occasionally at peak flow, discharge sand into pro-deltaic areas. Slumping and sediment gravity flows may be common down the

Figure 12.9   Grain size distribution of core materials sampled from the sea floor in front of the ice cliffs of Kongsvegen, Kongsfjorden, Svalbard (from Elverhoi *et al.*, 1980). Grain size scales apply to each ternary diagram.

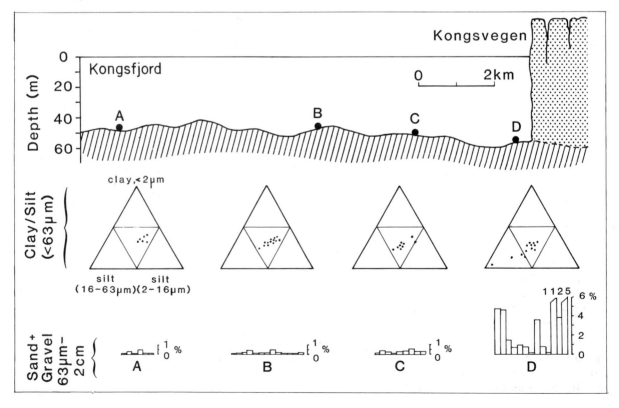

fore-slope of the ice-contact gravel banks. Such activity produces interbeds of sand with the mud produced by iceberg rafting. The latter may only locally reach high rates of deposition, for instance, close to fjord sills where bergs may cluster or queue, and release their entrained debris. In most cases, however, icebergs may be flushed too rapidly through the fjord to allow significant development of an ice-rafted facies (Molnia, 1983). Strong surface melting on the glacier may produce large lateral fan deltas prograding down-fjord, and composed of sand and gravel. This material will be found interbedded with iceberg mud both close to and at some distance from the ice front. Seasonal variations in the meltwater/sediment discharge of lateral streams produces characteristic laminations. Between outwash fans the ice-frontal zone may carry quantities of dumped superglacial debris.

12.4.2 Fjords and estuaries with stream input

Where glaciers have retreated inland, water input to the fjord is still largely controlled by glacial melt but the progradation of deltas or fans into the fjord or estuary head significantly alters sedimentation, buffering the estuary from coarse-grained deposition and some minor discharge fluctuations. Such deltas often constitute the distal portion of extensive sandar and braidplains made up of coarse glaci-fluvial outwash filling the inner parts of the fjord. Depositional processes are similar to those encountered in large proglacial lakes with the exception of the presence of sea water and some hydrographic effects (e.g. tides).

At the delta front coarse material is brought forward by streams to cover the finer-grained foreset sands. Slumping may cause minor turbidity currents resulting in graded bedding (cf. glaci-lacustrine activity) or more massive, but nevertheless well sorted, flow-induced sequences. Hoskin and Burrell (1972) and Powell (1981, 1983) indicate that such turbidity currents may also be responsible for erosion of channels on the fjord floor.

Away from the prograding fjord head fine-grained muds (of silt and clay-sized fractions) predominate. Turbid waters from outwash streams sometimes form underflows during peak summer discharges; at other times overflow and occasional interflow conditions may transfer fine-grained sediments down-fjord. Some of the fine sediment deposits may result from highly flocculated suspensions.

Ice-rafted debris will not usually represent a significant proportion of sediments in fjords where tidewater glaciers are absent. Nevertheless, debris released from melting sea ice or river ice brought down by influent streams, may give rise to occasional **dropstones**.

Relling and Nordseth (1979) report a grain size trend away from the fjord head with coarse components peaking a little way from the stream inlet and finer fractions becoming dominant further down the fjord. Gilbert (1982), however, finds little down-fjord trend in particle size in Maktak Fjord, Baffin Island which he ascribes to the effects of deposition from gravity and aeolian process, bioturbation, and widespread distribution of fine-grained sediments by active suspensions.

The rate of sedimentation in fjords and estuaries is well exemplified by the results in Gaupne Fjord from sediment trap studies (Figure 12.10). Results suggest an overall sedimentation rate of 0.1–0.2 m a^{-1} near the Josterdola delta and 0.01–0.02 m a^{-1} at the outermost sampling site some 2 km away. In glacierized fjords Gilbert reports a rate of 0.3–0.5 m a^{-1} in Coronation Fjord, Baffin Island while Elverhoi et al. (1980) suggest 0.05–0.1 m a^{-1} in the inner basin of Kongsfjorden, Svalbard and 0.4 mm a^{-1} in the outer basin with 0.5–1 mm a^{-1} as typical. Very high sedimentation rates have been reported from Alaska: >4.4 m a^{-1}, in Glacier Bay and 9 m a^{-1} in Muir Inlet (Powell, 1981; Molnia, 1983).

12.4.3 Biogenic sedimentation

Fjords, like many other large natural bodies of water, sustain varied and often abundant communities of marine organisms. Table 12.1

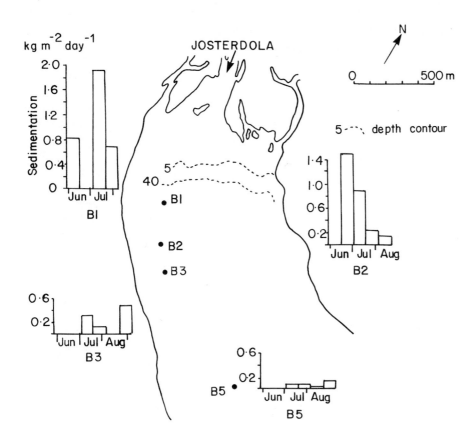

Figure 12.10 Sediment trap studies of the sedimentation rate on the bottom of Gaupnefjord, Norway (summer 1976). Note the pronounced decrease in sedimentation with distance from the Josterdøla infall (from Relling and Nordseth, 1979).

Table 12.1 General biological characteristics of fjord basins (from Pearson, 1980)

		Boreal	Arctic
Hydrodynamic energy input	High	High biomass High diversity High productivity NUTRIENT LIMITED	High biomass Medium diversity Low productivity CARBON LIMITED
	Low	Low biomass Low diversity Intermittently high productivity CARBON SINKS	Low biomass Low diversity Low productivity NUTRIENT SINKS

indicates the general biological characteristics of fjord communities.

The flora and fauna of fjords are important to sedimentation in several ways. Dead organisms, for instance, may settle to the fjord floor and contribute a significant fraction to bottom sediments. One of the more important biological aspects for sedimentation is the large nutrient supply associated with zones of upwelling in the coastal, outer fjord zone as well as towards the fjord head. Although the outer fjord region appears more productive, inner fjord basins are fed by glacial meltwater which acts as a major source of nutrients. Glaciers themselves may serve to enhance nutrient levels in fjords where tidewater glaciers are present. Apollonio (1973) demonstrated higher levels of silicate and nitrate in Southcape Fjord in Ellesmere Island which supports a glacier than in nearby Grise Fjord which is unglaciated. Phosphate levels were similar in both fjords. Appolonio ascribed these differences specifically to glacial inputs and suggested they may be significant in controlling phytoplankton growth. Pearson (1980) suggests that stirring of bottom sediments by grounding icebergs may also enhance fjord productivity by recycling nutrients. During the spring vertical mixing encourages high primary production and often **plankton** 'blooms' are encountered. Elverhoi *et al.* (1980) report production of abundant organic matter in Kongsfjorden. Sharma (1979) indicates that biannual **diatom** blooms in fjords of SE Alaska remove a considerable quantity of soluble silica from the **euphotic zone** which is converted into siliceous tests. When the diatoms die their tests sink and are partly exolved back into the sea water at greater depth, and eventually reach the sea bed where they contribute silica as diatomaceous sediment. A similar cycle is associated with other planktonic species such as foraminifera and **silicoflagellates**. The former favour warmer water conditions for growth of their calcium carbonate tests, however, and are, therefore, less abundant in the waters encountered in glacial fjord environments. Nevertheless Elverhoi *et al.* (1980) report small quantities of low-diversity benthonic and planktonic foraminifera in Kongsfjorden, Svalbard.

A further, important, biological activity in fjords is the bioturbation or mixing of sediments by the action of bottom-dwelling organisms. Such action destroys primary structures such as laminae and homogenizes the sediments particularly where sedimentation rates are low. In the fjord environment the principal organisms that affect glacial sediments are worms, molluscs and crustaceans, and Figure 12.11 shows some of these macro-benthic successions and their effects. Gilbert (1982) reports that bioturbation of bottom sediments in Coronation Fjord was least at the head of the fjord near Coronation Glacier where molluscs tended to be smaller and possibly younger. Bioturbation became more pronounced down-fjord. For those seeking additional detail of the processes involved a general discussion of bioturbation is given in Reineck and Singh (1980).

In those fjords where euxinic conditions occur

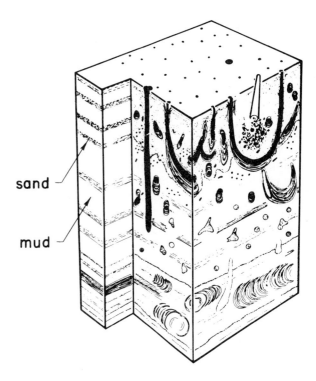

sand

mud

Figure 12.11 Modern muddy sediments with some sandy interbeds bioturbated by various organisms. Left part of the figure indicates the original bedding with burrowers absent. The right part of the diagram shows that burrowers have almost completely destroyed the bedding (from Reineck *et al.*, 1967).

(see Subsection 12.3.1) the lack of oxygenation of deep waters inhibits the oxidizing of organic matter which, as it decays, will generate H_2S by bacterial reduction of SO_4^{2-}:

$$SO_4^{2-} + \underbrace{2CH_2O}_{\text{organic material}} = 2HCO_3^- + H_2S \qquad (12.6)$$

This reaction produces typical black silts and clays with a high organic content (10–30% by weight in some fjords compared to 1–3% under normal oxidizing conditions). In addition Fe^{2+} exists in solution in fjord waters from the reduction of Fe^{3+} in sediments and reaction between H_2 and Fe^{2+} results in iron sulphides which form thin layers.

12.5 Facies models

Several authors have recently attempted to summarize or model the principal sedimentary regimes encountered in both the estuarine and open ocean glacial environment (e.g. Powell 1981, 1983, 1984; Elverhoi *et al.*, 1983; Molnia 1983). Although these workers have published detailed descriptions, both observed and inferred, of the patterns of sedimentation from a number of locations it is probably premature to provide generalizations of widespread application.

Elverhoi *et al.* (1983) recognize four main stratigraphic units in Kongsfjorden, Spitsbergen from 3.5 kHz **acoustic profiling**, and coring:
(i) mud with inter-bedded sand layers found close to the ice front (less than 0.5 km) and considered to be formed by sediment in glacial meltwater discharging from the terminus as underflows. Variations in the strength and position of the discharge are reflected in the inter-bedding.
(ii) homogeneous mud with low sand–gravel content, and found further out in the fjord and resulting from sediment settling.
(iii) sandy–gravelly mud located on slopes probably resulting from some ice reworking and possible iceberg overturning. (iv) mud and sandy mud, occasionally finely laminated with evidence of bioturbation and possessing a high content of monosulphides. This last group of sediments forms in the fjord basins again from progressive settling.

Powell (1981, 1983) has recognized five facies associations based upon detailed studies of sediments in Glacier Bay, Alaska obtained by shallow coring and grab sampling. Three principal facies associations are related to glaciers terminating in deep or shallow water, and two facies associations for deposition in fjords where. glaciers terminate close to or distant from the fjord head (Table 12.2). Powell's model takes account of many of the factors discussed in earlier sections of this chapter.
(a) Facies Association 1 is for a rapidly retreating glacier in deep water with calving. The sediments produced are summarized in Figure 12.12 and comprise subglacial till reworked by tidal currents, submarine streams or dense sediment-charged gravity flows. Subglacial streams issuing from portals in the submerged glacier front produce gravel, sand and mud, sometimes laminated, sometimes pseudo-esker type bodies (the mechanisms of sedimentation are similar to those discussed in Chapter 11 for glaci-lacustrine processes). Superglacial debris may be deposited directly at the front of the retreating glacier producing piles of coarse-grained clastics. In addition, terminal push moraines may be evident due to small oscillations at the ice front. At some distance from the glacier, mud from melting icebergs forms and may interbed with stream-produced sands and mud.
(b) Facies Association 2 derives from slowly retreating glaciers in shallow water and with active calving. Such situations arise at fjord constrictions due to bends or submarine sills. Close to the ice front considerable quantities of coarse-grained sediments often occur in banks or ridges, and sometimes they possess lag deposits from which the fines have not been winnowed. As the ice front slowly retreats submarine glacial streams build fans and deltas and occasionally at peak flow discharge sand into pro-deltaic areas. Slumping and sediment gravity flows may be common down the fore-slope of the ice-contact gravel bank. Such activity produces interbeds of sand and mud layers (the mud resulting from iceberg rafting) which may reach high rates of deposition ($4 \, \text{m a}^{-1}$).
(c) Facies Association 3 arises where the ice front is retreating or advancing slowly in protected shallow water with little calving. Strong surface melting on the glacier may produce large lateral fan deltas prograding down-fjord, and composed of sand and gravel. This material will be found interbedded with iceberg mud both close to and at some distance from the ice front. Seasonal variations in the meltwater/sediment discharge of the lateral streams produces characteristic laminations. Between outwash fans the ice-frontal zone may carry quantities of dumped supraglacial debris.
(d) Facies Associations 4 and 5 are formed once the glacier has retreated out of the water. Where the glacier is not far away (Association 4) sedimentation is characterized by substantial

Table 12.2 Glaci-marine facies associations according to Powell's scheme and indicating the relative development with respect to ice and marine conditions. Strong (S) and weak (W) categories are based upon occurence, distinctiveness and degree of preservation (from Powell, 1983)

Facies association	Glacier retreat	Calving state	Water depth (m)	Moraine bank	Push moraine	Gravel/rubble piles	Diamicton	Bergstone mud	Laminite	Turbidity current channels	Eskers	Central delta	Lateral delta	Outwash delta/braided stream	Fossils
1	Fast	Active	>150		S	S	S	S	S		S	W			
2	Slow	Active	<150	S	W	S	W	S	W	S		S			
3	Slow	Passive	Restricted <150		W	S	W	S	S			S	S		W
4	Slow	Very passive	0						S	S				S	S

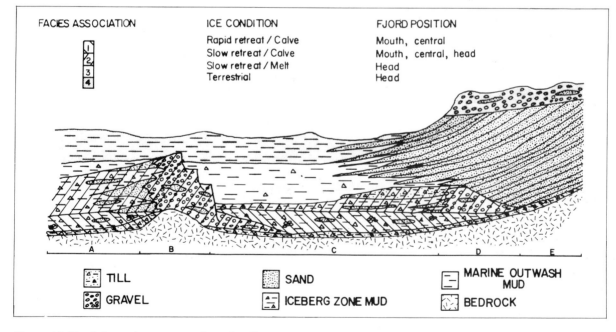

FACIES ASSOCIATION | ICE CONDITION | FJORD POSITION

1
2
3
4

Rapid retreat / Calve
Slow retreat / Calve
Slow retreat / Melt
Terrestrial

Mouth, central
Mouth, central, head
Head
Head

A   B   C   D   E

TILL   SAND   MARINE OUTWASH MUD
GRAVEL   ICEBERG ZONE MUD   BEDROCK

Figure 12.12 Schematic representation of sediment facies and facies associations in a glacial estuary for various ice conditions (from Powell, 1981).

outwash deltas prograding into the fjord and composed of coarse-grained glaci-fluvial deposits identical to braidplain sequences discussed in Chapter 10. The sands are interbedded with fine-grained marine outwash (silt and mud) which contain very little IRD. Where the glacier terminus is some distance away (Association 5) the marine environment is characterized by very shallow water conditions with extensive tidal flats of mud and braided stream channels.

Molnia (1983) has criticized Powell's facies associations on the grounds that they do not highlight the variations in time and space which result from fluctuations in the ice front with respect to the site of sediment deposition. He points out that active retreat of glaciers removes, in a progressive manner, any particular part of the fjord from the debris source, which complicates the resulting pattern of sedimentation. Molnia has investigated this effect using high-resolution seismic profiling and long gravity cores in Muir Inlet, Glacier Bay, Alaska where Muir Glacier has undergone 13 km of retreat since 1960. Sequences change at any one site from coarse ice-contact deposits to finer-grained and laminated fjord floor or basin deposits. These studies have led him to recognize three simple facies in fjords: ice-contact facies; beach and delta facies and a fjord floor facies.

12.5.1 Deposition by surging glaciers

Elverhoi *et al.* (1983) discuss a sediment assemblage resulting from the periodic surging (see Chapter 1.6) of Spitsbergen glaciers. Sequences were investigated in Kongsfjorden and Van Mijenfjorden and recognized on the basis of

acoustic opacity (on 3.5 kHz seismic records) combined with diffuse, parallel and discontinuous reflectors and hyperbolic returns. The units are typically found in depressions and proximal to moraine ridges. Cores reveal that the units comprise coarse, sandy-gravelly mud and show evidence of considerable rates of deposition, in part by meltwater and related to sliding conditions and production of basal water. Such sequences in Spitsbergen are associated with the known positions of surge lobes.

In the front of Bråsvellbreen, Nordaustlandet, Svalbard an ice cap lobe which surged in 1936–38 (see Figure 1.13) Solheim and Pfirman (1985) have identified distinctive sea floor morphologies created by the surge activity. The maximum extent of the surge is marked by a moraine ridge of asymmetrical cross-section 8–20 m in elevation and 0.5–1.7 km in width. The steepest slopes face in-shore towards the present day ice front which lies between 0.5 km (west side) to 5 km (east side) distant. Some small slumps are present on the distal slope. Solheim and Pfirman believe that sediments forming the ridge were deposited rapidly from meltwater; there is also some evidence of ice push. Between the ridge and the ice front a distinctive rhombohedral pattern is characteristic of the sea floor comprising linear semi-continuous ridges (5 m in elevation with a spacing of 20–50 m). Also present are irregularly distributed mounds (10 m high) and semi-continuous arcuate ridges nearer to the ice front. These features are believed to be related to the morphology of the ice-substratum interface at the termination of the surge (i.e. squeeze-up into bottom crevasses). The arcuate ridge may possibly represent subsequent minor oscillations of the Bråsvellbreen ice front.

13 Glaci-marine processes and sedimentation: ice shelves and ice tongues

13.1 Introduction

At the margin of grounded ice sheets and glaciers ice may enter the sea where it becomes buoyant and floats out as ice tongues or ice shelves (Figure 13.1). Sediments entrained in the ice are subsequently melted out close to shore depositing a characteristic assemblage of glacial material on the sea floor. These sediments were first identified and described by Philippi (1912) from material collected by the research vessel 'Gauss' off Antarctica. He proposed the name 'glazialmarine Ablagerungen' for crudely sorted terrigenous sediments possessing an abundant silt-sized fraction with scattered cobbles but poor in $CaCO_3$ and other organic matter.

Figure 13.1A Landsat image of Riiser-Larsenisen, Dronning Maud Land, Antarctica: a typical small ice shelf. The image is 180 km across and has a resolution determined by the pixel size (80 × 100 m). At the extreme top right open water is visible in a lead between the front of the ice shelf and sea ice and cloud in the Weddell Sea. A small ice rise (Blåenga) is shown towards the front of the ice shelf where ice is grounded on a sea bed high. Various zones of rifts and crevassing can also be discerned. To the right of the image the complex and irregular grounding line can be identified whilst further inland are some isolated nunataks (Vestfjella) (view, 18 January 1974 image E-1544-08325-7 0l courtesy Eros World Data Center).

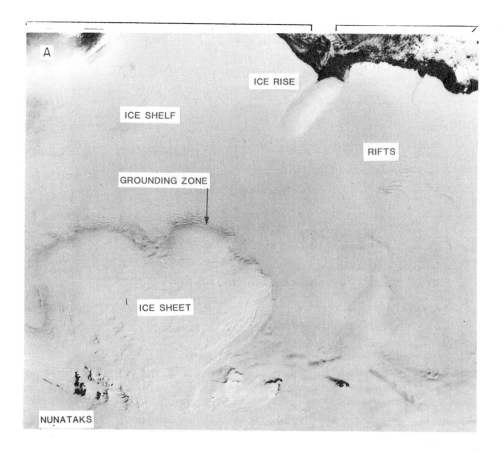

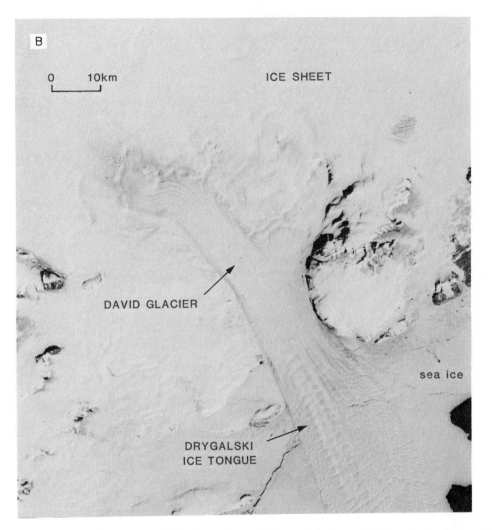

Figure 13.1B An ice stream at the margin of the Antarctic Ice Sheet (David Glacier) discharging into the Ross Sea where it forms a conspicuous floating ice tongue (Drygalski Ice Tongue).
Note sea ice (with leads) close in around the ice tongue. (Landsat image. No 1543-20305 Band 7, 17 January 1974. Courtesy of Eros World Data Center).

The main components of glaci-marine sediments are detrital (sands and gravels, silt- and clay-sized fractions) and biogenic material (including phyto- and zoo-plankton, sponge spicules, molluscs and occasional plant fragments). In addition **authigenic** or alteration products may be encountered: Fe- and Mg-rich silicates in the form of Fe-rich **sepiolite** developed as biogenic opal from **radiolaria** and diatoms, or **smectite** created by basalt–sea water interactions.

10% of the world's oceans and epicontinental seas have glaci-marine sediments forming in them today (some 36 M km^2 according to Lisitzin (1972) although some other sources place the figure much higher at 80 M km^2). Around Antarctica the belt of present-day glaci-marine sedimentation is several hundreds of kilometres in width (Figure 13.2) but during late **Cenozoic** glacial maxima this area was probably doubled. It should be recognized that deposition in the open ocean allows a substantially increased chance of survival of glacial events in the geological record. Terrestrial glacial deposits such as till and outwash are easily weathered, eroded

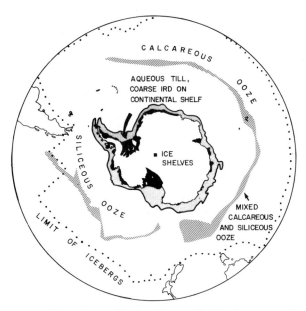

Figure 13.2 Generalized sediment distribution around Antarctica illustrating the inner girdle of diamicts and ice-rafted debris, principally on the continental shelf and slope. Further out in the Southern Ocean siliceous fine-grained pelagic sediments are characteristic of the annulus south of the Antarctic Convergence where mixing takes place with warmer-water calcareous oozes.

and reworked. As a result, palaeo-environmental interpretations relying solely on such evidence are often based upon a considerably fragmented record (Hambrey and Harland, 1981).

13.2 Sources of glacial sediment in sea water

Glacial sediments destined for deposition in the marine environment are contributed to sea water from a variety of sources. Floating ice shelves and outlet glaciers (including glacier tongues) may contain basal debris entrained up-flowline in their grounded segments and material contributed to their upper surfaces (as described in Chapter 7). Such sediments with distinctive concentrations, textures and lithologies are released into sea water by ice melt. Figure 13.3 summarizes several of these

Figure 13.3 Schematic depiction of the principal processes of glaci-marine sedimentation resulting from ice shelves and outlet glaciers. The sources of sediments are shown, the principal transport paths and various modes of deposition (from Drewry and Cooper, 1981).

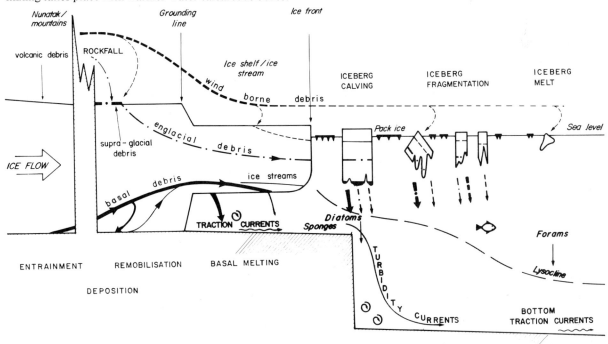

processes. Important differences exist between sediment contained in ice shelves and outlet glaciers as set out in Table 13.1. Differences in location are important once ice commences to melt in sea water. If, for example, basal debris layers are thin and discontinuous only minor sedimentation will occur, concentrated into the early stages of melting beneath an ice tongue or ice shelf. If the bulk of the sediment is located high in the ice mass, in englacial layers or at the ice surface, there may be a prolonged period before release, which will be delayed until the ice has calved as bergs into the open ocean. For **ice streams** issuing from adjacent ice sheets, and ice shelves little sediment is incorporated within the body of the ice from sub-aerial sources (except for volcanic dust, extra-terrestrial debris or fine material deflated from nearby ice-free terrain and only of local significance). The bulk of the sediments are entrained in ice layers close to the bed. For outlet and tidewater glaciers originating in and/or transecting mountainous terrain significant quantities of debris are supplied by rockfalls and avalanching off outcrops (either valley sides or **nunataks**) and incorporated into super- and

englacial positions. When grounded ice cliffs terminate in sea water sediments may be liberated by melt-back and minor calving of the ice front. In addition, where and when temperature conditions are favourable meltwater containing sediments may be discharged into the sea forming turbid plumes similar to those encountered in lakes and fjords. Such activity is usually restricted to the termini of sub-polar ice caps such as those in Svalbard and the Canadian Arctic or certain outlet glaciers of the Greenland ice sheet (see Figure 12.6).

Icebergs calving from ice shelves, outlet glaciers and ice cliffs may contain sediments not previously melted out which are transferred into the open sea. These processes, and those related to sea ice, are discussed in Chapter 14.

13.3 Melting and freezing of ice shelves and ice tongues

The release of sediments from glacier ice into sea water will ultimately be controlled by oceanographic and glaciological processes which aid or inhibit melting. Ice floating in sea water will usually melt due to the flux of heat into the ice from the water. Under certain circumstances, however, it is possible for freezing to occur. Doake (1976) has provided a useful general summary of these conditions:

$$\dot{M} = (Q_s{}^* - Q_i{}^*)/\rho_i L \qquad (13.1)$$

where $\dot{M}$ = melt rate

$Q_s{}^*$ = heat flux from the sea

$Q_i{}^*$ = heat flux upward through the ice

ρ_i = density of ice

L = latent heat of fusion.

The quantity $Q_i{}^*$ is the product of the thermal conductivity of ice and the temperature gradient. The latter will depend upon the upstream temperature regime, creep thinning as well as mean annual surface temperature (see Chapter 1). $Q_s{}^*$ results from water temperature, flow characteristics including velocity and **thermo-haline** behaviour (if any).

Table 13.1 Variations in the location and quantity of entrained sediments with type of ice mass

	Basal	Englacial	Super-glacial
Ice shelf			
from ice sheet	XX	X	X
via mountains	XXX	XX	X
Outlet glaciers			
via mountains	XXXX	XX	XX
ice stream	XXXX	X	X
Ice cliffs			
near mountains	XXX	XX	XX
ice sheet edge	XX	X	X

Sediment quantity
XXXX   substantial
XXX   moderate
XX   small
X   trace

N.B. englacial debris may result from either basal or superglacial locations. If superglacial inputs are small most englacial debris will be of basal origin.

13.3.1 Ice–water–salt system

If the sea water possesses a temperature above its freezing point basal ice will be melted in order to lower the water temperature towards the freezing point. Such a process and its reverse behaviour may be better understood in relation to a phase diagram of a salt, water and ice system shown in Figure 13.4A. If no salt were present (NaCl concentration = 0‰) a fall in temperature would result in the freezing of a packet of water at 0°C (point w on the diagram). Starting with pure water at (w) the addition of an increasing proportion of salt lowers the freezing point as described by the curve w–b. Similarly, the addition of water to salt produces the curve b–s. The point at which the two curves intersect, and where the whole system becomes solid, is called the **eutectic**.

Figure 13.4A Ice–water–salt system phase diagram. The liquidus is the equilibrium line w–b–s where ice and saline solution possess the same chemical potential. b = eutectic, the point at which the whole system becomes solid. The inset diagram **B** represents conditions for sea water containing ice such as an ice shelf. A particular ice–sea water combination (points A and B respectively) will attempt to equilibrate to point C on the liquidus (after Doake, 1976). See text for further discussion.

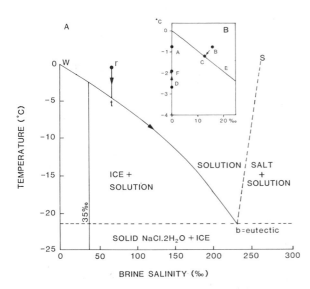

Given a solution represented in temperature and salt composition by point r, and subject to a drop in temperature no freezing will occur until the temperature reaches the **liquidus** at t. Thereafter, ice which is in excess of the eutectic proportion will commence to form. Removal of ice from the liquid increases the salt concentration with a continued fall in temperature towards b. At b both the salt and ice crystallize together until all the solution disappears. As the salinity of average sea water is of the order 35 g kg^{-1} (35‰) it is not necessary to examine details of the curve b–s.

For an ice mass floating in sea water Doake (1976) has described relations to the phase diagram (Figure 13.4B). Ice and surrounding sea water are assumed to be in a closed system at the same temperature (say − 0.8°C) and represented by points A and B. The sea water is above freezing temperature and according to the phase diagram is out of equilibrium. In order for the system to come into steady-state some ice must melt in order to reduce salinity and the temperature and hence achieve its position on the liquidus. Melting takes place because of an imbalance in the chemical **(free) energy** potential between the ice and sea water. In the mixture they react and the free energy decreases to tend to a constant and minimum value (Williams, 1982). The temperature of the sea water is thus lowered in order to provide the latent heat necessary to melt the ice. The final state of the system will be at point C on the liquidus.

For freezing-on to occur the water must already be at its freezing point, that is lying on the liquidus (say point E) with the ice at a lower temperature (point D). Freezing takes place at the ice–water interface and the latent heat produced is removed, not by raising the water temperature (which would merely move C up to the liquidus and hence only *increase* the temperature differences), but by conduction through the ice at a rate dependent upon its basal temperature gradient (basal ice will warm to F). At the same time the sea water would become more saline as water was removed by freezing.

13.3.2 Freezing and melting rates

Gill (1973) has given first-order estimates of the rates of freezing and melting at the base of a floating ice shelf or glacier tongue. For ice accretion, assuming all the heat lost from the water produces freezing and that the water is already at the freezing point, the rate is:

$$A_i = Q_s{}^*/\rho_i L \qquad (13.2)$$

where $Q_s{}^*$ is given by:

$$Q_s{}^* = \rho_i c_i T_b (k_i/\pi t')^{1/2} \qquad (13.3)$$

where c_i = specific heat of ice
　　　　k_i = thermal diffusivity of ice
　　　　T = temperature below freezing of the ice base
　　　　t' = time since ice comes into contact with water.

Typical values substituted into Equations (13.2) and (13.3) suggest that for an ice shelf of width 500 km the average freezing rate is quite small, in the order of 0.04 m a^{-1}.

　Once the temperature of the ice, raised by conduction of latent heat, reaches that of sea water freezing will cease. However, water temperatures are unlikely to be exactly at or below the freezing point. Water advected by currents from outside the ice shelf or ice tongue environment will be above the freezing point and will cause melting of the ice base. An upper limit to such melting has been calculated by Gill (1973) in the presence of a current:

$$\dot{M} = R_* \rho_w c_w \Delta T U_{tc} \qquad (13.4)$$

where R_* = coefficient depending, in part, upon interface and current velocity
　　　　c_w = specific heat of water
　　　　ΔT = temperature difference between ice and water
　　　　U_{tc} = current velocity.

Substituting typical values into Equation (13.4) yields an estimate for the melt rate of ~ 30 m a^{-1} which is clearly an upper bound.

　In addition to transfers of energy and mass from tidal and other currents as indicated above, the effective transfer of sensible and latent heat

between sea water and ice will also depend upon mixing coefficients and small-scale boundary layer dynamics, further discussed by Gill (1973) and Jacobs *et al.* (1981). Changes in pressure also affect melting and freezing regimes.

13.3.3 Effects of pressure on thermal regime

The effect of pressure has been discussed by Foldvik and Kvinge (1974, 1977), Doake (1976) and more recently by Robin (1979). A water mass (either a tidal current or a meltwater plume) moving along the base of an ice shelf will change depth due to bottom topography and orientation of the current. In general, ice thickens up-flowline. Two ice thickness profiles, typical of most Antarctic ice shelves, are shown in Figure 13.5A for flow lines on the Ross and Brunt Ice Shelves. It can be seen that in order to maintain hydrostatic equilibrium the bottom of the ice has to move down from its previous level. For the Ross Ice Shelf, the height of the ice–water interface is 100 m bsl at the ice front and 500 m bsl close to the grounding line. Maximum gradients are found at both these localities. Gradients in the more typical central region of an ice shelf are $1:2 \times 10^{-4}$ (Ross) and $1:2 \times 10^{-3}$ (Brunt). Within an ice shelf, transverse to the flowline, the bottom slope may also change in response to the presence of thick ice streams (Neal, 1979).

　If water flows up an ice flowline it will descend as the ice thickness increases and the ice base is lowered. The pressure will continue to depress the freezing point and hence the position of the water packet on the liquidus (Figure 13.4) by 7.64×10^{-5} C kPa^{-1}. The effect is for the descending water to reach equilibrium only by melting the overlying ice and lowering its temperature. Robin (1979) following Foldvik and Kvinge (1974) discusses in some detail the reverse process whereby a packet of water in a current *ascends* from a deeper level and freezes onto the base of an ice shelf. Robin shows that a 5 mm layer of ice will form if a metre cube of sea water is raised 500 m. Consider an irregular ice shelf base with thick ice stream lobes. If a current flows along the bottom perpendicular to the lobes

Figure 13.5   A Ice thickness profiles along centrally located flowlines of the Brunt Ice Shelf (left) and Ross Ice Shelf (right). Ice depths were determined by airborne radio echo sounding.
B Mass balance components for the Brunt and Ross Ice Shelves, Antarctica. The curves for accumulation, velocity and vertical strain rate were derived by fitting Chebyschev polynomials to raw data. Data for Brunt Ice Shelf from Thomas (1973) and for Ross Ice Shelf from Thomas (personal communication) and Thomas and McAyeal (1982).
C Calculated melt rates (Equation 13·5) and **D** derived sedimentation rates for the Brunt and Ross Ice Shelves. Quantities k_2 and C_0 derive from Equation (7.2) and refer to the thickness of the basal debris layer of constant debris concentration, and the initial concentration of the basal layer respectively (from Drewry and Cooper, 1981).

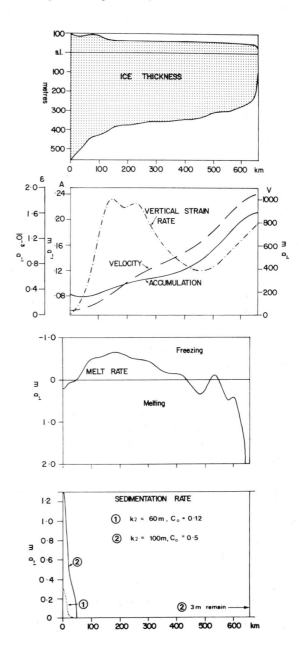

at 0.05 m s^{-1} an isothermal layer of 10 m thickness would melt off a layer 1.2 m depth across an ice stream of 25 km width and redeposit that amount as the current rises in the intervening hollow.

13.4 Measurements and estimates of melting and freezing

No direct measurements have yet been made of freezing or melting beneath ice shelves or ice tongues, nor have sufficient oceanographic observations been collected to allow accurate calculation of the heat fluxes. In the absence of such data a number of indirect glaciological and oceanographic methods have been applied to estimating these quantities (Thomas, 1979).

13.4.1 Ice shelves

At site J9 (82° 22.5'S, 168° 37.6'W) on the Ross Ice Shelf drilling to the ice–water interface in 1978 revealed a 6 m layer of frozen-on sea ice at the base of the 416 m core (Figure 13.6). Various estimates based upon temperature gradients and oceanographic conditions suggest an accretion rate, A_i, of between 0.01 and 0.035 m a^{-1} (Clough and Hansen, 1979; Jacobs *et al.* 1979; Zotikov *et al.* 1979), and compatible with the theoretical calculations of Gill (1973).

On the Amery Ice Shelf Morgan (1972) has interpreted another core as suggesting ad-freezing. Drilling revealed 270 m of glacier ice overlying 158 m of sea ice (crytallographically and isotopically distinctive). The inferred growth rate was 0.3 to 0.5 m a^{-1} – an order of magnitude greater than at J9. Numerical modelling of temperatures and ice flow of the Amery Ice Shelf by Neal Young (described by Budd *et al.*, 1982) suggests a similar basal accretion rate in the order of 0.6 m a^{-1}. Figure 13.7 shows the mass balance components for the Amery Ice Shelf.

Lennon *et al.* (1982) describe estimates of the basal melt rate beneath George VI Ice Shelf in the Antarctic Peninsula based upon identification of that fraction of the water outflow from George VI Channel supplied by ice melt. Salinity and temperature comparisons suggest this fraction is

Figure 13.6   Frozen-on ice from the bottom of the Ross Ice Shelf at J-9 discovered by deep-core drilling (courtesy US National Science Foundation).

1% yielding an average melt rate of 1.8 m a^{-1} and is probably a minimum value. Several workers (Thomas and Coslett 1970; Drewry and Cooper 1981; Jacobs *et al.*, 1981) have adopted the technique of estimating basal freezing/melting rates developed by Crary *et al.* (1962) which relies on the mass continuity equation and assumes the ice mass is in steady state. Melting or freezing at the base must be balanced by the vertical ice velocity which is assessed by measurement of accumulation and vertical strain rates. The melt rate (freezing rate is given as a negative value) is:

$$\dot{M} = -U(dh/dx) + \dot{A} + h\dot{\epsilon}_z \qquad (13.5)$$

where U   = horizontal velocity of the ice mass
dh/dx = thickness gradient
$\dot{A}$ = surface accumulation rate
$h\dot{\epsilon}_z$ = vertical strain rate.

Glaciological data from surface observations of $\dot{A}$ and $\dot{\epsilon}_z$ and U can be combined with ice thickness measurements from radar soundings to evaluate Equation (13.5).

Figure 13.5B, C shows the results of

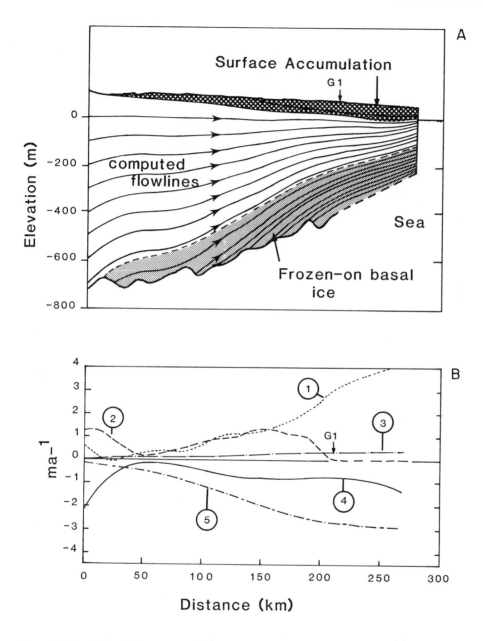

Figure 13.7 **A** Ice thickness profile along a central flowline of the Amery Ice Shelf with calculated particle paths and zones of freezing-on at the base (hatched) and surface accumulation.
B Mass balance components for the ice shelf.
1 = surface accumulation rate; 2 = basal ice accretion rate; 3 = horizontal advection; 4 = longitudinal strain thinning; 5 = transverse strain thinning (from Budd *et al.*, 1982).

calculations of melting and freezing according to Equation (13.5) for two typical Antarctic ice shelves (Ross and Brunt) and demonstrates important aspects of their behaviour. The larger ice shelves exhibit basal melting close to the grounding line (say within the first few tens of kilometres) resulting, in part, from convective, turbulent water-flow associated with tidal breathing and general circulation, and assisted by an ice base close to or at the pressure melting point. Further out beneath an ice shelf, freezing or melting is governed by the effects of active water circulation. Near the ice front, for instance, circulation is strongest and heat exchange greatest with considerable basal melting (several m a^{-1}). Further inland from the ice front heat exchange is weakened and a layer of cold **isothermal** water of low salinity may be present which encourages net basal freezing, enhanced where the ice base slopes strongly upwards (Robin, 1979). It is within this zone on the Ross Ice Shelf that bottom freezing was detected at J9.

13.4.2 Ice tongues

Ice tongues are of much more limited extent and thus strongly affected by the action of currents. Their importance is in their high basal debris content. It is rare to find extensive zones of basal freezing beneath them. Few ice tongues have been studied in any detail despite their importance to sedimentological processes (as discussed below). The Erebus Ice Tongue in McMurdo Sound, Antarctica has been investigated extensively by Holdsworth (1982) and Jacobs *et al.* (1981). Surface measurements allowed the use of a form of Equation (13.5) to estimate basal melting. Jacobs *et al.* (1981) suggest that 120 m of ice must have been melted off the base of the ice tongue in a period of 85 years yielding a melt rate of ~1.4 m a^{-1}. The rate is somewhat greater than for most ice shelves. Holdsworth (1982) has also calculated the sidewall melt rates by comparison of the measured width (or half-width) of the ice tongue with its theoretical width calculated from a consideration of lateral creep-spreading. Results are shown in Figure 13.8 and are of the order of 6 m a^{-1} for the south half and 1.5 m^{-1} for the north half.

13.5 Patterns of glaci-marine sedimentation

Sediments entrained in ice tongues and ice shelves will be released by melting. Coarse fractions fall rapidly to the sea floor while fine material will

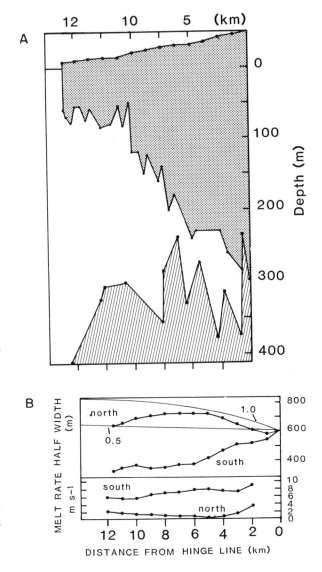

Figure 13.8   A Ice thickness profile for the Erebus Ice Tongue, McMurdo Sound, Antarctica (from Jacobs *et al.*, 1981).
B Melt rates (lower part) for the north and south edges derived from the comparison of ice tongue half-widths (upper part) (measured from the dynamic centreline) with spreading rates (calculated with transverse strain factor, $\alpha' = |\dot{\epsilon}_{yy}|/|\dot{\epsilon}_{xx}|$; $\dot{\epsilon}_{yy}$ and $\dot{\epsilon}_{xx}$ being the transverse and longitudinal strain rates, $\alpha' = 0.9$ is considered most likely but two cases 0.5 and 1.0 are shown (from Holdsworth, 1982).

remain in suspension for a considerable period being carried some distance by currents. Carter *et al.* (1981) have shown that suspended matter from the Antarctic is often dispersed on a global scale by movement of Antarctic Bottom Water.

Significant variations occur in sedimentation depending upon the type of ice mass contributing sediment, and it is important to remember the characteristics detailed in Table 13.1. Some early studies emphasized the role of ice shelves in the production of ice-rafted debris (e.g. Carey and Ahmed, 1961), but it is now recognized that sediment release from ice shelves takes place close to their grounding lines on the inner part of the continental shelf. Ice streams on the other hand are principally responsible for discharging large quantities of debris-rich icebergs into the open sea where melting liberates sediments either on the continental shelf or into deeper oceanic waters.

Drewry and Cooper (1981) investigated the release of debris from ice shelves in some detail. The quantity of coarse sediment relinquished to the sea floor was found to depend upon the vertical distribution of sediment in basal ice, the bottom melt rate at any point on the ice shelf and the integrated up-flowline sedimentation–melt history. Drewry and Cooper used Equation (7.2) to describe a suite of sediment concentrations in the ice base and then for two case studies (Ross and Brunt Ice Shelves) calculated melt rates using Equation (13.5). The results of this work are given in Figure 13.5C, D. It can be seen that if the melt rate and ice shelf horizontal velocity are constant deposition of sediments will mirror the debris concentration function.

Due to the relatively thin vertical extent of the basal debris-rich horizon sedimentation is highly sensitive to melt-rate. If there is vigorous melting close to the **grounding line** as suggested by Figure 13.5D all debris may be melted out within the first few tens of kilometres – a highly significant result. The effect of freezing prior to complete melt-release of sediments is to interrupt deposition and shift it further down-flowline where sedimentation will resume once the

additional ice layer has been stripped by melting. Several lines of evidence from the Ross, Filchner, Amery and Maudheim Ice Shelves, summarized by Drewry and Cooper (1981), support this notion of active sedimentation in the vicinity of the grounding line.

13.5.1 Suspended sediments

The very finest fractions, melting out at all stages from the grounding line outwards towards the ice front, are subject to considerable transport in the water column. Few studies have been made of suspended matter in the fully glaci-marine environment. Carter *et al.* (1981) report measurements from McMurdo Sound, Antarctica and in the water column beneath the Ross Ice Shelf at J9. In McMurdo Sound suspended particulate matter has a concentration of 2–4 mg l^{-1} with occasionally higher values when vertical mixing is particularly pronounced. At J9 the mean concentration was 0.7 mg l^{-1}. Off Coats Land in the Weddell Sea Elverhoi and Roalsted (1983) found concentrations of between 0.1 and 1.0 mg l^{-1}. Sharma (1979) has dealt in some detail with suspended sediment on the Alaskan Shelf. Typical values off-shore are between 0.5 and a few mg l^{-1}. These are close to the concentration of suspended matter found in the major oceans regardless of latitude and the presence of ice (Lisitzin, 1972, Elverhoi and Roalsted, 1983).

The material making up the suspended matter is usually composed of terrigenous, biogenic, aggregated and authigenic products. The clastics range in size from < 5 μm up to 100 μm, and the biogenic material from 10–50 μm up to 200 μm. Elverhoi and Roalsted (1983) note that the proportion of biogenic to clastic particles is usually greater in the higher levels of the water column than at depth. **Detrital** grains reach their greatest concentration near shore and towards the bottom. Carter *et al.* (1981) note, however, that in areas of greater vertical mixing zooplankton are moved to lower levels. In McMurdo Sound Hicks (1974) reports the greatest zooplankton **biomass** in surface waters near the seasonal ice edge whilst near the Ross Ice Shelf the greatest biomass was found in the bottom 200 m; terrigenous particles constitute at least 50%. The biogenic matter was essentially whole and fragmented diatom **frustules** with minor silicoflagellates, zooplankton and foraminiferal debris. Fecal pellets made up the bulk of the aggregated material (see also Subsection 11.5.2), the remainder being floccules. At J9, beneath the ice shelf and 600 km from the open ocean, terrigenous material was 75% of the total suspended matter and biogenic material and aggregated particles of fecal origin scarce due, presumably, to low **productivity**. Most of the diatom fragments exhibited considerable corrosion.

Suspended matter will settle out according to Stoke's Law (Equation 11.8), modified by the shape of the particles and local vertical mixing/turbulence. Fecal pellets in the size range 50–100 μm will settle at a velocity of between 0.12 and 0.46 mm s^{-1}.

13.5.2 Lithofacies

Glaciological and oceanographic processes and sea bed characteristics will control the pattern of glaci-marine sedimentation. To date there have been few investigations of contemporary glacial environments in Antarctica or in the Arctic to establish preliminary reliable depositional models and associated litho-facies (see Anderson et al., 1983; Molnia, 1983). A useful bibliography of papers published up until 1983, is provided by Andrews and Matsch (1983). Eyles et al. (1983) have suggested a lithofacies classification for glacial deposits, including glaci-marine sequences, based upon a suite of characteristics such as grain-size, fabric, internal structure and bed-contact relationships. These have already been discussed in Chapters 9 and 10. Miall (1983) has taken these ideas further with the interpretation of palaeo-glacial sequences (Table 13.2). Such studies are only exploratory and require more widespread application and testing, especially to the investigation of modern-day processes, if they are to be used as generalized sedimentary models. Nevertheless they show considerable merit and are worthy of

Table 13.2 Lithofacies types for glaci-marine sediments (from Miall, 1983)

Diamictite (D):	poorly sorted gravel-sand-mud
Dm-	matrix-supported
Dc-	clast-supported
D-m	massive
D-s	stratified on decimeter to meter scale
Sandstone (S):	very fine to very coarse, may be pebbly
Sr	ripples
Sh	horizontal lamination
Sm	massive
Sg	graded
Fine-grained units (F): mainly mud	
Fm	massive mud (argillite)
Fl	laminated mud (minor silt-sand laminae)
Flp	laminated fines with pseudonodule structures
F-d	dropstones present

development, refinement and vigorous testing. It has only proved possible to extend these ideas in a limited manner in detailing the patterns of sedimentation resulting from ice shelf and ice tongue activity. Figure 13.10 suggests the lithofacies characteristics, based upon Miall's scheme, which result principally from melt-out deposition beneath the inner part of an ice shelf. Inland of the grounding line basal melting of sliding ice streams will give rise to **diamictites**, deposited as lodgement or melt-out tills. If there is vigorous bottom melting, meltwater deposits may also be present (as suggested by Orheim and Elverhoi (1981) and reported from the Alaskan continental shelf by Molnia (1983)). Such sequences are characterized by stratified sands with laminated and rippled units similar to those discussed in Chapter 10 (10.4). At the grounding line when the ice commences to float and melt the entrained sediments are released and fresh water is liberated into the subjacent sea.

This water, although close to 0°C, will be less dense than the surrounding saline water. It is also unlikely that water released from the bottom will contain sufficient suspended matter to increase its density significantly and create a descending

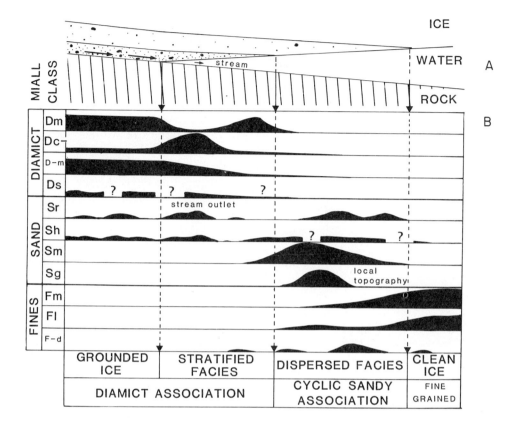

Figure 13.9 Lithofacies characteristics generated close to and across the grounding zone of a typical ice shelf. **A** An idealized ice base entering the sea and coming afloat. It contains a basal debris layer composed of a stratified lower horizon and a dispersed upper horizon (see Equation (7.2) and Figure 7.15). **B** Variations in principal lithofacies (according to the Miall (1983) scheme: Table 13.2), across the grounding zone (see text for details).

density current. The water will not be sufficiently turbulent to entrain the coarser solids and most particles melting out will fall freely to the sea bed according to Equation (11.8). This will give rise to a pulse of coarse diamictite. In addition **bottom traction currents** generated by deep, cold water flow or from large-scale circulation may be responsible for local scour and **winnowing**. The preferential erosion and transport of finer fractions by this process results in gravel lags or pavements.

Further out from the grounding line the rate of sedimentation, especially of coarser material originally entrained in basal, stratified layers, will fall off rapidly as melting proceeds. This is due to the negative exponential nature of the vertical concentration of debris in the ice (see Equation 7.2). Finer fractions from the dispersed ice facies increasingly dominate deposition: clast-supported diamictites wedge into massive, poorly graded pebbly to fine sands. These are also subject to bottom current winnowing, giving rise to occasional drapes or lenses of coarser debris as velocities fluctuate. Dropstones may be encountered frequently in this zone as large clasts in the dispersed ice facies melt out and penetrate finer-grained sea floor sediments. As the dispersed facies continues to be melt-stripped from the ice base even finer sediments are released to the sea bed, typically fine sand, silt and some clay-sized particles. These form massive to weakly laminated deposits often with sandy interbeds from migrating bottom currents. Rhythmic sequences and occasional dropstones may sometimes be present. As an example of the type of textures that may be expected from deposition of basal debris Figure 13.11 illustrates grain-size distributions from a study by Barrett *et al.* (1983) from McMurdo Sound, Antarctica.

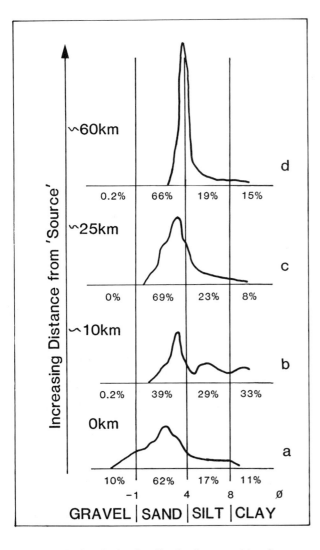

Figure 13.10 Grain-size distributions resulting from progressive deposition of basal debris in sea water. (After Barrett *et al.*, 1983).
At the bottom (a) is basal sediment from Taylor Glacier, southern Victoria Land, Antarctica. Curves b–d are for sediment samples at increasing distances

Effects of grounding line fluctuations

Since ice shelves and outlet glaciers respond rapidly to sea level, climatic and mass balance fluctuations, frequent migration of the grounding zone will be experienced. Such activity gives rise to a thick and complex sequence of strongly **diachronous** glaci-marine sediments based upon the units described above and shown in Figure 13.11. Molnia (1983) has developed a model of glaci-marine deposition for the sub-arctic based upon studies of the Quarternary history of the Gulf of Alaska. The most important factor in sedimentation is the location and fluctuation of the glacier terminus relative to the shoreline. It is implicit in Molnia's discussion that fluctuations of the grounding line are also relevant. At the point of flotation the entire weight of the ice is supported by the water column. Just upstream from the grounding line, however, a considerable fraction of the glacier's weight will still be borne by the water. The distance over which such support is maintained depends principally upon bulk rheological properties of the ice (Holdsworth, 1974b). Such support results in far less disruption of sediments from any grounding line oscillation. When major advances occur, however, considerable erosion and bulldozing of the upper sedimentary horizons will occur resulting in shear, accompanied by the development of **drag, slump** and **roll structures**, sharp erosional contacts and the **streaking-out** of less competent lithologies. Widespread **unconformities** are known from the Ross Sea, Antarctica attributed to expansion phases of the Ross Ice Shelf which grounded out over the continental shelf to the break-of-slope. The advance of Bråsvellbreen, a lobe of Austfonna in Nordaustlandet, Svalbard during a surge phase between 1936 and 1938 (Figure 1.14) bulldozed sediments in front of the semi-buoyant glacier snout. A distinctive ridge has been left marking the maximum extent of the ice – a sub-aqueous ice-push feature (Solheim and Pfirman, 1985).

offshore from Taylor valley, although it should be noted that suspended sediments are influenced by currents and do not follow a straight line offshore as described by the sample locations. Note the general features, however, of a reduction in coarse and fine grades and an increasing single modality.

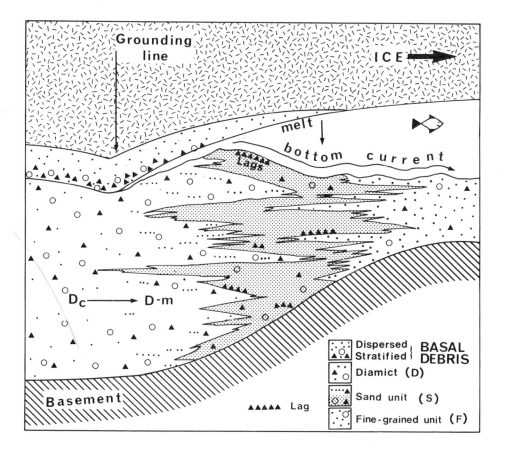

Figure 13.11   Sedimentary facies generated at the inner grounding zone of a floating ice shelf (or glacier tongue) undergoing grounding-line oscillations (due to sea level or upstream mass balance changes) but without experiencing major erosion of sea floor deposits. Compare with Figure 13.9.

13.5.3 Sediment gravity flows

Miall (1983) has drawn attention to the possible disruption and reworking of glaci-marine sediments by slumping, sliding and gravity flows. Miall considers debris flow activity may affect both coarse-grained diamictites as well as finer-grained sedimentary accumulations. Typically debris flows require relatively steep slopes with a slant-length sufficient for sustained flow (i.e. in the order of $10^3 - 10^4$ m). These factors suggest only local flow behaviour will characterize relatively low-relief

continental shelf areas and major flow processes will be focused around major submarine highs and primarily the continental margin. Wright and Anderson (1982) and Wright *et al.* (1983) have discussed in some detail the downslope, inter-canyon processes which affect glaci-marine sediments on the continental slope of Antarctica. They describe lithological logs and grain-size data along a track (at 10°W) off the northeastern continental slope of the Weddell Sea. On the continental shelf close to its edge sediments comprise homogeneous, poorly sorted gravelly-sandy muds typical of coarse clastic deposition from melting basal ice. A little way down the continental slope relatively homogeneous poorly sorted, lithic gravelly to muddy sand was encountered with 15–20% mud matrix, improved sand sorting up-core (absence of fractions finer than 2–2.5 ϕ). Further down still, the upper

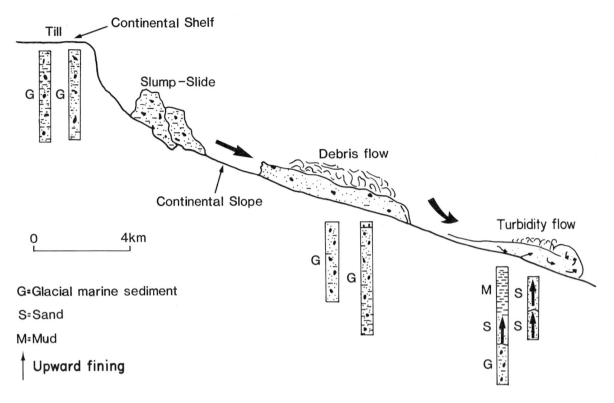

Figure 13.12   Model for inter-canyon gravity flow and probable corresponding core lithologies from the Weddell Sea. See text for details (from Wright and Anderson, 1982).

sections of cores display graded sands with upward fining and an up-core mineralogical sorting – characteristic of turbidite deposition. It is instructive that the coarsest debris in the **graded beds** is of 2ϕ – the size component missing from higher up the slope. On the abyssal plain the sediments are entirely graded sands with lithic material typically $< 5\%$. Wright and Anderson (1982) report that such sand covers an area of some 66 200 km². The presence, within this body of sand, of grain surfaces claimed to be diagnostic of glacial transport and the appearance of 'displaced' cold-water foraminifera characteristic of those found on the Weddell Sea continental shelf suggest an origin at higher elevation. Figure 13.12 provides an interpretation of the formation of this sequence of deposits. Outer shelf glaci-marine sediments are subject to downslope slumping and sliding. As increasing quantities of fluid are entrained during mass transport the sediments become partially suspended and fluidized, accompanied by size/density sorting. Fine-grained sands and muds may be sheared from the head region of the transitional debris flow and transported into a fine, turbulent suspension of solids above the flow. It is clear that auto-suspension may well occur. Motion will also be enhanced by dispersive stress caused by grain flow (Lowe, 1976b). With continued downslope transfer the finer grades are eventually winnowed out. The remainder of the debris flow becomes sluggish to form a separate turbulent suspension which continues downslope to yield fine-grained turbidites.

14 Glaci-marine processes and sedimentation: icebergs and sea ice

14.1 Introduction

Icebergs calve from the extremities of ice shelves, outlet glaciers and ice cliffs. They drift out into the open ocean where they gradually melt in warmer waters and release their sediment load. Sea ice, formed during winter **freeze-up**, may also transport a significant quantity of debris, frozen-in during formation or blown onto its surface. The importance of icebergs and sea ice to glaci-marine sedimentation lies in their capacity to carry debris considerable distances into the open ocean where they are released by melting and dumping and the disturbance to sea floor sediments resulting from the grounding and scour of bergs and sea ice keels. Sedimentational processes associated with these ice masses are, however, poorly understood. This chapter attempts to summarize current knowledge and indicate those areas worthy of further study.

14.2 Iceberg calving, fragmentation and toppling

The deposition of sediments entrained in icebergs will depend upon the mechanisms of calving, subsequent berg fragmentation, the location and quantity of debris, iceberg drift and deterioration (melt) histories.

Several mechanisms have been proposed to account for iceberg calving and are discussed by Thomas (1973b), Holdsworth (1978, 1982) and Robin (1979). A useful discussion is provided by Kristensen (1983) and only a summary is given here. The primary modes are:

1 Tidal bending and flexure
2 Calving due to an imbalance of hydrostatic forces on an unsupported ice cliff – the **Reeh calving** mechanism (Reeh, 1968)
3 Energy dissipation by **tsunamis**
4 Fracture by stress imposed from sea and ocean currents
5 Creep-failure due to lateral spreading
6 Impact of large, drifting icebergs (Swithinbank, 1969)

7 Vibrational cyclic bending and failure (Holdsworth, 1981)

Observations from tidewater glaciers in the northern hemishere (Ovenshine, 1970) and from ice streams around Antarctica (Bellair et al., 1964) show that for several hours and sometimes days after calving, icebergs fragment and turn over in order to attain equilibrium positions with low centres of gravity. This process is important as sediments exposed in the base, sides or surface may be moved into or out of the water thus affecting deposition on the sea bed. Figure 14.1 illustrates typical bergs.

If the total thickness of a typical tabular iceberg calving from an Antarctic ice shelf (Figure 14.2) is H_b Weeks and Mellor (1978) show that draft is $0.81 H_b$ and depth to centre of buoyancy (B') is $0.45 H_b$ below the waterline. The centre of gravity (G') is at $(H_b/2) + qi$ (where qi is a correction factor for the upper low-density layer of snow or firn, and taken as ~ 6 m). If the berg were tilted as in Figure 14.2 a new centre of buoyancy would be defined as b'. For a tilted berg the vertical line through b' intersects with the line through the centre of gravity, G', at the **metacentre** (M_c). In general terms if G' is below M_c the buoyancy force acting upwards through G' will rotate the iceberg back into an upright position. If it is above M_c the berg is unstable and may topple over. The height of M_c is given as:

$$M_c = (W_b/12[0.81 H_b] - B' - G')$$
$$= (W_b/9.72 H_b) - (0.095 H_b - 6) \quad (14.1)$$

where W_b = width of iceberg.

The minimum requirement for stability is that $M_c > 0$ or $M_c > 0.1 W_b$ (Weeks and Mellor, 1978). If an iceberg has a thickness of 200 m it will roll over if its width is < 220 m. For the berg to be considered reasonably stable from roll-over the width should be at least 320 m. A tiltmeter record obtained from a berg near to Clarence Island, South Shetland Islands of planimetric dimensions 1.2×0.48 km, with a freeboard between 39 and 57 m and a thickness of 273 m (Kristensen et al.,

Figure 14.1   A Typical Antarctic tabular iceberg from
the Weddell Sea (off James Ross Island). HMS
Endurance in the background (photo: V.S. Squire).
B Banded iceberg observed between South Georgia
and South Sandwich Islands (from Wordie and Kemp,
1933).

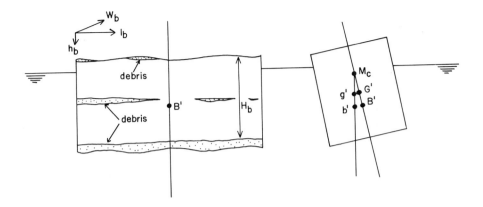

Figure 14.2 Iceberg characteristics and stability criteria.
Left: an iceberg showing possible locations of sediments (surface, internal and basal).
Right: definition of iceberg stability terms. G', g' = centres of gravity; B', b' = centres of buoyancy; M_c = metacentre. These terms are further explained in the text.

1982) exhibited rolling characterized by dominant periods of 14 s and 100 s. The former is introduced by ocean waves while the latter is the **natural period** or frequency of roll for the iceberg of the specified geometry: the iceberg was thus very close to the limit of the stability given above. Similar data have been reported for icebergs in the Weddell Sea by Wadhams *et al.* (1983).

14.3 Debris content of icebergs

The debris content of bergs will depend on whether they calve from ice shelves, ice streams and outlet glaciers or originate from ice cliffs. It has been shown in Chapter 13 that most *ice shelves* will not produce bergs containing significant amounts of basal debris. Any sediment that is present will probably consist of englacial ribbons from ice flowing into the ice shelf through exposed rock terrain. Such phenomena have been inferred from radar soundings of the Ross Ice Shelf (Bentley, 1980). Icebergs containing such debris bands (Figure 14.1B) have been observed off the Antarctic Peninsula (Wordie and Kemp, 1933). In the case of small ice shelves or those whose basal regime exceptionally favours extensive zones of

bottom freezing it is possible for a limited amount of basal debris to be present. *Outlet glaciers* appear to be the dominant source of debris-rich icebergs: they exhibit rapid calving and possess thick basal diamict layers. Figure 14.3 shows a view of an overturned iceberg that calved from the Astrolabe Glacier in Terre Adèlie. Exposed in the base of the berg is a thick layer of debris. Although Bellair *et al.* (1964) originally interpreted these sediment horizons as englacial, their distinctive grooving and composition leaves little doubt that this is basal till, and that Astrolabe Glacier was sliding just prior to decoupling. Anderson *et al.* (1980a) have also provided useful information on sediments in Antarctic icebergs.

Superglacial debris has been widely reported from icebergs in the Arctic and would appear to originate either as true superglacial accumulations from mechanisms described in Chapter 7 or by exposure and surface concentration of englacial or basal material after iceberg tilting. Detailed descriptions have been given from studies on two icebergs which were tracked and manned during the 1950s and 1960s (T–3 and Arliss II). These, and similar icebergs, have been given the inappropriate designation of 'ice islands'. Their drift in the Arctic Ocean of several thousands of kilometres during two or three decades hardly qualifies them for a term implying immobility! Crary (1960) reports a wide variety of sediment textures and rock types ranging from boulders (several m³ in size) to clay-sized particles on T–3, while Smith (1964) and Schindler (1968) describe extensive surface deposits on Arliss II.

Figure 14.3 Debris exposed in the base of an overturned iceberg newly calved from Astrolabe Glacier, Terre Adèlie, Antarctica. Note the large rocks that are exposed and the pronounced grooving of the debris which indicate that the glacier was actively sliding on its base prior to coming afloat and calving (from Bellair, 1960).

14.4   Iceberg trajectories

The drift of icebergs will depend principally upon the effect of ocean currents integrated through the upper 200–300 m of the water column. Additional factors are predominant winds, oceanic fronts, pack-ice conditions and sea bed topography. Until the advent of satellite surveillance it was difficult to continuously track iceberg movements with ease and reliability. Several icebergs were singled out for special study by Soviet and North American scientists and tracked in the Arctic Ocean during the 1950s and 1960s as a means of deducing large-scale oceanic circulation patterns, and in one or two cases as manned platforms for the prosecution of a wide range of oceanographic experiments. Even today few thorough studies have been undertaken to provide representative drift

information. Very large **tabular icebergs** originating from Antarctic ice shelves have received special attention (Swithinbank, 1969; Swithinbank *et al.*, 1977), but for the purposes of understanding ice-rafted sedimentation the average drift rate of ice stream-type bergs is important.

Tchernia (1971), Tchernia and Jeannin (1980, 1983) and Vinje (1980) tracked several icebergs around the coast of Antarctica by means of transponders attached to them, relaying to navigation satellites. Average speeds recorded over periods of several hundred days ranged from 0.11 to 0.2 m s^{-1}. Vinje (1980) found a maximum speed of 0.5 m s^{-1} over a 12-day period. Tchernia reports that such drift is between 60–75% of the ocean current velocity. The icebergs in question ranged in size from 0.4 to 3.8 km (length) with freeboards of 30–50 m. Parts of the trajectories of several bergs showed a pronounced east–west movement close to the Antarctica coastline (100–150 km offshore) in the zone dominated by the Eastwind Drift (Figure 14.4). Movement was at times irregular and exhibited twisting motion probably due to the constraints of pack ice, local currents and occasional grounding on shoals in the continental

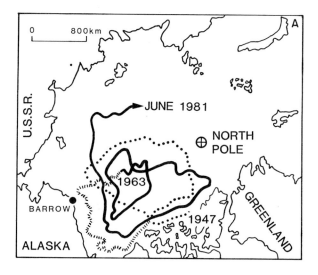

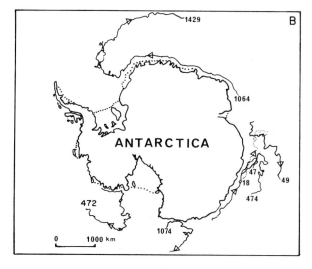

Figure 14.4 Selected iceberg drift trajectories. **A** drift of T-3 in the Arctic Ocean (after Clark and Hansen, 1983), **B** for several bergs around Antarctica (from Tchernia and Jeanin, 1983). Note the motion in the Eastwind Drift, close to the continental shelf, and recurvature into the Westwind Drift.

shelf. Four of the icebergs showed distinctive recurvature of their trajectories: they moved progressively further north from the coast and then turned to drift in an west–east direction. Such motion indicates transition from the influence of westward flow close to the continental slope to the

Westwind Drift – the primary circulation pattern of the Southern Ocean.

This twofold distinction in terms of drift tracks around Antarctica is important since many icebergs will attempt to circumnavigate the continent from east to west close in to the shore where waters are colder, constraining effects of thicker pack ice greater, and debris release consequently slower. Nevertheless the long periods spent in this zone probably result in most basal debris being released onto the continental shelf generating an extremely coarse inner girdle of IRD (see Figure 13.2). Bergs pushed towards the edge of the pack come under the influence of west–east currents and local **gyres**. Icebergs may then be swept both east and north into warmer waters where their melt-rates are enhanced. Longevity is thereafter a simple function of berg size at breakout from the Eastwind Drift.

In the northern hemisphere Robe (1980) has reviewed the drift paths of icebergs which calve almost exclusively from the fronts of major tidewater glaciers and occasionally from cliffs of small ice caps. A few icebergs find their way into the Polar Basin of the high Arctic. These are principally from Zemlya Franza Iosifa, Svernaya Zemlya and Svalbard archipelagos and from the Canadian Arctic Islands, especially the small Ward Hunt Ice Shelf in northern Ellesmere Island. The drift of a few icebergs in the central Arctic Ocean such as T–3 is well documented. The trajectory of T–3 reflected the presence of a major gyre in the Canadian Basin and Beaufort Sea (Figure 14.4A).

The south-flowing East Greenland Current draws off many icebergs from the Polar Basin, and major calving glaciers in northeast Greenland, channelling them between Svalbard and Greenland (the Fram Strait). These icebergs may find their way into the North Atlantic but around Cape Farvel (southern tip of Greenland) are more often entrained into the north-going West Greenland Current. They are thus carried into Baffin Bay. Here they mix with icebergs calving from the West Greenland outlet glaciers and from the Canadian Arctic Islands. Icebergs in the western part of Baffin Bay are discharged south into Davis Straits and beyond (to Newfoundland and the Grand Banks) by the Baffin Island–Labrador Currents.

Drift speeds vary typically between 0.08 and 0.25 m s^{-1}, although much higher velocities have been reported (0.75 m s^{-1}) in Baffin Bay.

14.5   Debris release from icebergs

14.5.1   Melt above the waterline

Debris cropping out at the surface (Figure 14.5) or sides of an iceberg is subject to incoming solar radiation and positive air temperatures which may be sufficient to melt the ice and release sediments, especially if their **albedo** is low. On the berg surface melt processes give rise to water saturation of sediments producing periodic debris-flow activity (see Chapter 9) as on Arliss II (Schindler, 1968). Meltwater rivulets, described by Ovenshine (1970), may also carry away fines in suspension or larger particles in traction. Such transporting agencies either redistribute sediments on the berg surface, carry them into crevasses and moulins or discharge them over the berg edge into the sea. Sediments melted out from iceberg sidewalls are transported

Figure 14.5   Debris on the surface of an iceberg (courtesy C-Core).

off the berg as either occasional debris falls, especially of coarse material, or as a semi-continuous rain of finer sediment. Close to the coast in such areas as Antarctica, northern and eastern Greenland and in the Arctic Ocean such release will be restricted to summer months, and even then may be of small magnitude due to low atmospheric temperatures. Orheim (1980) has noted that some Antarctic icebergs may even exhibit a positive mass balance!

14.5.2   Sidewall melt at and below the waterline

Wave action and the effects of strong currents give rise to both waterline melting and collapse of undercut cliffs. Unless substantial quantities of debris are exposed at these locations the sedimentological significance of waterline melting is not great. The submerged sidewalls of an Antarctic tabular iceberg can be investigated using side-scan sonar techniques as described by Klepsvik and Fossum (1980). Melting of submerged sidewalls will release debris and several studies of melting behaviour have been made. Huppert (1980) and Huppert and Joseberger (1980) show that

processes at the ice wall are complex due to vertical density gradients in the water but that the rates of melt are in the order of m a^{-1}. The net effect of melting in stratified sea water is to produce a thin upward-directed boundary layer close to the ice wall complemented by a thicker descending layer at a greater distance. The melt rate (in m day^{-1}) is greatest in the turbulent region and is given by:

$$M = 3.7 \times 10^{-2} (T_\infty - T_{ss})^{1.5} \qquad (14.2)$$

where T_∞ = far field temperature
T_{ss} = sea surface temperature.

The mean melt rate for the submerged ice wall is:

$$\overline{M} = 1/h_w \int_o^{hw} M(x/k)^{0.25} dx \qquad (14.3)$$

where h_w = height of submerged ice wall
x = coordinate in water flow direction
k = scale factor.

The descending layer eventually attains a level of zero net buoyancy (due to increasing salinity) and flows out forming a semi-horizontal lamella of several metres to tens of metres thickness. Such layers adjacent to the Erebus Ice Tongue form discontinuities of 0.1 °C in temperature, 0.04 ‰ in salinity and 3.5 $\times$ 10^{-4} g cm^{-3} in density. The layers extend to a distance of hundreds of metres and tend to coalesce into thicker bands (Jacobs et al., 1981). Such behaviour is important for the suspension–transport of very fine clay-sized fractions away from melting icebergs.

14.5.3 Melt at the iceberg base

Bottom melting of icebergs is not dissimilar to that of the base of ice tongues and ice shelves, the principal differences being the effects of spatial variations in oceanographic factors. Several approaches to the investigation of basal melting have been made – laboratory simulation, theoretical calculations and field experiments. No iceberg has been instrumented to directly measure bottom-melt. Laboratory studies were undertaken by Russell-Head (1980) who suspended ice blocks in water of varying salinity and temperature. His results yield a melt-relationship of the form:

$$M_b = 1.8 \times 10^{-2} (T_{sea} + 1.8)^{1.5} \qquad (14.4)$$

where M_b = iceberg melt rate
T_{sea} = sea water temperature.

These results are similar to the estimate made by Budd et al. (1980) from a consideration of the steady-state flux of icebergs into the Southern Ocean. The results of Martin and Kauffman (1977) give bottom-melt rates of between 0.5 and 1.0 m a^{-1} comparable to the theoretical predictions of both Gade (1979) and Weeks and Campbell (1973), but two orders of magnitude lower than the Russell-Head and Budd et al. conclusions. These differences (which are summarized in Table 14.1 for a berg of 1 km length and a relative velocity of 0.05 m s^{-1}) probably highlight more the serious shortcomings of laboratory experiments in simulating complex oceanographic and glaciological processes, particularly where scaling factors are ill-determined. Considerable caution is necessary in using and interpreting the melt rates shown in Table 14.1. It would appear from the several studies and the range of values suggested for bottom melting that a good understanding of the physical processes and determination of typical rates and controlling factors is still some way off. Perhaps the difficult task of real iceberg instrumentation might prove more effective in the long term. Notwithstanding the above comments the values for melting in Table 14.1 suggest that if layers of sediment a few metres in thickness are present in the basal portions of an iceberg (such as

Table 14.1 Calculated melt rates for the base of an iceberg (M_b, in m a^{-1}) from various studies (see text for further discussion)

ΔT °C	Weeks and Campbell*	Russell-Head**	Budd and others†
0.5	0.07	23	15–55
1.0	0.15	31	18–70
2.0	0.30	49	33–80
5.0	0.74	117	37–95

* $M_b = (L\, V_b^{0.8}\, \Delta T)/x_b^{0.2}$
where L is constant, V_b is relative free stream flow velocity (= 0.05 m s^{-1}), and x_b is iceberg length (= 1 km). ΔT is ice/water temperature difference.
** $M_b = 1.8 \times 10^{-2} (\Delta T + 1.8)^{1.5}$
† From Budd and others (1980, figure 15).

those calving from outlet glaciers) all debris would be melted out within a few years at the most. Vinje (1980) considers that around Antarctica icebergs are held close to the coast for up to five years so that the bulk of sediments would be melted out and deposited on the continental shelf. Only persistent fragmentation and overturning of icebergs would assist preservation of debris in the ice and continued rafting, and even this is not considered effective by Orheim and Elverhoi (1981).

A problem not yet worked out in detail is the effect of basal sediments on the melt–release process itself. The values quoted in Table 14.1 are based upon theoretical and experimental considerations for *pure* ice melting in sea water and their applicability to debris-charged ice may thus be questioned. Assume an iceberg as shown in Figure 14.6. The heat flux from the sea, Q_s^*, and temperatures at the top and base of the iceberg are taken as constant and known. If sediments are added to the ice base whose thermal conductivity $K_d \gg K_i$ their effect will be to drain away heat more rapidly from the interface and hence reduce the melt rate, thus slowing the melt–release process. An additional effect of basal sediments will be in changing interface roughness with an opposite effect on melting.

14.5.4 Mechanical modes of debris release

During the fragmentation, tilting and toppling of icebergs loose sediments located at the surface or sides of the berg may be dumped into the sea, giving rise to periodic sedimentation events. On the sea floor this process creates isolated carapaces of coarse detrital material.

14.6 Sedimentation and facies models

Figure 14.7 illustrates a hypothetical debris-release history for an outlet glacier iceberg. Dumping, mudflow and dirty meltwater activity are superimposed upon the more steady release of sediments by basal and sidewall melt. The result is a highly irregular, episodic and complex pattern of deposition. It is no wonder, therefore, that ice-rafted material is often extremely heterogeneous and difficult to categorize as suggested by accounts given in the literature (Anderson *et al.*, 1980b). Nevertheless the presence of dropstones (striated and faceted), till pellets and **conglomeratic** lenses are strong indications of ice rafting.

Clark *et al.* (1980) and Clark and Hansen (1983) recognize 13 lithographic units in the central Arctic Ocean spanning the last 5 M a. They exhibit a surprisingly uniform pattern brought about by the constant flux of sea ice and icebergs containing

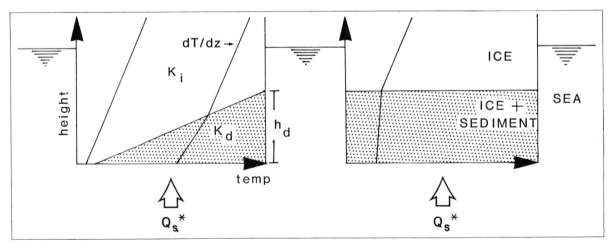

Figure 14.6 Effects of the presence of basal debris in icebergs on the melt-rate. The left-hand figure illustrates an iceberg with a variable basal sediment horizon (thickness h_d and thermal conductivity K_d less than the ice K_i). Since the heat flux from sea water (Q_s^*) cannot be rapidly conducted through the iceberg base melting of ice-debris mix may ensue with release of sediments to the sea. To the right $K_d \gg K_i$, the temperature gradient is steepened and results in less basal melt and sedimentation.

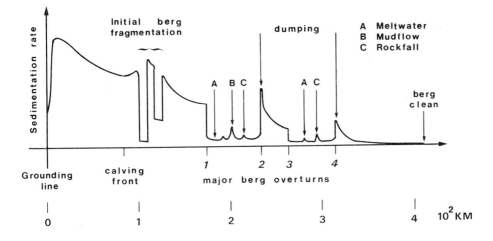

Figure 14.7 Debris-release history of a hypothetical ice-stream iceberg and resulting sea floor sedimentation rate. Note the various effects of calving and overturning, the contributions from meltwater, mudflow and rockfall activity (from Drewry and Cooper, 1981).

sediments, and the advection and mixing by mid-depth ocean currents of finer fractions. These authors identify four textural classes which suggest particular depositional processes and are described in Table 14.2. Of these Types I and IV owe their origin to sedimentation from icebergs.

14.7 Sedimentational processes associated with sea ice

Sea ice is an effective medium for the reworking, transport and deposition of clastic and biogenic matter, in both shallow, epi-continental seas and the central Arctic. Campbell and Collin (1958), for instance, have estimated that some 4–8 M tonnes of sediment become ice-borne in Foxe Basin in the Canadian Arctic archipelago during each annual cycle. Sea ice may account for between 1% and 10% of the total sediments released into the central Arctic Ocean according to Mullen et al. (1972). In the Beaufort Sea, along the north coast of Alaska, Barnes et al. (1982) calculate the total amount of sediment in sea ice during spring, 1978 to be in the order of 2.1 M tonnes. Although we shall deal with the effects of sea ice in the ocean several of the processes in this chapter will have importance for sedimentation in estuaries and fjords. In addition, debris-rafting by sea ice can be applied to the break-up and drift of ice in large lakes.

The effectiveness of sea ice as a geological agent depends upon entrainment of sediments,

Table 14.2 Textural types and origins for sediments in the size range 0.63–63.0 μm (fine clay to coarse silt) in the central Arctic Ocean (after Clark et al., 1980)

Sediment type	Texture characteristics					Inferred mode of deposition
	Mean grain size (μm)	Sorting (RMS μm)	% coarse	% silt	% clay	
I	11.6	13.9	14.2	46.4	39.4	Icebergs and sea ice ≡ 'background' deposition
II	14.1	15.4	16.0	48.4	35.6	Sea ice
III	6.1–9.6	10.2–13.4	2.0–15.2	31.2–37.0	47.7–66.8	Sea ice
IV	25.8–30.1	16.6–18.8	11.9–25.7	63.6–74.6	13.4–14.2	Icebergs

subsequent drift and the eventual deterioration of the ice releasing matter to the sea bed. A further process is the scouring action of *sea ice keels* on sea bed deposits – a topic dealt with separately in Section 14.11.

14.8 Entrainment of debris into sea ice

Several mechanisms may be responsible for the incorporation of sediments into ice and are summarized in Table 14.3. It is difficult, however, to assess the relative importance of each of these processes due to the lack of sufficient studies. The contribution from river flooding appears to be minor and geographically restricted (Collinson, 1971) – as are the effects of avalanching from coastal cliffs. Incorporation from the sea bed has been demonstrated to be important and widespread in shallow water.

Debris may also be incorporated into the surface of sea ice where sediments from coastal ice-free areas are blown out to sea and deposited. This appears to be an important process in McMurdo Sound (Barrett *et al.*, 1983) (Figure 14.8). Strong off-shore winds are common around glacierized coasts especially Antarctica and Greenland where katabatic drainage and surface topographic irregularities funnel cold dense air across ice-free terrain.

Table 14.3 Sources of debris entrained into sea ice (including fast ice)

Rivers	Flooding of sediment-laden water onto fast ice during springtime
Seabed	Freezing-in of sediments in suspension in sea water (due to turbulence generated by storms, tidal action, ice scour, etc.) Anchor ice and attached debris
Beaches	Freezing of sea ice onto beach materials
Coastal cliffs	Avalanching, slumping and sliding of debris onto sea ice
Wind	Deflation of fine-grained sediments from ice- and snow-free terrain onto adjacent sea ice

Figure 14.8 Debris on first-year sea ice in McMurdo Sound, Antarctica. Note streaked pattern from wind action and a substantial surface debris deposit in the centre of the photo which is a sediment-laden iceberg.

14.8.1 Frazil ice-debris suspensions

Sharma (1979) and Barnes *et al.* (1982) describe entrainment resulting from freezing of turbid water suspensions. The process takes place almost exclusively in shallow water during storm periods when waves roil the sea bed, cause vertical mixing and result in the resuspension of bottom sediments. Clay- and silt-sized materials are particularly susceptible and become quickly incorporated into growing **frazil ice** (Tsang, 1982). The increased viscosity of the water due to the presence of numerous ice crystals in suspension aids the retention of sediment particles. Frazil ice forms rapidly and hence possesses the capacity to entrain considerable quantities of debris. Frazil ice will solidify by interstitial development of congelation ice at a rate of ~ 100 mm d^{-1}. This process results in downward brine drainage, which may flush out some of the very finest sedimentary components.

14.8.2 Anchor ice

A second process involves the development of anchor ice on the sea bed. This particular formation is composed of individual or aggregations of ice platelets which vary in size from 100–150 mm diameter and 20–50 mm in thickness. They are found frozen to the sea floor usually at water depths of less than 30–40 m (Dayton *et al.*, 1969). Although attached to the substratum (hence the name), anchor ice may reach sufficient dimensions to lift loose material off the bottom due to its inherent buoyancy. Thus ice and debris (organic and inorganic) are carried up to the sub-ice platelet layer beneath sea ice where they are slowly frozen in.

Anchor ice forms when supercooled water is transferred to the sea bed (or a river bottom) where ice is nucleated around organic and inorganic artifacts. The adhesive bond between anchor ice and bed materials is strongest during the initial period of formation (due to the negative heat gain in the water) but heating of the water from incoming solar radiation, for instance, will later weaken the bond. Tsang (1982) observes, '. . . anchor ice forms mostly over boulders, stones, gravels, coarse sand and aquatic weeds, and almost never forms over bottoms of packed fine sand, silt

or clay.' This preference is attributed to the ease with which small particles can be lifted by anchor ice of relatively small dimensions, thus inhibiting the continued growth of substantial anchor ice. Secondly, large clasts on the sea bed extend upwards into supercooled waters whilst a compact finer-grained bed is under the influence of a laminar boundary layer and also of local subjacent heat flow which inhibits good anchor-ice adhesion and formation.

Using Archimedes principle it is possible to calculate the size of the clasts that can just be lifted off the bottom by ice platelets of specified dimensions (assuming no cohesion or frictional resistance). For buoyancy the upthrust on the ice and clast must be at least equal to their combined weight. The upthrust is the weight of the sea water that would fill the space occupied by both clast and ice:

$$\text{Total upthrust} = (\pi \, r_f^2 \, h_f)\rho_w + (1.33\pi \, r_c^3)\rho_w \tag{14.5}$$

where r_f = radius of ice platelet
   r_c = radius of clast (taken as sphere)
   h_f = thickness of ice platelet
   ρ_w = density of sea water.

$$\text{Combined mass} = (\pi \, r_f^2 \, h_f)\rho_i + (1.33\pi \, r_c^3)\rho_c \tag{14.6}$$

where ρ_i = density of ice
   ρ_c = density of clast.

For both ice and clast to float the volume of the platelet (V_p) must, therefore, be:

$$V_p \geqslant V_c \left(\frac{\rho_c - \rho_i}{\rho_w - \rho_i} \right) \tag{14.7}$$

where V_c = volume of clast.

Substituting into equations (14.5) and (14.6) the radius of the ice platelet of given thickness has to be:

$$r_f \geqslant \left[\frac{1.33 \, r_c^3}{h_f} \left(\frac{\rho_c - \rho_i}{\rho_w - \rho_i} \right) \right]^{1/2} \tag{14.8}$$

This solution is that derived by Drake and McCann (1982) for the movement of boulders on Arctic tidal flats. Figure 14.9 shows size of platelet vs. clast size for a thickness of 30 mm –

Figure 14.9 The size of rock particles that can be lifted by anchor ice platelets of specified dimensions as determined by Equation (14.8).

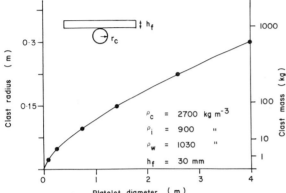

typical for anchor ice. Dayton *et al.* (1969) report observing masses of anchor ice up to 2 m² rising from the floor of McMurdo Sound, Antarctica, while Tsang (1982) records masses of 0.5–1.0 m diameter – these would have the potential of lifting particles with a radius of several centimetres and a mass of several kilograms. Usually the mass raised will be less than predicted by Equation (14.8) due to cohesion between clast and substratum. Using an extension of the above analysis it is possible to estimate the maximum size of clast that can be supported by sea ice floes and thus rafted into the open ocean. Drake and McCann (1982) undertook such a study and found that the bearing strength of a floe had also to be taken into account. Following Michel (1978) they consider the bearing strength as a function of floe thickness and tensile strength, with a value ~$10^5 h_f^2$. The ice floe will thus fail if:

$$V_c (\bar{\rho}_c - \rho_w) \geq 10^5 h_f \qquad (14.9)$$

where V_c = volume of the clast.

This can be expressed in terms of the critical floe size:

$$r_f \geq \left[\frac{10^5 h_f^2}{1.33 \pi (\bar{\rho}_c - \rho_w)} \right]^{1/3} \qquad (14.10)$$

An idealized floe will raft boulders up to several metres size with an upper limit constrained by floe thickness.

14.8.3 Concentration of sediments in sea ice

Few measurements of sediment concentration have been reported for sea ice. Early commentaries indicate patchy concentrations of a variety of macroscopic clastic and biogenic materials ranging in size from cobbles to clay-sized particles. Campbell and Collin (1958) collected sediments from ice floes in Foxe Basin, NWT which averaged approximately 0.5–2.0 g l⁻¹ for the surface and 0.05–0.1 g l⁻¹ at depth. Barnes *et al.* (1982) found sediment in sea ice along the coast of the Beaufort Sea up to 1.6 g l⁻¹, predominantly composed of finely disseminated clays.

The sediment concentration in a series of cores taken from Prudhoe Bay, Alaska shows a general pattern of an upper layer 100–150 mm deep of relatively clean ice overlying a sediment-rich zone of a few centimetres to over 1 m in thickness. Below this there is a **brine**-rich ice horizon.

14.9 Transport of sediments by sea ice and melt-out

The transport of sea ice and entrained sediments is highly variable in time and space. The bulk of sediments found in sea ice are incorporated in the nearshore zone dominated by the development of **fast ice** (i.e. that held to the shore and to points on the sea floor) which grows rapidly in the early winter primarily as a result of the presence of shallow water which reduces the convection depth needed to allow cooling of the water to its freezing point (Wadhams, 1980). Other factors such as lower salinity from river discharge, calmer surface conditions, effects of local bathymetry, shoreline geometry and offshore pack conditions play varying roles in fast-ice formation (Kovacs and Mellor, 1974). More than a few kilometres offshore, however, the sediment content of sea ice is extremely low and contributed almost exclusively by **deflation**.

The width of the fast-ice zone is principally controlled by water depth (and hence bathymetry) and the maximum likely draft of grounded pressure ridges. Off the north coast of Alaska this depth is in the order of 18 m but may occasionally rise to over 25 m. The resulting width of the fast-ice zone is several tens of kilometres and up to 70 km (Stringer, 1974). Fast ice displays little movement once formed. During the onset of the summer the fast ice lifts off the seabed and commences to melt. This activity continues for several weeks after which the ice rapidly breaks up and floats away. Reimnitz and Barnes (1974) and Walker (1974) demonstrate that melting takes place more or less *in situ*. Prior to break-up, therefore, melt releases up to 90% of the entrained sediments within a few hundred metres of where they were incorporated. Little sediment held in winter fast ice may thus remain to be rafted out at break-up and the sedimentary activity is essentially one of reworking rather than net transport. Furthermore, during the formation of sea ice brine drainage may flush out some of the fine-grained particulates held in the consolidated frazil ice. Only occasionally do major winter or spring storms break up the fast ice and allow sediment to be moved into the shear zone, the marginal ice zone or into the polar pack (Figure 14.10).

In some localities, however, there does appear to be more widespread movement of debris-charged sea ice – notably in the eastern part of the Canadian Arctic archipelago (Campbell and Collin, 1958). Barnes *et al*. (1982) consider that 10–20% of the ice involved in **ridges** does not melt completely during the first summer. Such ice may drift away from the continental shelf and be incorporated into the polar pack. Any sediments entrained in these ridges are eventually deposited in the Arctic Basin (Clark and Hansen, 1983). Considerable quantities of debris have been observed in ice in the polar pack north of Svalbard (Wadhams, personal communication). Maybe as much as 10% of ice floes exhibit discolourations from sediments. These are believed to originate from incorporation in fast ice along the Siberian coast, later broken out and moved westwards in the Transpolar Drift Stream

to eventually discharge through the Fram Strait into the East Greenland Current.

14.10 Sedimentary facies and models

Few investigations have been undertaken of the details of sediment facies resulting from sea-ice activity. Some of the more useful studies are those of Clark *et al*. (1980) and Clark and Hansen (1983) in the central Arctic Ocean and Barnes and Reimnitz (1974) in the Beaufort Sea. No detailed studies have been conducted in the Antarctic and most work to date (e.g. Anderson *et al*., 1983) stress the role of icebergs not sea ice. It is important to note that besides melt-release of sediments to the sea floor (i.e. in a manner similar to processes described for ice tongues, ice shelves and icebergs) the effects of river discharge, water circulation and wave activity have to be taken into account. The ice itself may also rework bottom sediments by bulldozing and gouging (see Section 14.11).

Important features of the sedimentary environment of sea ice-dominated continental shelves is the division into inner and outer shelf areas, usually demarcated by the shear zone separating fast ice from polar pack ice (Figure 14.10). In general, inshore of the shear zone sediment patterns are influenced by currents, tides, storm activity and discharge from rivers. Beyond the shear zone and in water depths of up to 20 m or so ice and ice–bottom interactions are the principal influences on sedimentation.

14.10.1 Inner shelf region

This zone lies out to a depth of no more than about 5 m and, depending upon the season, is affected by ice and wave action, the latter a key element in the summer. Barnes and Reimnitz (1974) consider that ice rafting and ice gouging are not so important as the influence of waves, currents and sometimes tides because the zone is dominated by fast ice which experiences minimal advection. The melting-out of sediments usually occurs *in situ*. Currents and wave ripple marks are frequently observed in addition to current

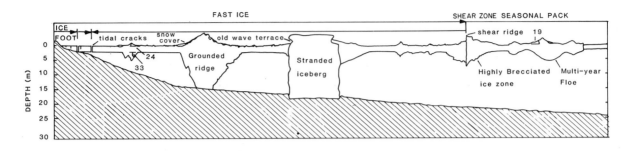

Figure 14.10 Association of seasonal pack, fast ice and shear (brecciated) zones as exemplified by ice conditions in the Beaufort Sea (from Kovacs and Mellor, 1974).

scours around grounded ice and ice-scoured features (see 14.10). Where rivers discharge into the shelf laminated, but rarely cross-laminated, sands and clays are found grading landwards into complex deltaic sequences (Walker, 1974). Bioturbation is common.

14.10.2 Central and outer shelf region

Further offshore, beyond the shear zone, sediments are typically massive muds with some sands and gravels. Sequences usually exhibit little structure and bioturbation. Barnes and Reimnitz (1974) also describe jumbled blocks of sediment with steeply dipping laminations. The lack of structure and disruption is attributed to the gouging action of drifting ice keels. Surficial clay and silt fractions may be derived from adjacent river influx or from ice melt-out. Some coarse-grained sediments are found on the Beaufort Shelf and may represent relict sequences of ice-rafted detritus.

14.10.3 Deepwater glaci-marine sedimentation

Clark *et al.* (1980) consider that sea ice has played the dominant role in sediment supply to the central Arctic Ocean during late Cenozoic times. The sediments are principally composed (80%) of silty **lutites** and arenaceous lutites with small quantities of clastic, carbonate-rich

sediments. The silty lutites are predominantly sea ice-rafted and correlated with periods of reduced ice flux in the Arctic Ocean. Sediments of the ice-rafted source are mixed with normal pelagic constituents (atmospheric and biogenic) and authigenic material. Some dropstones are encountered but are much more common in the arenaceous lutites which also exhibit poor sorting and coarse textures. They also represent contributions from iceberg processes. (Clark *et al.* (1980) and Goldstein (1983) report the presence of pellets (≤ 1 mm) within the silty–sandy lutites some of which may originate on sea ice. Traction currents, bioturbation processes and mid-depth ocean currents rework deep-water sediments. The latter are, possibly, the most significant and advect considerable quantities of the finer fractions changing the grain size distributions from place to place. These factors are shown schematically in Figure 14.11.

14.10.4 Seasonal variations

Barnes and Reimnitz (1974) provide a useful summary of the relative roles of ice and water in affecting sedimentation on a typical Arctic continental shelf. The separation of processes between summer and winter and with water depth (usually proportional to distance offshore) is pronounced. During the winter shore-fast ice grows quickly and usually ceases development by May in the northern hemisphere. Freezing-in of sediments from turbid suspensions occurs as previously described (14.8.1). Beach and nearshore sediments may be pushed and bulldozed during the early winter before the ice is fully stabilized. Once this condition is achieved

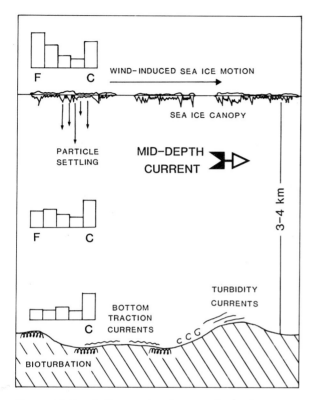

Figure 14.11 Sedimentation from sea ice in deep water. Three histograms show sediment textures (F = fine; C = coarse) in sea ice, notionally at mid-ocean depth and on sea floor. Note how the finer fractions are preferentially transported away by currents. On the sea floor sediments are bioturbated and reworked by traction and turbidity currents.

the influence of the ice decreases. Tidal currents are restricted in extent but intensified with the development of fast ice inducing sediment scour and transport to a shelf depth of up to 5 m. Seaward of the shear zone in 10–25 m of water ice action becomes dominant principally through scouring by ice keels. There is typically considerable ice transport, and current activity is greater due to fewer submarine constrictions. Further out, as shelf depth continues to increase, both hydrographic and ice processes decrease in intensity. Sediments mirror this change: there is better sorting, more pronounced lamination and bioturbation.

In the summer currents and waves become dominant near to shore as the winter ice is lost.

Littoral activity increases with beach sediments often being reworked by drift ice. Shelf currents become re-established and sediments display more pronounced lamination. Further out (20–60 km offshore) broken pack ice is usually present and dominates the sedimentary regime with ice scouring. The occasional grounding of keels may give rise to locally intensified currents with reworking and current scour of bottom sediments.

14.11 Sea bed scouring by floating ice

Gouging of the sea floor takes place when either sea-ice keels or the bottoms of icebergs run aground in shallow water. The result is considerable disruption and erosion of sea floor sediments. Ice plough marks on the sea bed are revealed on side-scan sonar records (Figure 14.12A) on seismic ('sparker') reflection records (Figure 14.12B), from bottom photography and diving inspections.

Plough marks are found on present-day polar continental shelves where they are currently being formed, and on shelves at lower latitudes where they are relict features from former glacial episodes. The terminology for these features is not clearly defined and numerous authors have variously used 'gouges', 'plough marks', 'scores', 'grooves', 'furrows' and 'scours'. In this section the term 'plough mark' is adopted as being suitably descriptive but other terms are used from time to time. The type of ploughing action by sea ice and icebergs is usually distinctive and geographically somewhat separate. Marks made by sea-ice keels are restricted to continental shelf areas shallower than 100 m and reach their maximum concentration between depths of 18 and 30 m. The forms produced by icebergs are found on the outer shelf and continental slope regions in water depths of 50–500 m.

14.11.1 Sea-ice plough marks

Grooves and furrows created by the ploughing of sea-ice keels are commonly observed on Arctic continental shelves. Weeks *et al.* (1983) provide detailed statistics of plough marks along the 190

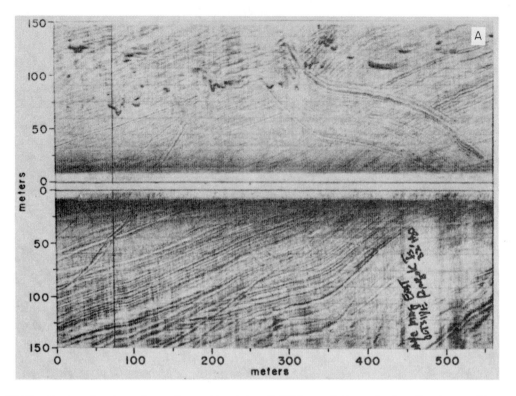

Figure 14.12   Sea ice plough marks on the continental shelf.
A Side-scan sonar record depicting plough marks on the sea floor (10 m deep) as a result of sea ice activity. Note that there are two stages of grooving. An initial pattern has been re-ploughed by the keel of a pressure ridge.

B High-resolution seismic reflection profile across the shelf of the Beaufort Sea off Kaparuk River, Alaska. Depth 31 m. The small-scale roughness results from sea ice ploughing (both **A** and **B** from Reimnitz and Barnes, 1974).

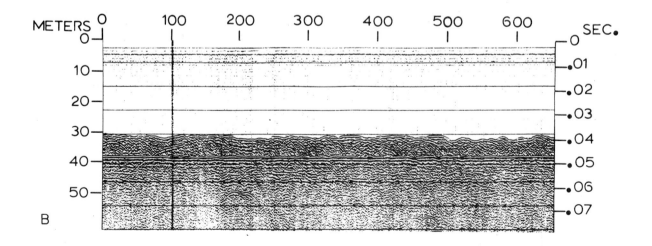

km stretch of the Alaskan shelf, and in the Beaufort Sea between Smith and Camden Bays (a distance of some 400 km). Their results are based on side-scan sonar and precision fathmometer measurements made between 1972 and 1979 in water depths up to 38 m. The morphology of these plough marks varies considerably depending upon geotechnical properties of the underlying sea floor materials, the geometry of the impinging ice keel, characteristics of the motion of the ice and nature of the impelling forces (Reimnitz and Barnes, 1974).

Some data on ploughing rates suggest values of about 5 km^{-1}a^{-1} (Weeks *et al.*, 1983).

In stiff, cohesive sediments ice keels may create irregular, rough grooves and associated flanking ridges. Blocks of sediment may be dislodged and pushed over the sea bed creating bands of broken-up sediment. Such plough marks are not readily modified and reworked by wave action (Figure 14.13A). In less cohesive, soft sediments plough marks are more regular and continuous but are rapidly modified by currents and wave action. Unconsolidated materials which are displaced into flanking levées and ridges are unstable and frequently subject to minor mass movements such as slumping. Sea-ice keels may be of various shapes, the simplest division being between single and multi-pronged forms. The resulting furrows are either single gouges or a complex sequence of semi-parallel troughs. The motion of the ice and keels is predominantly governed by wind-coupling to the upper surface, currents, and thrusting action from adjacent ice. The trajectory of a ploughing

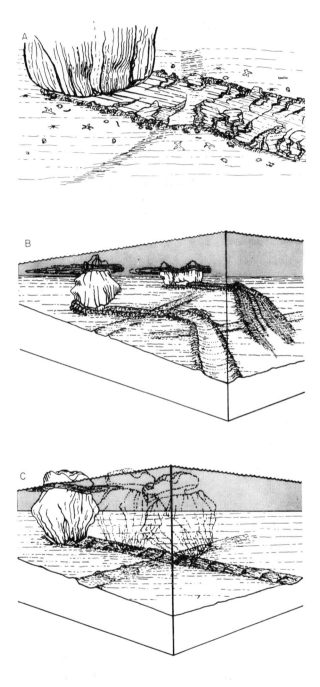

Figure 14.13 Formation of sea ice plough marks.
A Base of an ice keel dragging and grooving stiff, cohesive sea floor sediments. Note how the materials form massive, fractured blocks.
B The different plough directions may be due to either reorientation of a single keel by constraints of pack ice or successive passes by keels. Rotation of floes with more than one grooving keel cause complex patterns.
C In shallow water with incomplete buoyancy keels may exhibit unstable, rocking motions which gives rise to sprag or jigger marks (from Reimnitz and Barnes, 1974).

keel may be steady and uni-directional but more frequently gouges exhibit rapid changes of direction reflecting constraining effects of pack, and changes in the driving forces (Figure 14.13B). Rotation of keels modify the shape of plough marks whilst 'wobbling' motion of solitary, unstable ice blocks gives rise to unequally spaced 'sprag' or 'jigger' marks (Reimnitz and Barnes, 1974) (Figure 14.13C).

Whilst the above conditions make it difficult to generalize about plough marks they are for the most part a few metres wide but range from a few centimetres up to 10 m or more. Depths are highly variable, typically 1–3 m and only very rarely more than 10 m deep. Gouge lengths are from a few metres to several kilometres. *Fresh* plough marks exhibit a sharp, steep-sided relief often with sub-parallel minor grooves, lying at the angle of repose of the sediment. Associated with these features is a lack of biological activity such as the presence of burrows, trails or tracks although active bioturbation may rapidly alter plough marks.

Kovacs (1972) and Kovacs and Mellor (1974) recognize three zones with distinctive ice plough characteristics as defined from a study in the Beaufort Sea. In the nearshore coastal zone where water depths are <7 m there is frequent scouring of the sea floor giving rise to typically shallow grooves (0.5 m) which are smoothed by wave and current action. The bulk of the ploughing takes place during break-up and ice formation as fast ice is an inhibiting presence during the winter (up to 8 months). Most of present-day ice ploughing takes place in the mid-shelf zone, with water depths between 7 and 30 m. Plough mark densities (measured normal to their trend) of 10–15 km^{-1} are typical and often >100 km^{-1} are observed (see also Weeks *et al.*, 1983). This region, within and oceanward of the shear zone, possesses optimum drift rates and keel frequencies for scouring. Wadhams (1980) suggests that the rate of ploughing at any point on the sea floor (defined by x, y and h$_z$ coordinates) may be given by:

$$E_f = \overline{N_t}\, \overline{U_f} \int_{h_z}^{\infty} P(Z)\, dz \qquad (14.11)$$

where P(h) = probability density function of keel draft

U_f = annual drift rate
N_t = number of keels per km.

In the outer shelf zone plough mark density, and hence plough rate, falls off rapidly where water depths are >30 m since the frequency of ice keels of such depth is small (Figure 14.14). The deepest keel measured to date is ~ 47 m. Wadhams and Horne (1980) have shown that the distribution of keel drafts per kilometre per metre of draft increment follows a negative exponential function:

$$N_t'\, dh = k_1 \exp(-k_2 h_k)\, dh \qquad (14.12)$$

where N_t' = number of keels per kilometre per metre of draft increment
h_k = keel draft
k_1, k_2 = empirically determined parameters expressed in terms of mean keel draft and number of keels per kilometre.

Ploughing mechanisms

Ploughing may be considered analogous to abrasional grooving discussed in Chapter 4 and a similar analysis employed. Kovacs and Mellor (1974), however, have discussed sea-ice ploughing mechanisms in some detail, evaluating resistance to ploughing by sea bed sediments, the strength and integrity of ice keels and the magnitudes of forces and energy potentials necessary for sustained grooving. Total resistance to ploughing (R_t') is given by the general expression:

$$R_t' = R_1 + R_2 + R_3 \qquad (14.13)$$

where R_1 = bulldozing resistance
R_2 = resistance along near-vertical groove sides
R_3 = resistance along groove bottom.

Kovacs and Mellor (1974) consider an order of magnitude estimate of R_t' is ~ 14 MPa. Most keels appear to possess adequate strength to score soft sediments to moderate depth (~ 1 m), except perhaps for newly formed first-year pressure ridge keels impinging abruptly on the sea bed. The force propelling keels and hence responsible for ploughing is wind shear on the upper surface

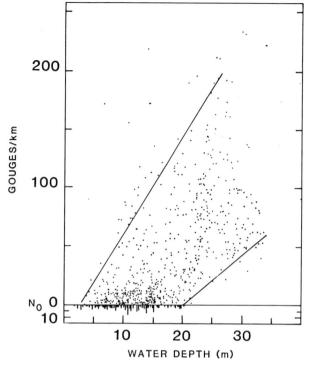

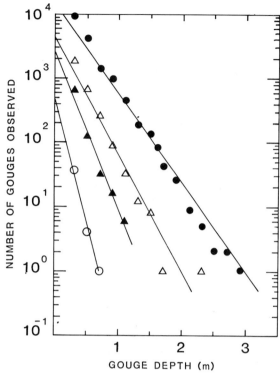

of sea ice which takes place over distances of 50–200 km. Pressure for gouging can also be transmitted to particular nearshore pressure ridges by first-year sea ice.

14.11.2 Iceberg plough marks

Iceberg plough marks occur in considerably deeper water than those resulting from sea-ice keels (Figure 14.15). They are frequently encountered in sediments of the outer continental shelf and upper continental slope as well as around shoal areas of the deep sea, and primarily at depths of 50–500 m (Belderson *et al.*, 1973). The dimensions of these plough marks are also highly variable, from 20–100 m in width with depths of 2–10 m (Harris and Jollymore 1974). Lien (1983) reports plough marks up to 250 m in width and 20 m depth on the Norwegian continental shelf. Lengths are several kilometres. Where small icebergs are responsible for ploughing the resulting grooves may be broad

Figure 14.14 Statistics on plough marks in the Beaufort Sea. Left-hand graph shows the number of plough marks per kilometre of sea floor as a function of water depth. The envelope shows no plough marks in water less than ~3.5 m deep and that greater frequencies of ploughing require increasing water depths. Right hand figure shows the exponential relationship between plough mark depth and number of plough marks for four separate coastal sub-regions of the Beaufort Sea Shelf indicating different regimes of ocean and ice characteristics. (from Weeks *et al.*, 1983).

and flat reflecting a relatively uniform ice base. It should be remembered that the bulk of observations of iceberg scour marks come from the northern hemisphere and the present-day Arctic where icebergs may not approach the dimensions of Antarctic tabular bergs. The characteristic sizes and form of plough marks from bergs several kilometres or tens of kilometres in size is

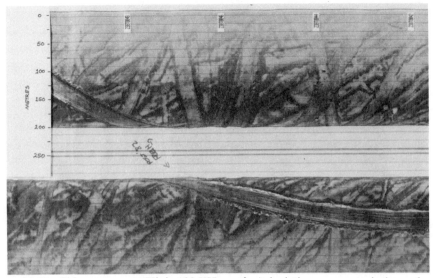

Figure 14.15   Side-scan sonar (Kloiv 100 KHz system) depicting numerous iceberg plough marks on the Saglek Bank, northern Labrador Shelf in a water depth of 120 m. Data were gathered in August 1985 from M.V. Pandora II operated by Canadian Federal Department of Fisheries and Oceans. Note very recent scouring by an iceberg base which occurred post 1982 and has obliterated previous marks. Bottom sediments have been pushed and squeezed out to form the lateral berms and the intervening seabed has been grooved by the passage of the berg. The berms are up to 2–3 m in elevations as determined from submersible dives. (Courtesy Fisheries and Oceans, C-CORE, and Atlantic Geoscience Centre, Bedford Institute of Oceanography.)

Figure 14.16   Depth of iceberg ploughing as a function of iceberg kinetic energy for given sea floor sediment shear strengths (τ) and drag coefficients (after Chari *et al.*, 1980).

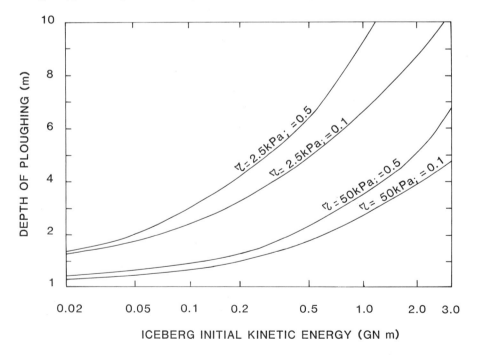

virtually unknown. Sonargraphs recorded close to the coast of Dronning Maud Land, by Klepsvik and Fossum (1980) indicate that major grooving occurs to depths of 20 m or more. Lien (1983) found that all areas of the Norwegian continental shelf shallower than 100 m and deeper than 500 m recorded no iceberg activity. It is now generally accepted that iceberg plough marks in areas such as the Norwegian Shelf were produced during periods of expanded continental ice and mostly during the Late Wisconsin. Since sea level at this time was on average 100 m lower than today the maximum likely draft of icebergs is of the order of 400 m.

As a result of the Dynamics of Iceberg Grounding and Scouring (DIGS-85) programme undertaken off the coast of Labrador during summer 1985 considerable data have been acquired on iceberg-sea floor processes (Josenhans et al., 1985; Barrie and Collins, 1985). Direct observations were made of fresh and relict iceberg impact and ploughing sites using a manned submersible (PISCES IV) and side-scan sonar (Figure 14.15). In particular resurvey of scours showed much new sea bed ploughing and substantial modification of scours by currents and macrobenthos, but dependent upon the geotechnical properties of the sea floor sediments.

Iceberg ploughing mechanisms

In a similar manner to ploughing by sea-ice keels iceberg scouring depends upon berg dimensions, drift velocity and drag, geotechnical properties and depth and configuration of the sea floor. As an iceberg grounds on a shallowing sea floor its very considerable kinetic energy (resulting from its large mass) is either converted into potential energy as it moves forward, or is dissipated by ploughing soft sediments. Of course, in reality both conditions apply. Chari (1979) and Chari et al. (1980) have investigated the mechanics of iceberg scour in some detail and have developed a model to predict depth of plough marks. Figure 14.16 illustrates the depth of ploughing according to Chari et al. (1980) for various geotechnical properties of the sea floor sediments and iceberg drag coefficient. Although the iceberg they consider is of small dimensions (30 m) the plough mark depths of a few metres are not unrealistic.

As a result of the DIGS-85 programme data were gathered on three groundings of icebergs off the Labrador coast. Wind, current and tidal fluctuation (Lewis et al., 1985) were recorded in order to calculate the driving forces on each iceberg. A complementary suite of observations was made on the icebergs themselves with deployment of instruments to measure accelerations and attitude and orientation variations. The net result has been to estimate forces exerted on the sea floor during the grounding of icebergs. Direct observations were made of grounding and pitting events, although no ploughing activity (i.e. with linear motion on the sea floor) was observed.

Such detailed observations of currently active processes will do much to improve knowledge of the driving forces on icebergs, the resulting motion and interactions with sea bed sediments and assist in developing phenomenological and quantitative models of iceberg ploughing.

References

AARSETH, I., HOLTEDAHL, H., KJELDSEN, O., LIESTØL, O., ØSTREM, G. and SOLLID, J.L., 1980: Glaciation and deglaciation in central Norway. *Field guide to excursion 31 Aug. – 3 Sep. 1980 organized in conjunction with the 'Symposium on Processes of Glacier Erosion and Sedimentation'*, Geilo, Norway. Norsk Polarinstitutt (58pp).

ALESTALO, J. and HAIKIO, J., 1979: Forms created by the thermal movement of lake ice in Finland in winter 1972–73. *Fennia* **157** (2), 51–92.

ALLEN, J.R.L., 1970: *Physical processes of sedimentation*. Earth Science Series 1, Allen & Unwin Ltd, London (248pp).

—— 1971: A theoretical and experimental study of climbing-ripple cross-lamination, with a field application to the Uppsala esker. *Geogrf. Annlr.* **53A** (3–4), 157–87.

—— 1984: *Sedimentary structures: Their character and physical basis*. Amsterdam, Elsevier (663pp).

ALLISON, I., 1976: Glacier regimes and dynamics. In Hope, G.S.; Peterson, J.A.; Radok, U.; Allison, I. (eds.), *The equatorial glaciers of New Guinea*, A.A. Balkema, Rotterdam. 39–59.

AMBACH, W., BEHRENS, H., BERGMANN, H. and MOSER, H., 1972: Markierungsversuche am inneren Abfluss-system des Hintereisferners (Otztaler Alpen). *Zeit. Gletscherk. Glazialgeol.* **8** (1–2), 137–45.

ANDERSEN, J.L. and SOLLID, J.L., 1971: Glacial chronology and glacial geomorphology in the marginal zones of the glaciers Mitdalsbreen and Nigardsbreen, south Norway. *Norsk geogr. Tidsskr.* **25**, 1–38.

ANDERSON, J.B., DOMACK, E.W. and KURTZ, D.D., 1980a: Observations of sediment-laden icebergs in Antarctic waters, implications to glacial erosion and transport. *J. Glaciol.* **25**(93), 387–96.

ANDERSON, J.B., KURTZ, D.D., DOMACK, E.W. and BALSHAW, K.M., 1980b: Glacial and glacial marine sediments of the Antarctic Continental Shelf. *J. Geol.* **88**, 399–414.

ANDERSON, J.B., BRAKE, C., DOMACK, E., MYERS, N. and WRIGHT, R., 1983: Development of a polar glacial-marine sedimentation model from Antarctic Quaternary deposits and glaciological information. In Molnia, B.F. (ed.), *Glacial-marine sedimentation*, New York, Plenum Press, 233–64.

ANDERSON, R.S., HALLET, B., WALDER J. and AUBRY, B.F., 1982: Observations in a cavity beneath Grinnel Glacier. *Earth Surf. Proc. and Landfms.* **7**, 63–70.

ANDREWS, J.T. and MATSCH, C.L., 1983: *Glacial marine sediments and sedimentation. An annotated bibliography*. Geo Abstracts Ltd, Norwich, (277pp).

APOLLONIO, S., 1973: Glaciers and nutrients in Arctic seas. *Science* **180** (4085), 491–3.

ARNDT, R.E.A., 1981: Recent advances in cavitation research. *Adv. Hydrosci.* **12**, 1–78.

ASHBY, M.F. and JONES, D.R.H., 1980: *Engineering materials – an introduction to their properties and applications*. Pergamon Press, Oxford (278pp).

BAGNOLD, R.A.,1954: Experiments on a gravity free dispersion of large solid spheres in a Newtonian fluid under shear. *Proc. Roy. Soc. (London) Ser. A* **225**, 49–63.

—— 1966: An approach to the sediment transport problem from general physics. *US Geol. Surv. Prof. Pap.* **422–I**, (37pp).

—— 1968: Deposition in the process of hydraulic transport. *Sedimentology* **10** (1), 45–56.

BANERJEE, I. and McDONALD, B.C., 1975: Nature of esker sedimentation. In Jopling, A.V. and McDonald, B.C. (eds.), *Glaciofluvial and glaciolacustrine sedimentation*, Tulsa, Oklahoma. Society of Economic Palaeontologists and Mineralogists, Spec. Publ. 23, 132–54.

BARNES, H.L. 1956: Cavitation as a geological agent. *Am. J. Sci.* **254**, 493–505.

BARNES, P., TABOR, D. and WALKER, J.C.F., 1971: The friction and creep of polycrystalline ice. *Proc. Roy. Soc. (London). Ser. A* **324** (1557), 127–55.

BARNES, P.W. and REIMNITZ, E., 1974: Sedimentary processes on arctic shelves off the northern coasts of Alaska. In Reed, J.C. and Sater, J.E., *The coast and shelf of the Beaufort Sea*, Arctic Inst. of N. America, Arlington (Va), 439–76.

BARNES, P.W., REIMNITZ, E. and FOX, D., 1982: Ice rafting of fine-grained sediment, a sorting and transport mechanism, Beaufort Sea, Alaska. *J. Sed. Petrol.* **52** (2), 493–502.

BARNETT, D.M. and HOLDSWORTH, G., 1974: Origin, morphology and chronology of sublacustrine moraines, Generator Lake, Baffin Island, Northwest Territories, Canada. *Can. J. Earth Sci.* **11**, 380–408.

BARRETT, P.J., 1980: The shape of rock particles, a critical review. *Sedimentology* **27**, 291–303.

BARRETT, P.J., PYNE, A.R. and WARD, B.L., 1983: Modern sedimentation in McMurdo Sound, Antarctica. In

Oliver, R.L., James, P.R. and Jago, J.B. (eds.), *Antarctic earth science*, Canberra, Australian Academy of Science, 550-40.

BARRIE, J.V. and COLLINS, W.T., 1985: Iceberg pits; description and postulated origin (abstract) in *14th Arctic Workshop: Arctic Land-Sea Interaction.* Bedford Institute of Oceanography, 90.

BATES, C.C., 1953: Rational theory of delta formation. *Am. Assoc. Petrol. Geol.* **37** (9), 2119-62.

BEECROFT, I., 1983: Sediment transport during an outburst from glacier Tsidjiore Nuove, Switzerland, 16-19 June 1981. *J. Glaciol.* **29** (101), 185-90.

BEHRENS, H., BERGMANN, H., MOSER, H., AMBACH, W. and JOCHUM, O., 1975: On the water channels of the internal drainage system of the Hintereisferner, Otztal Alps, Austria. *J. Glaciol.* **14** (72), 375-82.

BEHRENS, H., BERGMANN, H. MOSER, H. RAUERT, W., STICHLER, W., AMBACH, W., EISNER, H. and PESSL, K., 1971: Study of the discharge of alpine glaciers by means of environmental isotopes and dye tracers. *Zeit. Gletscherk. Glazialgeol.* **7** (1-2), 79-102.

BEHRENS, H., MOSER, H., OERTER, H., RAUERT, W., STICHLER, W., AMBACH, W. and KIRCHLECHNER, P., 1978: Models for the runoff from a glaciated catchment area using measurements of environmental isotope contents. *Int. Ass. Sci. Hydrol. Publ. No.,* 829-46.

BEHRENS, H., MOSER, H., OERTER, H., BERGMANN, H., AMBACH, W., EISNER, H., KIRCHLECHNER, P. and SCHNEIDER, H., 1979: Neue Ergebnisse zur Bewegung des Schmelzwassers im Firnkorper des Akkumulationsgebietes eines Alpengletschers (Kesselwandferner- Otztaler Alpen). *Zeit. Gletscherk. Glazialgeol.* **15** (2), 219-28.

BELDERSON, R.H. and WILSON, J.B., 1973: Iceberg plough marks in the vicinity of the Norwegian Trough. *Norsk Geol. Tidsskr.* **53** (3), 323-8.

BELDERSON, R.H., KENYON, N.H. and WILSON, J.B., 1973: Iceberg plough marks in the Northeast Atlantic. *Palaeogeog., Palaeoclim., Palaeoecol.* **13**, 215-24.

BELLAIR, P., 1960: La bordure côtière de la Terre Adèlie. In *Territoire des Terres Australes et Antarctiques Françaises. Terre Adèlie, Campagne d'Eté 1960.* Expeditions Polaires Françaises. Missions Paul-Emile Victor 218, 7-40.

BELLAIR, P., TOURENQ, J. and VERNHET, S., 1964: Un echantillon de moraine interne du Glacier de l'Astrolabe (Terre Adèlie). *Rev. Geogr. Phys. Geol. Dynam.* **6** (2) 115-21.

BENEDICT, P.C., BONDURANT, D.C., McKEE, J.E., PIEST, R.F., SMALLSHAW, J. and VANONI, V.A., 1971: Sediment transportation. *ASCE* **12**, 2010-7.

BENTLEY, C.R., 1971: Seismic evidence for moraine within the basal Antarctic Ice Sheet. In Crary, A.P. (ed.), *Ice and snow studies II,* American Geophysical Union, Washington, DC, Antarctic Research Series 16, 89-129.

—— 1980: Variations in valley glacier activity in the Transantarctic Mountains as indicated by associated flow bands in the Ross Ice Shelf. *Int. Ass. Sci. Hydrol. Publ.* **131**, 247-52.

BERNER, W., STAUFFER, B. and OESCHGER, H., 1977: Dynamic glacier flow model and the production of internal meltwater. *Zeit. Gletscherk. Glazialgeol.* **13** (1/2), 209-17.

BERTONE, M., 1972: *Aspectos glaciologicos de la zona del hielo continental Patagonico.* Instituto Nacional del Hielo Continental Patagonico, Buenos Aires (126pp).

BINDSCHADLER, R., HARRISON, W.D., RAYMOND, C.F. and CROSSON, R., 1977: Geometry and dynamics of a surge-type glacier. *J. Glaciol.* **18** (79), 181-94.

BJÖRNSSON, H., 1975: Subglacial water reservoirs, jökulhlaups and volcanic eruptions. *Jökull* **25**, 1-14.

—— 1976: Marginal and supraglacial lakes in Iceland. *Jökull* **26**, 40-51.

—— 1974: Explanation of jökulhlaups from Grimsvotn, Vatnajökull, Iceland. *Jökull* **24**, 1-26.

—— 1977: The cause of jökulhlaups in the Skafta River, Vatnajökull. *Jökull* **27**, 71-8.

BLATT, H., MIDDLETON, G. and MURRAY, R., 1972: *Origin of sedimentary rocks.* Prentice-Hall, New Jersey (634pp).

BLUCK, B.J., 1974: Structure and directional properties of some valley sandur deposits in southern Iceland. *Sedimentology* **21**, 533-54.

BOND, F.C., 1952: Third theory of comminution. *Trans. Am. Inst. Mech. Engnr.* **193**, 484.

BOOTHROYD, J.C., 1976: Braided streams and alluvial fans. In Hayes, M.D. and Kana, T.W. (eds.), *Terrigenous clastic depositional environments,* AAPG field course notes, 17-26.

BOOTHROYD, J.C. and ASHLEY, G.M., 1975: Processes, bar morphology and sedimentary structures on braided outwash fans, northeastern Gulf of Alaska. In Jopling, A.V. and McDonald, B.C. (eds.), *Glaciofluvial and glaciolacustrine sedimentation,* Society of Economic Paleontologists and Mineralogists, Tulsa, Oklahoma. Spec. Publ. 23, 193-222.

BOOTHROYD, J.C. and NUMMEDAL, D., 1978: Proglacial braided outwash: a model for humid alluvial-fan deposits. In Miall, A.D. (ed.), *Fluvial sedimentology*, Can. Soc. Petrol. Geol. Mem. **5**, 641–68.

BORLAND, W.M., 1961: Sediment transport of glacier-fed streams in Alaska. *J. Geophys. Res.* **66** (10), 3347–50.

BOULTON, G.S., 1970: On the origin and transport of englacial debris in Svalbard glaciers. *J. Glaciol.* **9** (56), 213–29.

—— 1971: Till genesis and fabric in Svalbard, Spitsbergen. In Goldthwait, R.P. (ed.), *Till, a symposium*, Ohio State Univ. Press, Columbus, Ohio, 41–72.

—— 1972: The role of thermal regime in glacial sedimentation. In Price, R.J. and Sugden, D.E. (eds.), *Polar geomorphology*, Inst. Br. Geogr. Spec. Pub. 4, 1–19.

—— 1974: Processes and patterns of glacial erosion. In Coates, D.R. (ed.), *Glacial geomorphology*, State Univ. New York, New York, 41–87.

—— 1975: Processes and patterns of subglacial sedimentation: a theoretical approach. In Wright, A.E. and Moseley, F. (eds.), *Ice ages: ancient and modern*, Seel House Press, Liverpool, 7–42.

—— 1976: The development of geotechnical properties in glacial tills. In Legget, R.F. (ed.), *Glacial till*, Royal Society of Canada, Ottawa, 292–303.

—— 1978: Boulder shapes and grain-size distributions of debris as indicators of transport paths through a glacier and till genesis. *Sedimentology* **25**, 773–99.

—— 1979: Processes of glacier erosion on different substrata. *J. Glaciol.* **23** (89), 15–38.

—— 1982: Subglacial processes and the development of glacial bedforms. In Davidson-Arnott, R., Nickling, W. and Fahey, B.D. (eds.), *Research in glacial, glacio-fluvial and glacio-lacustrine systems*, Proc. 6th Guelph Symposium on Geomorphology, 1980, Geo Books, Norwich, 1–31.

BOULTON, G.S. and EYLES, N., 1979: Sedimentation by valley glaciers; a model and genetic classification. In Schlüchter, C. (ed.), *Moraines and varves*, Balkema, Rotterdam, 11–23.

BOULTON, G.S. and PAUL, M.A., 1976: The influence of genetic processes on some geotechnical properties of glacial tills. *Q. J. Engng. Geol.* **9**, 159–94.

BOULTON, G.S. and VIVIAN, R.A., 1973: Underneath the glaciers. *Geogr. Mag.* **45** (4), 311–16.

BOULTON, G.S., MORRIS, E.M., ARMSTRONG, A.A. and

THOMAS, A., 1979: Direct measurements of stress at the base of a glacier. *J. Glaciol.* **22** (86), 3–24.

BOUTRON, C., 1979: Trace element content of Greenland snows along an east–west transect. *Geochim. Cosmochim. Acta* **43** (8), 1253–8.

BOUTRON, C., ECHEVIN, M. and LORIUS, C., 1972: Chemistry of polar snows: estimation of rates of deposition in Antarctica. *Geochim. Cosmochim. Acta* **36**, 1029–41.

BOWDEN, F.P. and TABOR, D., 1964: *The friction and lubrication of solids, Parts 1 and 2*. Clarendon Press, Oxford (544pp).

—— 1968: *Friction and lubrication*. Methuen, London (150pp).

—— 1973: *Friction: an introduction to tribology*. Heinemann, London (178pp).

BOYÉ, M., 1950: *Glaciaire et periglaciaire de l'Ata Sund nord-oriental Groenland. Expeditions Polaires Françaises, Missions Paul-Emile Victor 1*. Hermann and Co., Paris (176pp).

BRUNSDEN, D., 1979: Weathering. In Embleton, C. and Thornes, J. (eds.), *Process in geomorphology*, Arnold, London, 72–129.

BUDD, W.F., 1969: *The Dynamics of Ice Masses*. ANARE Sci. Rpt. 108, Melbourne, Australia (216pp).

BUDD, W.F. and McINNES, B.J., 1974: Modelling periodically surging glaciers. *Science* **186**, 925–27.

BUDD, W.F. and RADOK, U., 1971: Glaciers and other large ice masses. *Progress in Physics* **34** (1), 1–70.

BUDD, W.F., JACKA, T.H. and MORGAN, V.I., 1980: Antarctic iceberg melt rates derived from size distributions and movement rates. *Ann. Glaciol.* **1**, 103–12.

BUDD, W.F., KEAGE, P.L. and BLUNDY, N.A., 1979: Empirical studies of ice sliding. *J. Glaciol.* **23** (89), 157–70.

BUDD, W.F., CORRY, M.J. and JACKA, T.H., 1982: Results of the Amery Ice Shelf Project. *Annals of Glaciol.* **3**, 36–41.

CAMBELL, N.J. and COLLIN, A.E., 1958: The discoloration of Foxe Basin ice. *J. Fish. Res. Bd. Canada* **15** (6), 1175–88.

CAMPBELL, W.J. 1973: Structure and inferred circulation of South Cascade Lake, Washington, USA. *Int. Ass. Sci. Hydrol. Publ.* **95**, 259–62.

CANT, D.J., 1978: Development of a facies model for sandy braided river sedimentation: comparison of the south Saskatchewan River and the Battery Point Formation. In Miall, A.D. (ed.), *Fluvial sedimentology*, Can. Soc. Petrol. Geol. Mem. 5, 627-39.

CAREY, S.W. and AHMAD, N., 1961: Glacial marine · sedimentation. In Raasch, G.O. (ed.), *Geology of the Arctic vol. 2*, University of Toronto Press, Toronto, 865-94.

CARTER, L., MITCHELL, J.S. and DAY, N.J., 1981: Suspended sediment beneath permanent and seasonal ice, Ross Ice Shelf Antarctica. *NZ J. Geol. Geophys.* 24, 249-62.

CARY, P.W., CLARKE, G.K.C. and PELTIER, W.R., 1979: A creep instability analysis of the Antarctic and Greenland ice sheets. *Can. J. Earth Sci.* 16, 182-8.

CHAMBERLIN, 1888: The rock scorings of the great ice invasions. *US Geol. Survey 7 Annual Rpt.* (1888), 155-248.

CHARI, T.R., 1979: Geotechnical aspects of iceberg scours on ocean floors. *Can. Geotech. J.* 16, 379-90.

CHARI, T.R., PETERS, G.R. and MUTHUKRISHNAIAH, K., 1980: Environmental factors affecting iceberg scour estimates. *Cold Reg. Sci. and Tech.* 1, 223-30.

CHERNOVA, L.P., 1981: Influence of mass balance and run-off on relief-forming activity of mountain glaciers. *Ann. Glaciol.* 2, 69-70:

CHURCH, M., 1972: *Baffin Island sandurs: a study of arctic fluvial processes*. Geol. Surv. Canada Bull. 216 (208pp).

CHURCH, M. and GILBERT, R., 1975: Proglacial fluvial and lacustrine environments. In Jopling, A.V. and McDonald, B.C. (eds.), *Glaciofluvial and glaciolacustrine sedimentation*, Tulsa, Oklahoma. Society of Economic Paleontologists and Mineralogists, Spec. Publ. 23, 22-100.

CLARK, D.L. and HANSON, A., 1983: Central Arctic Ocean sediment texture: A key to ice transport mechanisms. In Molnia, B.F. (ed.), *Glacial-marine sedimentation*, Plenum Press, New York, 301-30.

CLARK, D.L., WHITMAN, P.R., MORGAN, K.A. and MACKEY, S.D., 1980: Stratigraphy and glacial-marine sediments of the Amerasian Basin, Central Arctic Ocean. *Geol. Soc. Am. Spec. Paper* 181, 57pp.

CLARKE, G.K.C., 1982: Glacier outburst floods from Hazard Lake, Yukon Territory, and the problem of flood magnitude prediction. *J. Glaciol.* 28 (98), 3-21.

CLARKE, G.K.C. and MATHEWS, W.H., 1981: Estimates of the magnitude of glacier outburst floods from Lake Donjek, Yukon Territory, Canada. *Can. J. Earth Sci.* 18, 1452-63.

CLARKE, G.K.C., COLLINS, S.G., and THOMPSON, D.F.,

1984: Flow, thermal structure and subglacial conditions of a surge-type glacier. *Can. J. Earth Sci.* 21, 232-40.

CLOUGH, J.W. and HANSEN, B.L., 1979: The Ross Ice Shelf Project. *Science* 203 (4379), 433-4.

COLBECK, S.C., 1973: Effects of stratigraphic layers on water flow through snow. *CRREL Res. Rpt.* 311.

—— (ed.), 1980: *Dynamics of snow and ice masses*. Academic Press, New York (468pp).

COLBECK, S.C. and EVANS, R.J., 1973: a flow law for temperate glacier ice. *J. Glaciol.* 12 (64), 71-86.

COLLINS, D.N., 1977: Hydrology of an alpine glacier as indicated by the chemical composition of meltwater. *Zeit. Gletscherk. Glazialgeol.* 13 (1/2), 219-38.

—— 1979a: Sediment concentration in melt waters as an indicator of erosion processes beneath an Alpine glacier. *J. Glaciol.* 23 (89), 247-57.

—— 1979b: Hydrochemistry of meltwater draining from an alpine glacier. *Arctic and Alpine Res.* 11 (3), 307-24.

—— 1979c: Quantitative determination of the subglacial hydrology of two Alpine glaciers. *J. Glaciol.* 23 (89), 347-62.

—— 1981: Seasonal variation of solute concentration in meltwaters draining from an Alpine glacier. *Ann. Glaciol.* 2, 11-16.

COLLINSON, J.D. 1970: Bedforms of the Tana River, Norway. *Geogrf. Annlr.* 52A, 31-55.

—— 1971: Some effects of ice on a river bed. *J. Sed. Petrol.* 41 (2), 557-64.

—— 1978: Alluvial sediments. In Reading, H.G. (ed.), *Sedimentary environments and facies*, Blackwell, Oxford, 15-60.

COLLINSON, J.D. and THOMPSON, D.B., 1982: *Sedimentary structures*. Allen & Unwin, London (194pp).

CONNELL, D.C. and TOMBS, J.M.C., 1971: The crystallization pressure of ice – a simple experiment. *J. Glaciol.* 10 (59), 312-15.

COOKE, R.U. and SMALLEY, I.J., 1968: Salt weathering in deserts. *Nature* (London) 220, 1226-7.

COOPER, A.P.R., McINTYRE, N.F. and ROBIN, G. de Q., 1982: Driving stress in the Antarctic ice sheet. *Ann. Glaciol.* 3, 59-64.

CORBEL, J., 1962: *Neiges et glaciers*. Paris.

CORDON, W.A., 1966: Freezing and thawing of concrete – mechanisms and control. *Am. Concrete Inst. Mono.* 3, 23-41.

COX, G.F.N. and WEEKS, W.F., 1974: Salinity variations in sea ice. *J. Glaciol.* 13 (67), 109-20.

CRARY, A.P., 1960: Arctic ice island and ice shelf

studies. In Bushnell, V. (ed.), *Scientific studies at Fletchers' Ice Island, T-3, 1952-1955 vol. III*; Terrestrial Sci. Lab., Geophys. Res. Direct. Airforce Cambridge Res. Cent., Air Res. and Dev. Command. USAF Bedford mans. Geophys. Res. Paper 63, 1-37.

—— 1966: Mechanism for fjord formation indicated by studies of an ice-covered inlet. *Geol. Soc. Am. Bull.* **77**, 911-29.

CRARY, A.P., ROBINSON, E.S., BENNETT, H.F. and BOYD, W.W. Jr., 1962: Glaciological regime of the Ross Ice Shelf. *J. Geophys. Res.* **67** (7), 2791-807.

DAKE, J.M.K., 1972: *Essentials of Engineering Hydraulics.* Macmillan, London.

DAVIDISON, G.P. and NYE, J.F., 1985: A photoelastic study of ice pressure in rock cracks. *Cold Reg. Res. Tech.* **11**, 141-53.

DAYTON, P.K., ROBILLIARD, C.A. and DEVRIES, A.L., 1969: Anchor ice formation in McMurdo Sound, Antarctica, and its biological effects. *Science* **163** (3864), 273-4.

DELMAS, R., BRIAT, M. and LEGRAND, M., 1982:Chemistry of South Polar snow. *J. Geophys. Res.* **87** (C6), 4314-18.

DIONNE, J.C. (ed.), 1976: *First international symposium on the geological action of drift ice (Le glaciel).* La revue de géographie de Montréal, 30 (1-2) (236pp).

DOAKE, C.S.M., 1976: Thermodynamics of the interaction between ice shelves and the sea. *Polar Record* **18** (112), 37-41.

DOMACK, E.W. 1982: Sedimentology of glacial and marine glacial deposits on the George V - Adèlie Continental Shelf, East Antarctica. *Boreas* **11**, 79-97.

DORRER, E., 1971: Movement of the Ward Hunt Ice Shelf, Ellesmere Island, NWT, Canada. *J. Glaciol.* **10** (59), 211-25.

DORT, W., 1967: Internal structure of Sandy Glacier, Southern Victoria Land, Antarctica. *J. Glaciol.* **6** (46), 529-40.

DOWDESWELL, J.A., 1982: Supraglacial re-sedimentation from melt-water streams on to snow overlying glacier ice, Sylgjujökull, West Vatnajökull, Iceland. *J. Glaciol.* **28** (99), 365-75.

DOWDESWELL, J.A., HAMBREY, M.J. and WU, R., 1985: A comparison of clast fabric and shape in Late Precambrian and modern glaci-genic sediments. *J. Sed. Petrol.* **55** (5), 691-704.

DRAKE, J.J. and McCANN, S.B., 1982: The movement of

isolated boulders on tidal flats by ice floes. *Can. J. Earth Sci.* **19**, 748-54.

DRAKE, L.D., 1972: Mechanisms of clast attribution in basal till. *Geol. Soc. Am. Bull.* **83**, 2159-65.

—— 1974: Till fabric control by clast shape. *Geol. Soc. Am. Bull.* **85** (2), 247-50.

DREIMANIS, A., 1976: Tills: their origins and properties. In Legget, R.T. (ed.), *Glacial till*, Royal Society of Canada, Ottawa, 11-49.

DREIMANIS, A. and VAGNERS, U.J., 1969: Lithologic relationship of till to bedrock. In Wright, H.E.Jr (ed.), *Quaternary geology and climate,* Nat. Acad. Sci., Publ. **1701**, 93-8.

—— 1971: Bimodal distribution of rock and mineral fragments in basal tills. In Goldthwait, R.P. (ed.), *Till - a symposium*, Ohio State Univ. Press, Columbus, Ohio, 237-50.

—— 1972: The effect of lithology upon texture of till. In Yatsu, E. and Falconer, A. (eds.), *Research methods in Pleistocene geomorphology*, Proc. 2nd Guelph Symp. Geomorph, 1971, Geo Abstracts, Norwich, 66-82.

DREWRY, D.J., 1972: A quantitative assessment of dirt-cone dynamics. *J. Glaciol.* **11** (63), 431-46.

—— 1981: Radio echo sounding of ice masses: principles and applications. In Cracknell, A.P. (ed.), *Remote sensing in meteorology, oceanography and hydrology*, Ellis Horwood, Chichester, 270-84.

—— 1983: *Antarctica: glaciological and geophysical folio.* Scott Polar Res. Inst., Cambridge.

DREWRY, D.J. and COOPER, A.P.R., 1981: Processes and models of Antarctic glaciomarine sedimentation. *Ann. Glaciol.* **2**, 117-22.

DREWRY, D.J. and LIESTØL, O., 1985: Glaciological investigations of surging ice caps in Nordaustlandet, Svalbard, 1983. *Polar Record 22*, 359-78.

DUNBAR, M.J., 1973: Glaciers and nutrients in arctic fjords – comment on Apollonio, 1973. *Science* **182** (4110), 398.

DUVAL, P., 1979: Creep and recrystallization of polycrystalline ice. *Bull. de Mineralogie* **102** (2-3), 80-5.

DYER, K.R., 1973: *Estuaries: a physical introduction.* John Wiley & Sons, London (140pp).

DYSON, J.L., (1962): *The world of ice.* Knopf, New York (292pp).

EDWARDS, M.B., 1978: Glacial environments. In Reading, H.G. (ed.), *Sedimentary environments and facies*, Blackwell, Oxford, 416-38.

ELLISTON, G.R., 1973: Water movement through the Gornergletscher. *Int. Ass. Sci. Hydrol. Publ.* **95**, 79–84.

ELVERHOI, A. and ROALDSET, E., 1983: Glaciomarine sediments and suspended particulate matter, Weddell Sea Shelf, Antarctica. *Polar Res.* **1**(1), 1–21.

ELVERHOI, A., LIESTØL, O. and NAGY, J., 1980: Glacial erosion, sedimentation and microfauna in the inner part of Kongsfjorden, Spitsbergen. *Norsk Polarinstitutt Skrifter* **172**, 33–61.

ELVERHOI, A., LONNE, O. and SELAND, R., 1983: Glaciomarine sedimentation in a modern fjord environment. *Polar Res.* **1** (2), 127–49.

EMBLETON, C. and KING, C.A.M., 1975: *Glacial geomorphology*. Arnold, London (583pp).

ENGELHARDT, H.F., HARRISON, W.D. and KAMB, W.B., 1978: Basal sliding and conditions at the glacier bed as revealed by bore-hole photography. *J. Glaciol.* **20** (84), 469–508.

EVANS, I.S., 1970: Salt crystallization and rock weathering: a review. *Revue geomorph. dyn.* **19**, 153–77.

EVENSON, E.B., DREIMANIS, A. and NEWSOME, J.W., 1977: Subaquatic flow tills: a new interpretation for the genesis of some laminated till deposits. *Boreas* **6**, 115–33.

EYLES, N., EYLES, C.H. and MIALL, A.D., 1983: Lithofacies types and vertical profile models; an alternative approach to the description and environmental interpretation of glacial diamict and diamictite sequences. *Sedimentology* **30**, 393–410.

EYLES N., SASSEVILLE, D.R., SLATT, R.M. and ROGERSON, R.J., 1982: Geochemical denudation rates and solute transport mechanisms in a maritime temperate glacier basin. *Can. J. Earth Sci.* **19**, 1570–81.

FADDICK, R.R., 1975: Pipeline wear from abrasive studies. *First Int. Conf. Internal and External Protection of Pipes G3*, Durham, 29–37.

FAHEY, B.D., 1983: Frost action and hydration as rock weathering mechanisms on schist: a laboratory study. *Earth Surf. Proc. and Landfms.* **8** (6), 535–45.

—— 1985: Salt weathering as a mechanism of rock break up in cold climates: an experimental approach. *Zeit. Geomorph.* **29** (1), 99–111.

FAHNSTOCK, R.K., 1963: Morphology and hydrology of a glacial stream – White River, Mount Rainier, Washington. *US Geol. Soc. Prof. Paper* **422–A**.

FAIRHURST, C., 1964: Discussion of Eurby, J., Strain wave behaviour in the percussive drilling processes. *Trans. Inst. Min. Metall.* **73**, 671–82.

FICKER, E. and WEBER, E., 1981: Auf den Spuren der Gletscher – oder – Untersuchungen zum Problem der Rissentstehung auf Felsoberflächen. *VDI-Berichte* **399**, 89–96.

FICKER, E., SONNTAG, G. and WEBER, E., 1980: Ansätze zur mechanischen Deutung der Rissentstehung bei Parabelrissen und Sichelbrüchen auf glazial-geformten Felsoberflächen. *Zeit. Gletscherk. Glazialgeol.* **16** (1), 25–43.

FISHER, W.L., BROWN, L.F.Jr, SCOTT, A.J. and McGOWEN, J.H., 1969: *Delta systems in the exploration for oil and gas.* Bur. Econ. Geol., Austin, Texas (78pp).

FLINT, R.F., 1971: *Glacial and Quaternary geology.* Wiley, New York (892pp).

FOLDVIK, A. and KVINGE, T., 1974: Conditional instability of sea water at the freezing point. *Deep Sea Res.* **21** (3), 169–74.

—— 1977: Thermohaline convection in the vicinity of an ice shelf. In Dunbar, M.J. (ed.), *Polar oceans*, Montreal, 247–55.

FORBES, D.L., 1983: Morphology and sedimentology of a sinuous gravel-bed channel system: lower Babbage River, Yukon coastal plain, Canada. *Spec. Pub. Int. Ass. Sediment* **6**, 195–206.

FORD, D.C., FULLER, P.G. and DRAKE, J.J., 1970: Calcite precipitates at the soles of temperate glaciers. *Nature* (London) **226** (5244), 441–2.

FRANCIS, J.R.D., 1969, *Fluid mechanics.* Arnold, London.

FRANK, D., POST, A. and FRIEDMAN, J.D., 1975: Recurrent geothermally induced debris avalanches on Boulder Glacier, Mount Baker, Washington. *US Geol. Survey J. Res.* **3** (1), 77–87.

FRASER, J.Z., 1982: Derivation of a summary facies sequence based on Markov chain analysis of the Caledon outwash: a Pleistocene braided glacial fluvial deposit. In Davidson-Arnott, R., Nickling, W. and Fahey, B.D., *Research in glacial, glacio-fluvial and glacio-lacustrine systems*, Proc. 6th Guelph Symposium on Geomorphology, 1980, Geo Books, Norwich, 175–99.

FREEZE, R.A. and CHERRY, J.A., 1979: *Groundwater.* Prentice-Hall, New Jersey.

FRIEDMAN, G.M. and SANDERS, J.E., 1978: *Principles of sedimentology.* Wiley, New York (792pp).

GADE, H.G., 1979: Melting of ice in sea water: a primitive model with application to the Antarctic ice shelf and icebergs. *J. Phys. Oceanog.* **9** (1), 189-98.

GIBBARD, P., 1980: The origin of stratified Catfish Creek till by basal melting. *Boreas* **9** (1), 71-85.

GILBERT, G.K., 1906: Crescentic gouges on glaciated surfaces. *Geol. Soc. Am. Bull.* **17**, 303-16.

GILBERT, R. and SHAW, J., 1981: Sedimentation in proglacial Sunwapta Lake, Alberta. *Can. J. Earth Sci.* **18**, 81-93.

GILBERT, R., 1972a: Draining of ice-dammed Summit Lake, British Columbia. *Scientific Series No. 20, Water Resources Branch*, Inland Waters Directorate, Environment Canada, Ottawa (17pp).

—— 1972b: Observations on sedimentation at Lillooet Delta, British Columbia. In Slaymaker, H.O. and McPherson, H.J. (eds.), *Mountain geomorphology*, Tantalus Press, Vancouver, 187-94.

—— 1973: Processes of underflow and sediment transport in a British Columbia mountain lake. Natl. Res. Council of Canada, Assoc. Comm. Geodesy and Geophys. Subcomm. Hydrol., *Proc. Hydrol. Symposium 9*, 493-507.

—— 1975: Sedimentation in Lillooet Lake, British Columbia. *Can. J. Earth Sci.* **12** (10) 1697-711.

—— 1982: Contemporary sedimentary environments on Baffin Island, NWT, Canada: glaciomarine processes in fjords of Eastern Cumberland Peninsula. *Arctic and Alpine Res.* **14** (1), 1-12.

GILL, A.F., 1973: Circulation and bottom water information in the Weddell Sea. *Deep Sea Res.* **20** 111-40.

GLAZYRIN, G.E., 1975: The formation of ablation moraines as a function of the climatological environment. *Int. Ass. Sci. Hydrol. Publ.* **104**, 106-10.

GLEN, J.W., 1955: The creep of polycrystalline ice. *Proc. Roy. Soc. (London) Ser. A* **228**, 519-38.

GOLD, L.W., 1973: Ice – a challenge to the engineer. *Proc. Fourth Can. Congress of Applied Mech.,* 19-36.

GOLDSTEIN, R.H., 1983: Stratigraphy and sedimentology of ice-rafted and turbidite sediment, Canada Basin, Arctic Ocean. In Molina, B.F. (ed.), *Glacial-marine sedimentation*, Plenum Press, New York, 367-400.

GOLDTHWAIT, R.P., 1960: Study of ice cliff in Nanatarssuaq, Greenland. *US Snow, Ice & Permafrost Res. Estab. Tech. Rep.* **39** (108pp).

—— Introduction to till, today. In Goldthwait, R.P. (ed.), *Till, a symposium*, Ohio State Univ. Press, Columbus, Ohio, 3-26.

GOODMAN, D.J., FROST, H.J. and ASHBY, M.F., 1977: The effect of impurities on the creep of ice I_h and its illustration by the construction of deformation maps. *Int. Ass. Sci. Hydrol. Publ.* **118**, 17-22.

GOODMAN, R.E., 1980: *Introduction to rock mechanics*. Wiley & Sons, New York.

GOUDIE, A. 1974: Further experimental investigation of rock weathering by salt and other mechanical processes. *Zeit. Geomorph. Suppl.* **21**, 1-12.

GOW, A.J. and WILLIAMSON, T., 1976: Rheological implications of the internal structure and crystal fabrics of the West Antarctic ice sheet as revealed by deep core drilling at Byrd Station. *Geol. Soc. Am. Bull.* **87**, 1665-77.

GRAF, W.H., 1981: *Hydraulics of sediment transport*. McGraw-Hill, New York (513pp).

GRESZCZUK, L.B., 1982: Damage in composite materials due to low velocity impact. In Zukas, J.A., Nicholas, T., Swift, H.F., Greszczuk, L.B. and Curran, D.R., *Impact dynamics*, Wiley, New York, 55-94.

GRIFFITH, A.A., 1924: The theory of rupture. *First Int. Congr. Appl. Mech. Proc.,* Delft, 55-63.

GRIGGS, D.T. and HANDIN, J., (1960): Observations on fracture and an hypothesis of earthquakes. In *Rock deformation,* Geol. Soc. America Mem. 79, 347-64.

GRIPP, K., 1929: Glaciologische und geologische Ergebnisse der Hamburgischen Spitzbergen Expedition 1927. Naturwiss. Verein in Hamburg. *Abh. Geb. Naturw.* **22** (2-4), 146-249.

GURNELL, A.M., 1982: The dynamics of suspended sediment concentration in an Alpine pro-glacial stream network. *Int. Assoc. Sci. Hydrol. Publ.* **138**, 319-30.

GUSTAVSON, T.C., 1975: Sedimentation and physical limnology in proglacial Malaspina Lake, southeastern Alaska. In Jopling, A.V. and McDonald, B.C. (eds.), *Glaciofluvial and glaciolacustrine sedimentation*, Tulsa, Okla., Society of Economic Paleontologists and Mineralogists, Spec. Publ. 23, 249-63.

GUSTAVSON, T.C., ASHLEY, G.M. and BOOTHROYD, J.C., 1975: Depositional sequences in glaciolacustrine deltas. In Jopling, A.V. and McDonald, B.C. (eds.), *Glaciofluvial and glaciolacustrine sedimentation*, Tulsa, Okla., Society of Economic Paleontologists and Mineralogists, Spec. Publ. 23, 264-80.

HAAKENSEN, N. and WOLD, B. (eds.), 1981: *Glasiologiske undersokelser i Norge 1979*. Norges Vassdrags- og elektrisitetsvesen. Vassdrågsdirektoratet. Hydrologisk Avdeling 3 (80pp).

HAAKENSEN, N., OLSEN, H.C. and ØSTREM, G., 1975: Materialtransport i norska glaciaralvar 1973 Stokholms Universitet. *Naturgeografiska Inst. Forskringsrapport* 20 (107pp).

HAAKENSEN, N., LIESTØL, O., MESSEL, S., TREDE, A. and ØSTREM, G., 1982: *Glaciologiske Undersokelser i Norge 1980*. Hydrologisk Avdeling, Norges Vassdrogs- og Elektrisitetsvesen, Rapport NV.1–82.

HAGEN, J.O., WOLD, B., LIESTØL, O., ØSTREM, G. and SOLLID, J.L., 1983: Subglacial processes at Bondhusbreen Glacier, Norway: Preliminary results. *Ann. Glaciol.* 4, 91–8.

HALDORSEN, S., 1981: Grain-size distribution of subglacial till and its relation to glacial crushing and abrasion. *Boreas* 10, 91–105.

HALLET, B., 1975: Subglacial silica deposits. *Nature* (London) 254 (5502), 682–3.

—— 1976: Deposits formed by subglacial precipitation of $CaCO_3$. *Geol. Soc. Am. Bull.* 87 (7), 1003–15.

—— 1979a: Subglacial regelation water film. *J. Glaciol.* 23 (89), 321–34.

—— 1979b: A theoretical model of glacial abrasion. *J. Glaciol.* 23 (89), 39–50.

—— 1981: Glacial abrasion and sliding: their dependence on the debris concentration in basal ice. *Ann. Glaciol.* 2, 23–8.

HALLET, B. and ANDERSON, R.S., 1982: Detailed glacial geomorphology of a proglacial bedrock area at Castleguard Glacier. *Zeit. Gletscherk. Glazialgeol.* 6 (2), 171–84.

HALLET, B., LORRAIN, R.D. and SOUCHEZ, R.A., 1978: The composition of basal ice from a glacier sliding over limestones. *Geol. Soc. Am. Bull.* 89, 314–20.

HALLING, J., 1976: *Introduction to tribology*. Wykeham Pub. Ltd., London (157pp).

HALLING, J. and NURI, K.A., 1975: Contact of surfaces. In Halling, J. (ed.), *Principles of tribology*, Macmillan, London, 40–71.

HAMBLIN, P.F. and CARMACK, E.C., 1978: River-induced currents in a fjord lake. *J. Geophys. Res.* 83 (C2), 885–99.

HAMBREY, M.J. and HARLAND, W.B., (eds.), 1981: *Earth's pre-Pleistocene glacial record*. CUP, Cambridge (1004pp).

HAMILTON, W.L., 1970: Atmospheric dust records in permanent snowfields: implications to marine sedimentation: discussion. *Geol. Soc. Am. Bull.* 81, 3175–6.

HARLAND, W.B., HEROD, K.N. and KRINSLEY, D.H., 1966: The definition and identification of tills and tillites. *Earth Sci. Rev.* 2, 225–56.

HARMS, J.C., SOUTHARD, J.B., SPEARING, D.R. and WALKER, R.G., 1975: *Depositional environments as interpreted from primary sedimentary structures and the stratification sequences*. Lecture notes, course 2, Soc. Econ. Paleontologists and Mineralogists, Dallas, 1975.

HARRIS, I.M. and JOLLYMORE, P.G., 1974: Iceberg furrow marks on the Continental Shelf northeast of Belle Isle, Newfoundland. *Can. J. Earth Sci.* 11 (1), 43–52.

HARRISON, W.D. and RAYMOND, C.F., 1976: Impurities and their distribution in temperate glacier ice. *J. Glaciol.* 16 (74), 173–81.

HARTSHORN, J.H., 1958: Flowtill in southeastern Massachusetts. *Geol. Soc. Am. Bull.* 69, 477–82.

HARTSHORN, J.H. and ASHLEY, G.M., 1972: *Glacial environment and processes in Southeastern Alaska*. Tech. Rpt. 4-CRC University of Massachusetts, Amherst, Coastal Research Center, (69pp).

HEM, 1982: Conductance as a collective measure of dissolved ions. In Minear R.A. and Keith C.H. (eds.), *Water analysis vol. 1*, Academic Press, New York, 137–61.

HEYWOOD, R.B., 1972: Antarctic limnology: a review. *Br. Ant. Surv. Bull.* 29, 35–65.

HICKS, G.R.F., 1974: Variation in zooplankton biomass with hydrological regime beneath the seasonal ice, McMurdo Sound, Antarctica. *NZ J. Marine and Freshwater Res.* 8 67–78.

HJULSTRÖM, F., 1935: Studies of the morphological activities of rivers as illustrated by the River Fyris. *Bull. Geol. Inst. Univ. Uppsala* 25, 221–527.

—— 1957: Glacialhydrologiska Studies och delta-undersokningar pa Island. *Medd. Uppsala Univ. Geograf Inst. Ser. A* 122 (24pp).

HNATIUK, J. and WRIGHT, B.D., 1983: Sea bottom scouring in the Canadian Beaufort Sea. In *Fifteenth Annual Offshore Technology Conference*. 1983 Proceedings, Houston, Texas, 3, 35–40.

HOBBS, B.E., MEANS, W.D. and WILLIAMS, P.F., 1976: *An outline of structural geology*. Wiley & Sons, New York (571pp).

HOBBS, P.V., 1974: *Ice physics*, Clarendon Press, Oxford (837pp).

HODGE, S.M., 1974: Variations in the sliding of a

temperate glacier. *J. Glaciol.* **13** (69), 349–69.

—— 1976: Direct measurement of basal water pressures: a pilot study. *J. Glaciol.* **16** (74), 205–18.

HOLDSWORTH, G., 1974a: Meserve Glacier, Wright Valley, Antarctica: Part 1, Basal processes. *Inst. Polar Studies Rep.* **37** (104pp).

—— 1974b: Erebus Glacier tongue, McMurdo Sound, Antarctic, *J. Glaciol.* **13** (67), 27–35.

—— 1978: Some mechanisms for the calving of icebergs. In Husseiny, A.A. (ed.), *Iceberg utilization*, Pergamon Press, Oxford, 160–75.

—— 1982: Dynamics of Erebus Glacier tongue. *Ann. Glaciol.* **3**, 131–7.

HOLDSWORTH, G. and GLYNN, J.E., 1981: A mechanism for the formation of large icebergs. *J. Geophys. Res.* **86** (C4), 3210–22.

HOLMES, C.D., 1941: Till fabric. *Geol. Soc. Am. Bull.* **52**, 1301–52.

HOOKE, R.B., 1973: Flow near the margin of the Barnes Ice Cap, and the development of ice-cored moraine. *Geol. Soc. Am. Bull.* **84** (12), 3929–48.

—— 1977: Basal temperatures in polar ice sheets: a qualitative review. *Quat. Res.* **7**, 1–13.

HOOKE, R.Le B., DAHLIN, B.B. and KAUPER, M.T., 1972: Creep of ice containing dispersed fine sand. *J. Glaciol.* **11** (63), 327–36.

HOSKIN, C.M. and BURRELL, D.C., 1972: Sediment transport and accumulation in a fjord basin, Glacier Bay, Alaska, *J. Geol.* **80**, 539–51.

HOUBOLDT, J.J.H.C. and JONKER, J.B.M., 1968: Recent sediments in the eastern part of the Lake of Geneva (Lac Leman). *Geol. en Mijn.* **47**, 131–48.

HOUGHTON, D.L., BORGE, D.E. and PAXTON, J.A., 1978: Cavitation resistance of some special concretes. *J. Am. Concrete Inst.* **75** (68), 664–7.

HUDDLESTONE, P.J., 1973: An analysis of 'single layer' folds developed experimentally in viscous media. *Tectonophysics* **16**, 198–214.

HUGHES, T.J., 1981: Numerical reconstruction of paleo-ice sheets. In Denton, G.H. and Hughes, T.J. (eds.), *The last great ice sheets*, Wiley & Sons, New York, 221–61.

HUKKI, R.T., 1961: Proposal for a solomonic settlement between the theories of Von Rittinger, Kick and Bond. *Trans. AIME/SME* **220**, 403–8.

HUMLUM, O., 1981: Observations on debris in the basal transport zone of Myrdalsjökull, Iceland. *Ann. Glaciol.* **2**, 71–7.

HUPPERT, H.E., 1980: The physical processes involved in the melting of icebergs. *Ann. Glaciol.* **1** 97–101.

HUPPERT, H.E. and JOSBERGER, E.G., 1980: The melting of ice in cold stratified water. *J. Phys. Oceanogr.* **10** (6), 953–60.

HUTCHINSON, G.E., 1957: *A treatise on limnology. Geography, physics and chemistry 1.* Wiley, New York.

HUTTER, K., 1982a: Dynamics of glaciers and large ice masses. *Annual Rev. Fluid Mech.* **14**, 87–130.

—— 1982b: Glacier flow. *Am. Scient.* **70** (1), 26–34.

—— 1983: *Theoretical glaciology.* Reidel. Rotterdam (510pp).

IKEN, A., 1981: The effect of the subglacial water pressure on the sliding velocity of a glacier in an idealized numerical model. *J. Glaciol.* **27** (97), 407–21.

Journal of Strain Analysis, Monograph, 1978: *A general introduction to fracture mechanics*, Mech. Engrg. Pubs. Ltd., London (178pp).

JACOBS, S.S., GORDON. A.L. and ARDAI, J.L., 1979: Circulation and melting beneath the Ross Ice Shelf. *Science* **203** (4379), 439–43.

JACOBS, S.S., HUPPERT, H.E., HOLDSWORTH, G. and DREWRY, D.J., 1981: Thermohaline steps induced by melting of the Erebus Glacier tongue. *J. Geophys. Res.* **86** (C7), 6547–55.

JAEGER, J.C. and COOK, N.G.W., 1979: *Fundamentals of rock mechanics*, 3rd edn. Chapman-Hall, London (593pp).

JELLINEK, H.H.G., 1959: Adhesive properties of ice. *J. Colloidal Sci.* **14**, 268–80.

JEWTUCHOWICZ, S., 1965: Description of eskers and kames in Gåshamnöyra and on Bungebreen, south of Hornsund, Vestspitsbergen. *J. Glaciol.* **5** (41), 719–25.

—— 1969: Kame structure west of Zieleniew. In Goldthwait, R.P. (ed.), 1975, *Glacial deposits*, Benchmark Papers in Geology 21, Halsted Press, New York, 325–33. (Translated from *Folia Quaternai* **30**, 59–68.)

JOHNSON, A.M., 1970: *Physical processes in geology.* Freeman, Cooper & Co., San Francisco (577pp).

JOHNSON, P.G., 1972: The morphological effects of surges of the Donjek glacier, St Elias Mountains, Yukon Territory, Canada. *J. Glaciol.* **11** (62), 227–34.

JONES, B.F. and BOWSER, C.J., 1978: The mineralogy and related chemistry of lake sediments. In Lerman, A. (ed.), *Lakes: chemistry, geology, physics*, Springer-Verlag, New York, 179–235.

JONES, S.J. and GLEN, J.W., 1969: The effects of dissolved impurities on the mechanical properties of ice crystals. *Philosophical Mag.* **19** (157), 13–24.

JOPLING, A.V. and WALKER, R.G., 1968: Morphology and origin of ripple-drift cross lamination, with examples from the Pleistocene of Massachusetts. *J. Sed. Petrol.* **38**, 971–84.

JOSBERGER, E.G. and MARTIN, S., 1981: A laboratory and theoretical study of the boundary layer adjacent to a vertical melting ice wall in salt water. *J. Fluid Mech.* **111**, 439–73.

JOSENHANS, H.W., BARRIE, J.W., WOODWORTH-LYNAS, C.M.T. and PARROTT, D.R., 1985: DIGS-85 Dynamics of iceberg grounding and scouring: observations of iceberg scour marks by manned submersible (abstract) in *14th Arctic Workshop Arctic Land-Sea Interaction*. Bedford Institute of Oceanography.

KAMB, W.B., 1959: Ice petrofabric observations from Blue Glacier, Washington, in relation to theory and experiment. *J. Geophys. Res.* **64** (11), 1891–909.

KAMB, B., 1970: Sliding motion of glaciers: theory and observation. *Rev. Geophys. Space Phys.* **8** (4), 673–728.

KAMB, B. and La CHAPELLE, E., 1964: Direct observations of the mechanism of glacier sliding over bedrock. *J. Glaciol.* **5** (38), 159–72.

KAMB, B., POLLARD, D. and JOHNSON, C.B., 1976: Rock-frictional resistance to glacier sliding. *EOS Trans. Am. Geophys. Union* **57**, 325.

KAMB, B., RAYMOND, C.F., HARRISON, W.D., ENGLEHARDT, H., ECHELMAYER, K.A., HUMPHREY, N., BRUGMAN, M.M. and PFEFFER, T., 1985: Glacier surge mechanism 1981–1983 surge of Variegated Glacier, Alaska. *Science* **227**, 469–479.

KARABELAS, A.J., 1978: An experimental study of pipe erosion by turbulent slurry flow. *Hydrotransport* **5**, E2–E15.

KAWASHIMA, T., YAGI, T., ISE, T., SATO, E., WASHIMI, H. and YOKOGAWA, A., 1978: Wear of pipes for hydraulic transport of solids. *Hydrotransport* **5** (E3), 25–44.

KAYE, G.W.C. and LABY, T.H., 1975: *Tables of physical and chemical constants* (14th edn). Longmans, London (386pp).

KELLER, W.D. and REESMAN, A.L., 1963: Glacial milks and their laboratory-simulated counterparts. *Geol. Soc. Am. Bull.* **74**, 61–76.

KELLY, E.G. and SPOTTISWOOD, D.J., 1982: *Introduction to mineral processing*. John Wiley & Sons, Inc., New York.

KELLY, W.C. and ZUMBERGE, J.H., 1961: Weathering of a quartz diorite at Marble Point, McMurdo Sound, Antarctica. *J. Geol.* **69** (4), 433–46.

KENN, M.J., 1966: Cavitating eddies and their incipient damage to concrete. *Civ. Engng. Publ. Wks. Rev., 1404–5.*

KICK, F., 1885: *Das Gesetz der proportionalen Widerstände und seine Anwendung.* Leipzig.

KIZAKI, K., 1969: Ice-fabric study of the Mawson region, East Antarctica. *J. Glaciol.* **8** (53), 253–76.

KJELDSEN, O. (ed.), 1981: *Material transport undersokelser i Norske Breelver 1980.* Norges Vassdrags- og Elektrisitetsvesen; Vassdragondirektoretet Hydrologisk Avedeling Rapport 1981 N14, 41pp.

KJELDSEN, O. and ØSTREM, G., 1980 in Aarseth, I. *et al.*: Glaciation and deglaciation in central Norway, *Fieldguide to excursion 31 August-3 September 1980 organized in conjunction with 'Symposium on Processes of Glacier Erosion and Sedimentation'.* Geilo, Norway. Norsk Polarinstitutt, 27–33.

KLEPSVIK, J.O. and FOSSUM, B.A., 1980: Studies of icebergs, ice fronts and ice walls using side-scanning sonar. *Ann. Glaciol.* **1**, 31–6.

KNAPP, R.T., DAILY, J.W. and HAMMITT, F.G., 1970: *Cavitation.* McGraw Hill, New York (578pp).

KOVACS, A., 1972: Ice scoring marks floor of the Arctic Shelf. *Oil and Gas Jnl.* **70**, 92–106.

KOVACS, A. and MELLOR, M., 1974: Sea ice morphology and ice as a geologic agent in the southern Beaufort Sea. In Reed, J.C. and Sater, J.E., *The coast and shelf of the Beaufort Sea,* Arctic Inst. of N. America, Arlington (Va), 113–64.

KRIMMEL, R.M., TANGBORN, W.V. and MEIER, M.F., 1972: Water flow through a temperate glacier. *Int. Ass. Sci. Hydrol. Publ.* **107**, 406–16.

KRISTENSEN, M., SQUIRE, V.A. and MOORE, S.C., 1982: Tabular icebergs in ocean waves. *Nature* (London) **297** (5868), 669–71.

KRISTENSEN, M., 1983: Iceberg calving and deterioration in Antarctica. *Prog. in Phys. Geogr.* **7** (3), 313–28.

KRÜGER, J., 1976: Structures and textures in till indicating subglacial deposition. *Boreas* **8**, 323–40.

KUMAI, M., 1977: Electron microscope analysis of aerosols in snow and deep ice cores from Greenland. *Int. Ass. Sci. Hydrol. Publ.* **118**, 341–50.

KURTZ, D.D. and ANDERSON, J.B., 1979: Recognition and sedimentologic description of recent debris flow deposits from the Ross Sea and Weddell Sea, Antarctica. *J. Sed. Petrol.* **49**, 1159–70.

KYLE, P.R. and JEZEK, P.A., 1978: Compositions of three tephra layers from the Byrd Station ice core, Antarctica. *J. Volcan. Geothermal Res.* **4**, 225–32.

KYLE, P.R., JEZEK, P.A., MOSLEY-THOMPSON, E. and THOMPSON, L.G., 1981: Tephra layers in the Byrd Station ice core and the Dome C ice core, Antarctica and their climatic importance. *J. Volcan. Geotherm. Res.* **11**, 29–39.

LAKE, R.A. and WALKER, E.R., 1976: A Canadian arctic fjord with some comparisons to fjords of the western Americas. *J. Fish. Res. Bd. Can.* **33** (10), 2272–85.

LAMBE, T.W. and WHITMAN, R.V., 1969: *Soil mechanics.* Wiley, New York (553pp).

LANE, E.W., CARLSON, E.J. and HANSON, O.S., 1949: How temperature increases sediment transportation in Colorado River. In *Civil Engineering, American Society Civil Engineers* **79**, 303–13.

LANG, H. SCHADLER, B. and DAVIDSON, G., 1977: Hydroglaciological investigations on the Ewigschneefeld – Gr. Aletschgletscher. *Zeit. Gletscherk. Glazialgeol.* **XII** (2), 109–24.

LANG, H., LEIBUNDGUT, C. and FESTEL, E., 1979: Results from tracer experiments on the water flow through the Aletschgletscher. *Zeit. Gletscherk. Glazialgeol.* **15** (2), 209–18.

LARSON, G.J., 1978: Meltwater storage in a temperate glacier, Burroughs Glacier, Southeast Alaska. *Inst. Polar Stud. Rep.* **66** (56pp).

LAUTRIDOU, J.F. and OZOUF, J.C., 1982: Experimental frost shattering: 15 years of research at the Centre de Géomorphologie du CNRS. *Prog. in Phys. Geogr.*, 215–32.

LAWSON, D.E., 1979: Sedimentological analysis of the western terminus region of the Matanuska Glacier, Alaska. *CRREL Rep.* **79–9** (122pp).

—— 1981a: Distinguishing characteristics of diamictons at the margin of Matanuska Glacier, Alaska. *Ann. Glaciol.* **2**, 78–84.

—— 1981b: Sedimentological characteristics and classification of depositional processes and deposits in the glacial environment. *CRREL Rep.* **81–27** (22pp).

—— 1982: Mobilization, movement and deposition of active subaerial sediment flows, Matanuska Glacier, Alaska. *J. Geol.* **90**, 279–300.

LEEDER, M.R., 1982: *Sedimentology: process and product.* Allen & Unwin, London (344pp).

LEMMENS, M. and ROGER, M., 1978: Influence of ion exchange on dissolved load of alpine waters. *Earth Surf. Processes* **3**, 179–87.

LENNON, P.W., LOYNES, J., PAREN, J.G. and POTTER, J.R., 1982: Oceanographic observations from George VI Ice Shelf, Antarctic Peninsula. *Ann. Glaciol.* **3**, 178–83.

LERMAN, A., 1979: *Geochemical processes, water and sediment environments.* Wiley, New York (418pp).

LEWIS, C.F.M., PARROT, D.R., LEVER, J.H., DIEMAND, D., DYKE, M., CARTER, W.J. and STIRBYS, A.F., 1985: DIGS-85 Dynamics of iceberg grounding and scouring: iceberg and seabed mapping operations (abstract) in *14th Arctic Workshop: Iceberg Land-Sea Interaction.* Bedford Institute of Oceanography, 93.

LIEN, R., 1983: Iceberg scouring on the Norwegian Continental Shelf. In *Fifteenth Annual Offshore Technological Conference. 1983 Proceedings,* Houston, Texas, 3, 41–8.

LIESTØL, O., 1967: Storbreen glacier in Jotunheimen, Norway. *Norsk Polarinst. Skrifter* **141** (63pp).

LIESTØL, O., 1969: Glacier surges in West Spitsbergen. *Can. J. Earth Sci.* **6**, 895–7.

LIESTØL, O., 1956: Glacier-dammed lakes in Norway. *Norsk Geogr. Tidsskr.* **15**, 122–49.

LINDSAY, J.F., 1970: Clast fabric of till and its development. *J. Sed. Petrol.* **40** (2), 629–41.

LISITZIN, A.P. (ed.), 1972: *Sedimentation in the world oceans.* Soc. Econ. Palaeontologists Mineralogists Spec. Publ. 17, 1–218.

LIU, T.C., 1981: Abrasion resistance of concrete. *J. Am. Concrete. Inst. Tech. Pap.* **78–29**, 341–50.

LIVINGSTONE, D.A., 1963: Chemical compositions of rivers and lakes. *US Geol. Survey Prof. Pap.* **440–G**.

LJUNGGREN, D. and SUNDBORG, A., 1968: Some aspects of fluvial sediments and fluvial morphology II: a study of some heavy mineral deposits in the valley of the Lule Alv. *Geogrf. Annlr.* **50A**, 121–35.

LLIBOUTRY, L., 1962: L'erosion glaciaire. *Int. Ass. Sci. Hydrol. Publ.* **59**, 219–25.

—— 1964 and 1965: *Traité de glaciologie,* Masson, Paris (2 vols., 428pp and 162pp).

—— 1968: General theory of subglacial cavitation and sliding of temperate glaciers. *J. Glaciol.* **7** (49), 21–58.

—— 1975: Loi de glissement d'un glacier sans cavitation. *Ann. Geophys.* **31** (2), 207–26.

—— 1976: Physical processes in temperate glaciers. *J. Glaciol.* **16** (74), 151–8.

—— 1979: Local friction laws for glaciers: a critical review and new openings. *J. Glaciol.* **23** (89), 67–95.

LOOMIS, S.R., 1970: Morphology and ablation processes

on glacier ice. *Assoc. Am. Geogr. Proc.* **2**, 88–92.

LORIUS, C., BAUDIN, G., CITTANOVA, J. and PLATZER, R., 1969: Impuretés solubles contenues dans la glace de l'Antarctique. *Tellus* **21** (1), 136–48.

LORRAIN, R.D. and SOUCHEZ, R.A., 1972: Sorption as a factor in the transport of major cations by meltwaters from an alpine glacier. *Quat. Res.* **2** (2), 253–6.

LOWE, D.R., 1976a: Grain flow and grain flow deposits. *J. Sed. Petrol.* **46** (1), 188–99.

—— 1976b: Subaqueous liquefied and fluidized sediment flows and their deposits. *Sedimentology* **23**, 285–308.

LOWRISON, G.C., 1974: *Crushing and grinding.* Butterworths, London (286pp).

LUTSCHG, O., 1926: Beobachtungen über das Verhalten des vorstrassenden Allalingletschers im Wallis. *Zeit. Gletscherk.* **14**, 257–65.

LUTSCHG, O., HUBER, P., HUBER, H., QUERVAIN, M. de, 1950: Zum Wasserhaushalt des schweizer Hochgebirges. *Beit. Geol. Schweiz, Geotech. Series, Hydraulogie* **4**.

LYONS, W.B. and MAYEWSKI, P.A., 1984: Glaciochemical investigation as a tool in the historical delineation of the acid precipitation problem. In *The acid deposition phenomenon and its effects. Vol. 1 Atmospheric science* A.P. Altshuller and R.P. Linthurst (eds.), 8.71–8.81.

McCALL, J.G., 1960: The flow characteristics of a cirque glacier and their effects on glacial structure and cirque formation. In Lewis, W.V. (ed.), *Norwegian cirque glaciers, Roy. Geogr. Soc. (London) Res. Ser.* **4**, 39–62.

McCLINTOCK, F.A. and WALSH, J., 1962: Friction on Griffith cracks in rocks under pressure. *Proc. Fourth US Nat. Cong. Appl. Mech*, Berkeley.

McDONALD, B.C. and VINCENT, J.S., 1972: Fluvial sedimentary structures formed experimentally in a pipe and their implications for interpretation of subglacial sedimentary environments. *Geol. Surv. Can.* **72-27** (30pp).

McDOWELL, D.M. and O'CONNER, B.A., 1977: *Hydraulic behaviour of estuaries.* Methuen, London (292pp).

MAAG, H., 1969: *Ice dammed lakes and marginal glacial drainage on Axel Heiberg Island.* Axel Heiberg Island Res. Rept. McGill Univ., Montreal (147pp).

MAIZELS, J.K., 1979: Proglacial aggradation and changes in braided channel patterns during a period of glacier advance: an Alpine example. *Geogrf. Annlr.* **61A** (1–2), 87–101.

—— 1983: Proglacial channel systems: change and thresholds for change over long, intermediate and short time-scales. *Spec. Publs. Int. Ass. Sediment* **6**, 251–66.

MALE, D.H., 1980: The seasonal snowcover. In Colbeck, S.C. (ed.), *Dynamics of snow and ice masses*, Academic Press, New York, 305–96.

MARK, D.M., 1973: Analysis of axial orientation data, including till fabrics. *Geol. Soc. Am. Bull.* **84**, 1369–74.

MARTIN, S. and KAUFFMAN, P., 1977: An experimental and theoretical study of the turbulent and laminar convection generated under a horizontal ice sheet floating in warm salty water. *J. Phys. Oceanog.* **7** (2), 272–83.

MATTHES, F.E., 1930: Geological history of the Yosemite valley. *US Geol. Surv. Prof. Pap.* **160**.

MATTHEWS, J.B. and QUINLAN, A.V., 1975: Seasonal characteristics of water masses in Muir Inlet, a fjord with tidewater glaciers. *J. Fish. Res. Bd. Can.* **32**, 1693–1703.

MATHEWS, W.H., 1956: Physical limnology and sedimentation in a glacial lake. *Geol. Soc. Am. Bull.* **67**, 537–52.

—— 1973: Record on two jökulhlaups. *Int. Ass. Sci. Hydrol. Publ.* **95**, 99–110.

MATTHEWS, J.B.L. and HEIMDAL, B.R., 1980: Pelagic productivity and food chains in fjord systems. In Freeland, H.J., Farmer, D.M. and Levings, C.D. (eds.), *Fjord oceanography*, Plenum Press, New York, 375–98.

MEIER, M., 1972: Hydraulics and hydrology of glaciers. *Int. Ass. Sci. Hydrol. Publ.* **107**, 353–70.

MEIER, M.F., 1974: Ice sheets and glaciers. *Encyclopedia Britannica* 15th edition, Helen Hemmingway, Benton, New York, 175–86.

MEIER, M.F. and POST, A., 1969: What are glacier surges? *Can. J. Earth Sci.* **6** (4), 807–17.

MEIER, M.F., KAMB, W.B. ALLEN, C.R. and SHARP, R.P., 1974: Flow of Blue Glacier, Olympic Mountains, Washington, USA. *J. Glaciol.* **13** (68), 187–212.

MELLOR, M., 1973: Mechanical properties of rocks at low temperatures. *Permafrost: North American Contribution, Second International Conference, Washington, DC;* National Academy of Sciences, 334–44.

MESRI, G. and GIBALA, R., 1972: Engineering properties of a Pennsylvanian shale. *Proceedings 13th Symposium on Rock Mechanics (ASCE)*, 57–75.

METCALF, R.C., 1979: Energy dissipation during subglacial abrasion at Nisqually Glacier, Washington, USA. *J. Glaciol.* **23** (89), 233–46.

MIALL, A.D., 1977: A review of the braided-river

depositional environment. *Earth Sci. Rev.* **13**, 1–62.

—— 1978: Lithofacies types and vertical profile models in braided river deposits: a summary. In Miall, A.D. (ed.), *Fluvial sedimentology,* Can. Soc. Petrol. Geol. Mem. **5**, 597–60.

—— 1979: Deltas. In Walker, R.G. (ed.), *Facies models*, Geoscience Canada Reprint Series I, Geol. Assoc. Can., Ontario, 43–56.

—— 1983: Glaciomarine sedimentation in the Gowganda formation (Huronian), Northern Ontario. *J. Sed. Petrol.* **53**(2), 477–91.

—— 1984: *The principles of sedimentary basin analysis*. Springer-Verlag, Berlin (390pp).

MICHEL, B., 1978: *Ice mechanics*. Les Presses de l'Université Laval, Quebec (499pp).

MIDDLETON, G.V., 1966, 1967: Experiments on density and turbidity currents, I, II, and III. *Can. J. Earth Sci.* **3**, 523–46, 627–37; **4**, 475–505.

MIDDLETON, G.V., 1970: Experimental studies related to flysch sedimentation. In Lajoie, J. (ed.), *Flysch sedimentology in North America,* Spec. Pap. Geol. Assoc. Can. **7**, 253–72.

MIDDLETON, G.V. and BOUMA, A.H. (eds.), 1973: *Turbidites and deep water sedimentation: Pacific section.* Soc. Econ. Paleont. Min., Short course notes, Los Angeles.

MILLS, D. and MASON, J.S., 1975: Learning to live with erosion of bends. *First International Conference on the internal and external protection of pipes.* G1–G19.

MOLNIA, B.F., 1983: Subarctic glacial-marine sedimentation: a model. In Molnia, B.F. (ed.), *Glacial marine sedimentation*, Plenum Press, New York, 95–144.

MORAN, S.R., 1971: Glaciotectonic structures in drift. In Goldthwait, R.P. (ed.), *Till, a symposium.* Columbus, Ohio, Ohio State Univ. Press, 127–48.

MORAN, S.R., CLAYTON, L., HOOKE, R.L., FENTON, M.M. and ANDRIASHEK, L.D., 1980: Glacier-bed landforms of the prairie region of North America. *J. Glaciol.* **25** (93), 457–76.

MORGAN, V.I., 1972: Oxygen isotope evidence for bottom freezing on the Amery Ice Shelf. *Nature* (London) **238** (5364), 393–4.

MORGAN, V.I. and BUDD, W.F., 1978: The distribution, movement and melt rates of Antarctic icebergs. In Husseiny, A.A. (ed.), *Iceberg utilisation*, Pergamon Press, New York, 220–8.

MORLAND, L.W., 1976: Glacier sliding down an inclined wavy bed with friction. *J. Glaciol.* **17** (77), 463–77.

MORLAND, L.W. and BOULTON, G.S., 1975: Stress in an elastic hump: the effects of glacier flow over elastic bedrock. *Proc. Roy. Soc. (London) Ser. A* **344**, 157–73.

MORRIS, E.M., 1976: An experimental study of the motion of ice past obstacles by the process of regelation. *J. Glaciol.* **17** (75), 79–98.

—— 1979: The flow of ice, treated as a Newtonian viscous liquid, around a cylindrical obstacle near the bed of a glacier. *J. Glaciol.* **23** (89), 117–30.

MOSER, H. and AMBACH, W., 1977: Glaci-hydrological investigations in the Oetztal Alps made between 1968 and 1975. *Zeit. Gletscherk. Glazialgeol.* **13** (1–2), 167–79.

MULLEN, R.E., DARBY, D.A. and CLARK, D.L., 1972: Significance of atmospheric dust and ice rafting for Arctic Ocean sediment. *Geol. Soc. Am. Bull.* **83**, 205–12.

MULLER, L., 1963: *Der Felsbau.* Enke, Stuttgart (624pp).

MOROZUMI, M., NAKAMURA, S. and YOSHIDA, Y., 1978: Chemical constituents in the surface snow in Mizuho Plateau. *Memoirs National Inst. Polar Res. Spec. Issue* **7**, 255–63.

MURRELL, S.A.F., 1958: Strength of coal under triaxial compression. In *Mechanical properties of non-metallic brittle materials*, Butterworths, London.

—— 1963: A criterion for brittle fracture of rocks and concrete under triaxial stress, and the effect of pore pressure on the criterion. In *Rock mechanics*, Pergamon Press, Oxford.

NAKAMURA, T. and JONES, S.J., 1973: Mechanical properties of impure ice crystals. In Whalley, E., Jones, S.J. and Gold L.W. (eds.), *Physics and chemistry of ice*, Royal Soc. Canada, Ottawa, 365–9.

NAKA(W)O, M., 1979: Supraglacial debris of G2 glacier, Hidden Valley, Mukut Himal, Nepal. *J. Glaciol.* **22** (87), 273–84.

NAKA(W)O, M. and YOUNG, G.J., 1981: Field experiments to determine the effect of a debris layer on ablation of glacier ice. *Ann. Glaciol.* **2**, 85–91.

NEAL, C.S., 1979: The dynamics of the Ross Ice Shelf revealed by radio echo sounding. *J. Glaciol.* **24** (90), 295–307.

NEWITT, D.M., RICHARDSON, J.F., ABBOTT, M. and TURTLE, R.B., 1955; Hydraulic conveying of solids in horizontal pipes. *Trans. Inst. Chem. Engnrs.* **33**, 93–190.

NICKLING, W.G. and BENNETT, L., 1984: The shear

strength characteristics of frozen coarse granular debris. *J. Glaciol.* **30** (106), 348–57.

NOBLES, L.H. and WEERTMAN, J., 1971: Influence of irregularities of the bed of an ice sheet on deposition rate of till. In Goldthwait, R.P. (ed.), *Till, a symposium*, Ohio State Univ. Press, Columbus, Ohio, 117–26.

NYE, J.F., 1952: The mechanics of glacier flow. *J. Glaciol.* **2** (12), 82–93.

—— 1953: The flow rate of ice from measurements in glacier tunnels, laboratory experiments, and the Jungfraufirn borehole experiment. *Proc. Roy. Soc.* **A219** (1139), 477–89.

—— 1969a: The effect of longitudinal stress on the shear stress at the base of an ice sheet. *J. Glaciol.* **8** (53), 207–13.

—— 1969b: A calculation on the sliding of ice over a wavy surface using a Newtonian viscous approximation. *Proc. Roy. Soc. (London) Ser. A* **311**, 445–67.

—— 1970: Glacier sliding without cavitation in a linear viscous approximation. *Proc. Roy. Soc. (London) Ser. A* **315**, 381–403.

—— 1973: Water at the bed of a glacier. *Int. Ass. Sci. Hydrol. Publ.* **95**, 189–94.

—— 1976: Water flow in glaciers: jökulhlaups, tunnels and veins. *J. Glaciol.* **17** (76), 181–207.

NYE, J.F. and FRANK, F.C., 1973: Hydrology of the intergranular veins in a temperate glacier. *Int. Ass. Sci. Hydrol. Publ.* **95**, 157–61.

OLSEN, E.J., 1981: Estimates of the total quantity of meteorities in the East Antarctic Ice Cap. *Nature* (London) **292**, 516–18.

ORHEIM, O., 1980: Physical characteristics and life expectancy of tabular Antarctic icebergs. *Ann. Glaciol.* **1**, 11–18.

ORHEIM, O. and ELVERHOI, A., 1981: Model for submarine glacial deposition. *Ann. Glaciol.* **2**, 123–8.

ØSTREM, G., 1959: Ice melting under a thin layer of moraine, and the existence of ice cores in moraine ridges. *Geogrf. Annlr.* **41**, 228–30.

ØSTREM, G., 1964: A method of measuring water discharge in streams. *Geographical Bull.* **21**, 21–43.

ØSTREM, G., 1974: Studier av glaciarers massbalans och av materialtransporten i glaciaralvar som grundval for planering av vattenkraftverk i Norge. *Uppsala Univ. Naturgeografiska Inst. Rapport* **34**, 511–31.

ØSTREM, G., 1975: Sediment transport in glacial meltwater streams. In Jopling, A.V. and McDonald, B.C. (eds.), *Glaciofluvial and glaciolacustrine sedimentation*, Tulsa, Oklahoma, Society of Economic Paleontologists and Mineralogists, Spec. Publ. 23, 101–22.

ØSTREM, G., LIESTOL, O. and WOLD, B., 1976: Glaciological investigations at Nigardsbreen, Norway. *Norsk Geogr. Tidsskr.* **30**, 187–209.

ØSTREM, G., BRIDGE, C.W. and RANNIE, W.F., 1967: Glacio-hydrology, discharge, and sediment transport in the Decade Glacier area, Baffin Island, NWT. *Geogrf. Annlr.* **49A** (3–4), 268–82.

OSWALD, G.K.A. and ROBIN, G.de Q., 1973: Lakes beneath the Antarctic Ice Sheet. *Nature* (London) **245** (5423), 251–4.

OVENSHINE, G., 1970: Observations of iceberg rafting in Glacier Bay, Alaska, and the identification of ancient ice-rafted deposits. *Geol. Soc. Am. Bull.* **81** (3), 891–4.

PATERSON, W.S.B., 1977: Secondary and tertiary creep of glacier ice as measured by borehole closure rates. *Rev. Geophys. Space Phys.* **15** (1), 47–55.

—— 1981: *The physics of glaciers* 2nd edn. Pergamon Press, Oxford (380pp).

PEARSALL, I.S., 1972: *Cavitation.* Mills & Boon, London (80pp).

PEARSON, T.H., 1980: Macrobenthos of fjords. In Freeland, H.J., Farmer, D.M. and Levings, C.D. (eds.), *Fjord oceanography*, Plenum Press, New York, 569–602.

PERLA, R.I., 1980: Avalanche release, motion and impact. In Colbeck, S.C. (ed.), *Dynamics of snow and ice masses*, Academic Press, New York, 397–462.

PESSL, F. and FREDERICK, J.E., 1981: Sediment source for melt-water deposits. *Ann. Glaciol.* **2**, 92–6.

PETERSON, D.N., 1970: Glaciological investigations on the Casement Glacier, southeastern Alaska. Ohio State Univ. Inst. of Polar Studies Rep. 36 (161pp).

PETTS, G. and FOSTER, I., 1985: *Rivers and landscape.* Arnold, London (274pp).

PHILIPPI, H., 1912: Über ein rezentes alpines Os und seiner Bedeutung für die Bildung der Diluvialen Osar. *Z. Deut. Geol. Ges.* **64**, 68–102.

PLATT, C.M., 1966: Some observations on the climate of Lewis Glacier, Mount Kenya, during the rainy season. *J. Glaciol.* **6** (44), 267–88.

POWELL, R.D., 1981: A model for sedimentation by tidewater glaciers. *Ann. Glaciol.* **2**, 129–34.

—— 1983: Glacial-marine sedimentation process and lithofacies of temperate tidewater glaciers, Glacier Bay, Alaska. In Molnia, B.F., (ed.), *Glacial-*

marine sedimentation, Plenum Press, New York, 185–232.

—— 1984: Glacimarine processes and lithofacies modelling of ice shelf and tidewater glacier sediments based on Quaternary examples. *Marine Geology* **57**, 1–52.

PRICE, N.J., 1966: *Fault and joint development in brittle and semi-brittle rock.* Pergamon Press, Oxford (176pp).

PRICE, R.J., 1973: *Glacial and fluvioglacial landforms.* Oliver & Boyd, Edinburgh (242pp).

PRICE, W.H., 1947: Erosion of concrete by cavitation and sands in flowing water. *J. Am. Concrete Inst.,* 1009–23.

PRICE, W.H. and WALLACE, G.B., 1949: Resistance of concrete and protective coatings to forces of cavitation. *J. Am. Concrete Inst.* **21** (2), 109–20.

PRICE, W.H., CLARKE, R.R., CRESKOFF, J.J., McCLENAHAN, W.T., PECKWORTH, H.F., RAWN, A.M., WALTER, D.S., WASTLUND, G., YOUNG, R.B., 1955: Erosion resistance of concrete in hydraulic structures. *J. Am. Concrete Inst.* **27** (3), 259–71.

PRIOR, D.B. and WISEMAN Wm.J. Jr., 1981: Submarine chutes on the slopes of fjord deltas. *Nature* (London) **290** (5804), 326–8.

PYOKARI, M., 1981: Ice action on lakeshores near Schefferville, central Quebec–Labrador, Canada. *Can. J. Earth Sci.* **18**, 1629–34.

QUERVAIN, A. de, 1919: Über Wirkungen eines vorstossenden Gletschers. *Vjschr. Naturf. Ges. Zurich* **64**, 336–49.

RAGOTZKIE, R.A., 1978: Heat budget in lakes. In Lerman, A. (ed.), *Lakes: chemistry, geology, physics*, Springer-Verlag, New York, 1–19.

RAINWATER, F.H. and GUY, H.P., 1961: Some observations on the hydrochemistry and sedimentation of the Chamberlain Glacier area, Alaska. *US Geol. Surv. Prof. Pap.* **414-C**, (14pp).

RAITHBY, G.D., 1976: *Predictions of dispersion by surface discharge*, Canadian Center For Inland Waters.

RAUDKIVI, A.J., 1976: *Loose boundary hydraulics*, 2nd edn. Pergamon Press, Oxford.

RAPP, A., 1960: Recent development of mountain slopes in Karkevagge and surroundings, northern Scandinavia. *Geogrf. Annlr.* **42**, 65–200.

RAYMOND, C.F., 1980: Temperate valley glaciers. In Colbeck, S.C. (ed.), *Dynamics of snow and ice masses*, Academic Press, New York, 79–139.

RAYMOND, C.F. and HARRISON, W.D., 1975: Some observations on the behaviour of the liquid and gas phases in temperate glacier ice. *J. Glaciol.* **14** (71), 213–33.

REEH, N. 1968: On the calving of ice from floating glaciers and ice shelves. *J. Glaciol.* **7** (50), 215–32.

REID, H.F., 1892: Studies of Muir Glacier, Alaska. *Natnl. Geogr. Mag.* **4**, 19–84.

REIMNITZ, E. and BARNES, P.W., 1974: Sea ice was a geologic agent on the Beaufort Sea Shelf of Alaska. In Reed and J.C. Sater, J.E., *The coast and shelf of the Beaufort Sea*, Arctic Inst. of N. America, Arlington (Va), 301–53.

REINECK, H.E. and SINGH, I.B., 1980: *Depositional sedimentary environments*, 2nd edn. Springer-Verlag, Berlin (549pp).

REINECK, H.E., GUTMANN, W.F. and HERTWECK, G., 1967: Das Schlickgebiet südlich Helgoland als Beispiel rezenter Schelfablagerungen. *Senkenbergiana Lethaea* **48**, 219–75.

REKSTAD, J.B., (1911–12): Die Ausfüllung eines Sees vor dem Ergabrae, dem grössten Ausläufer des Svartisen, als Mass der Gletschererosion. *Zeit. Gletscherkunde* **6**, 212–14.

RELLING, O. and NORDSETH, K., 1979: Sedimentation of a river suspension into a fjord basin: Gaupnefjord in Western Norway. *Norsk Geogr. Tidsskr.* **33**, 187–203.

REMENYI, K., 1974: *The theory of grindability and the comminution of binary mixtures.* Akademiai Kiado, Budapest (144pp).

REYNOLDS, R.C. and JOHNSON, N.M., 1972: Chemical weathering in the temperate glacial environment of the Northern Cascade Mountains. *Geochim. Cosmochim. Acta* **36**, 537–54.

RICE, R.J., 1982: The hydraulic geometry of the lower portion of the Sunwapta River valley train, Jasper National Park, Alberta. In Davidson-Arnott, R., Nickling, W. and Fahey, B.D. (eds.), *Research in glacial, glacio-fluvial and glacio-lacustrine systems*, Proc. 6th Guelph Symposium on Geomorphology, 1980, Geo Books, Norwich, 151–73.

RILEY, N.W., 1979: Discussion comments in 'Symposium on glacier beds: the ice-rock interface', *J. Glaciol.* **23** (89), 384–5, 391.

—— 1982: *Rock wear by sliding ice.* PhD dissertation, Univ. Newcastle upon Tyne.

RIST, S., 1955: Skeidararhlaup 1954. The hlaup of Skeidara 1954. *Jökull* **5**, 30–6.

von RITTINGER, P.R., 1867: *Lehrbuch der Aufbereitungskunde.* Ernst & Korn, Berlin.

ROBE, R.Q., 1980: Iceberg drift and deterioration. In Colbeck, S.C. (ed.), *Dynamics of snow and ice masses*, New York, Academic Press, 211–59.

ROBERTS, A., 1977: *Geotechnology*. Pergamon Press, Oxford.

—— 1981: *Applied geotechnology*. Pergamon Press, Oxford (344pp).

ROBERTS, M.C. and ROOD, K.M., 1984: The role of the ice contributing area in the morphology of transverse fjords, British Columbia. *Geogrf. Annlr.* **60A** (4), 381–93.

ROBIN, G. de Q. and MILLAR, D.H.M., 1982: Flow of ice sheets in the vicinity of subglacial peaks. *Ann. Glaciol.* **3**, 290–4.

—— 1976: Is the basal ice of a temperate glacier at the pressure melting point? *J. Glaciol.* **16** (74), 183–96.

—— 1979: Formation, flow and disintegration of ice shelves. *J. Glaciol.* **24** (90), 259–71.

ROBIN, G. de Q., DREWRY, D.J. and MELDRUM, D.T., 1977: International studies of ice sheet and bedrock. *Phil. Trans. Roy. Soc. (London) Ser. B* **279**, 185–96.

RODINE, J.D. and JOHNSON, A.M., 1976: The ability of debris, heavily freighted with coarse clastic materials, to flow on gentle slopes. *Sedimentology* **23**, 213–34.

RÖTHLISBERGER, H., 1968: Erosive processes which are likely to accentuate or reduce the bottom relief of valley glaciers. *Intl. Ass. Sci. Hydrol. Publ.* **79**, 87–97.

—— 1972: Water pressure in intra- and subglacial channels. *J. Glaciol.* **11** (62), 177–203.

—— 1979: Comments in *J. Glaciol.* **23** (89), 385.

RÖTHLISBERGER, H. and IKEN, A., 1981: Plucking as an effect of water-pressure variations at the glacier bed. *Ann. Glaciol.* **2**, 57–62.

RUMPF, H., 1973: Physical aspect of comminution and new formulation of a Law of Comminution. *Powder Technology* **7**, 145–59.

RUSSELL-HEAD, D.S., 1980: The melting of free-drifting icebergs. *Ann. Glaciol.* **1**, 119–22.

RUST, B.R., 1972: Structure and process in a braided river. *Sedimentology* **18**, 221–45.

—— 1975: Fabric and structure in glaciofluvial gravels. In Jopling, A.V. and McDonald, B.C. (eds.), *Glaciofluvial and glaciolacustrine sedimentation*. Tulsa, Oklahoma. Society of Economic Paleontologists and Mineralogists, Spec. Publ. 23, 238–48.

—— 1978a: A classification of alluvial channel systems. In Miall, A.D. (ed.), *Fluvial sedimentology*, Can. Soc. Petrol. Geol. Mem. 5, 187–98.

—— 1978b: Depositional models for braided alluvium. In Miall, A.D. (ed.), *Fluvial sedimentology*, Can Soc. Petrol. Geol. Mem. 5, 605–25.

—— 1979: Coarse alluvial deposits. In Walker, R.G. (ed.), *Facies models*, Geoscience Reprint Series 1, Geological Assoc. Canada, Ontario, 9–21.

RUST, BR.. and ROMANELLI, R., 1975: Late Quaternary subaqueous outwash deposits near Ottawa, Canada. In Jopling, A.V. and McDonald, B.C. (eds.), *Glaciofluvial and glaciolacustrine sedimentation*. Tulsa, Oklahoma. Society of Economic Paleontologists and Mineralogists, Spec. Publ. 23, 177–92.

Symposium on glacier beds: the ice-rock interface. *J. Glaciol.* **23** (89).

SARKAR, A.D., 1980: *Friction and wear*. Academic Press, London (423pp).

SAUNDERSON, H.C., 1975: Sedimentology of the Brampton esker and its associated deposits: an empirical test of theory. In Jopling, A.V. and McDonald, B.C. (ed.), *Glaciofluvial and glaciolacustrine sedimentation*, Tulsa, Oklahoma. Society of Economic Paleontologists and Mineralogists, Spec. Publ. 23, 155–76.

—— 1977: The sliding bed facies in esker sands and gravels: a criterion for full-pipe (tunnel) flow? *Sedimentology* **24**, 623–38.

—— 1982: Bedform digrams and the interpretation of eskers. In Davidson-Arnott, R., Nickling, W. and Fahey, B.D., (eds.), *Research in glacial, glacio-fluvial and glacio-lacustrine systems*, Proc. 6th Guelph Symposium on Geomorphology, 1980, Geo Books, Norwich, 139–50.

SAUNDERSON, H.C. and JOPLING, A.V., 1980: Palaeohydraulics of a tabular, cross-stratified sand in the Brampton esker, Ontario. *Sed. Geology* **25**, 169–88.

SCHEI, B., EILERTSEN, H.G., FALK-LARSEN, S., GULLIKSON, B. and TAASEN, J.P., 1979: Marinbiologiske undersøkelser i Van Mijenfjorden (Vest Spitsbergen) etter objesøllekasje ved Sveagruva 1978. *Tromsø Museums Rapportserie, Naturvitenshap. Nr.* **2**, Universitetet i Tromsø (50pp).

SCHINDLER, J.F., 1968: The impact of ice islands: The story of Arlis II and Fletcher's Ice Island T-3, since 1962, in Sater, J.E., (ed.), *Arctic drifting stations*, Washington, DC, AINA, 49–80.

SCHNEIDER, B., 1967: Moyens nouveaux de

reconaissance des massifs rocheux. *Supp. Annals de l'Inst. Tech. de Batiment et des Travaux Publics* **20** (235–236), 1055–93.

SCHOMMER, Von, P., 1977: Wasserspiegelmessungen im firn des Ewigschneefeldes (Schweizer Alpen) 1976. *Zeit. Gletscherk. Glazialgeol.* **12** (2), 125–41.

SELBY, MJ., 1980: A rock mass strength classification for geomorphic purposes: with tests from Antarctica and New Zealand. *Zeit. Geomorph.* **24** (1), 31–51.

—— 1982: *Hillslope materials and processes.* OUP, Oxford (264pp).

SHARMA, G.D., 1979: *The Alaskan Shelf.* Springer-Verlag, New York (498pp).

SHARP, R.P., 1949: Studies of superglacial debris on valley glaciers. *Am. J. Sci.* **247**, 289–315.

—— 1960: *Glaciers.* Condon Lectures, Oregon State System of Higher Education, Oregon (78pp).

SHAW, J., 1977a: Till deposited in arid polar environments. *Can. J. Earth Sci.* **14**, 1239–45.

—— 1977b: Sedimentation in an alpine lake during deglaciation, Okanagan Valley, British Columbia, Canada. *Geogrf. Annlr.* **59A** (3–4), 221–40.

SHAW, J., GILBERT, R. and ARCHER, J.J.J., 1978: Proglacial lacustrine sedimentation during winter. *Arctic and Alpine Res.* **10** (4), 689–99.

SHERWIN, J.A. and CHAPPLE, W.M., 1968: Wavelengths of single layer folds: a comparison between theory and observation. *Am. J. Sci.* **266**, 167–79.

SHREVE, R.L., 1972: Movement of water in glaciers. *J. Glaciol.* **11** (62), 205–14.

—— 1984: Glacier sliding at subfreezing temperature. *J. Glaciol.* **30** (106) 341–7.

—— 1985: Esker characteristics in terms of glacier physics. Katahdin Esker System, Maine. *Geol. Soc. Am. Bull.* **95** (5), 639–46.

SINHA, N.K. and NAKAWO, M., 1981: Growth of first-year sea ice, Eclipse Sound, Baffin Island, Canada. *Can. Geotech. J.* **18** (1), 17–23.

SLATT, R.M., 1972: Geochemistry of meltwater streams from nine Alaskan glaciers. *Geol. Soc. Am. Bull.* **83** (4), 1125–32.

SLATT, R.M. and EYLES, N., 1981: Petrology of glacial sand: implications for the origin and mechanical durability of lithic fragments. *Sedimentology* **28** (2), 171–83.

SMITH, D.D., 1964: Ice lithologies and structure of ice island Arlis II. *J. Glaciol.* **5** (37), 17–38.

SMITH, N.D., 1978: Sedimentation processes and patterns in a glacier-fed lake with low sediment input. *Can. J. Earth Sci.* **15**, 741–56.

—— 1981: The effect of changing sediment supply on sedimentation in a glacier-fed lake. *Arctic and Alpine Res.* **13** (1), 75–82.

SMITH, N.D., and SYVITSKI, J.P.M., 1982: Sedimentation in a glacier-fed lake: the role of pelletization on deposition of fine-grained suspensates. *J. Sed. Petrol.* **52** (2), 503–13.

SMITH, N.D., VENOL, M.A. and KENNEDY, S.K., 1982: Comparison of sedimentation regimes in four glacier-fed lakes in western Alberta. In Davidson-Arnott, R., Nickling, W. and Fahey, B.D. (eds.), *Research in glacial, glacio-fluvial and glacio-lacustrine systems*, Proc. 6th Guelph Symposium on Geomorphology 1980. Geo Books, Norwich, 203–38.

SOLHEIM, A. and PFIRMAN, S.L., 1985: Sea-floor morphology outside a grounded surging glacier, Bråsvellbreen, Svalbard. *Marine Geology.* **65** (1–2), 127–43.

SOUCHEZ, R.A. and LORRAIN, R.D., 1975: Chemical sorting effect at the base of an alpine glacier. *J. Glaciol.* **14** (71), 261–5.

—— 1978: Origin of the basal ice layer from alpine glaciers indicated by its chemistry. *J. Glaciol.* **20** (83), 319–28.

SOUCHEZ, R.A. and TISON, J.L., 1981: Basal freezing of squeezed water: its influence on glacier erosion. *Ann. Glaciol.* **2**, 63–6.

SOUCHEZ, R.A., LORRAIN, R.D. and LEMMENS, M.M., 1973: Refreezing of interstitial water in a subglacial cavity on an Alpine glacier as indicated by the chemical composition of ice. *J. Glaciol.* **12** (66), 453–9.

SOUCHEZ, R.A., LEMMENS, M., LORRAIN, R.D. and TISON, J.L., 1978: Pressure melting within a glacier indicated by the chemistry of re-gelation ice. *Nature* (London) **273** (5662), 454–6.

SOUCHEZ, R.A., LEMMENS, M., LORRAIN, R.D. and TISON, J.L., 1979: Discussion in *J. Glaciol.* **23** (89), 421–2.

SPRING, U. and HUTTER, K., 1981: Numerical studies of jökulhlaups. *Cold Reg. Sci. Tech.* **4**, 227–44.

—— 1982: Conduit flow of a fluid through its solid phase and its application to intraglacial channel flow. *Int. J. Eng. Sci.* **20** (2), 327–63.

STEINEMANN, I.S., 1958 Experimentelle Untersuchungen zur Plastizität von Eis. *Beitr. Geol. Schweiz, Geotech.* **10**, 1–72.

STEINTHORSSON, S., 1977: Tephra layers in a drill zone from the Vatnajökull Ice Cap. *Jökull* **27**, 2–27.

STENBORG, T., 1968: Glacier drainage connected with ice structures. *Geogrf. Annlr.* **50A** (1), 25–53.

—— 1969: Studies of the internal drainage of glaciers. *Geogrf. Annlr.* **51A** (1–2), 13–41.

—— 1970: Delay of runoff from a glacier basin. *Geogrf. Annlr.* **52A** (1), 1–30.

STRAKHOV, N.M., 1967: *Principles of lithogenesis 1.* Oliver & Boyd, London (245pp).

STRINGER, W.J., 1974: Morphology of the Beaufort Sea shorefast ice. In Reed, J.C. and Sater, J.E. (eds.), *The coast and shelf of the Beaufort Sea*, Arlington (Va), AINA, 165–72.

STURM, M., 1979: Origin and composition of clastic varves. In Schlüchter, C. (ed.), *Moraines and varves*, Balkema, Rotterdam, 281–5.

SUGDEN, D.E. and JOHN, B.S., 1976: *Glaciers and landscape.* Arnold, London (376pp).

SUNDBORG, A., 1967: Some aspects on fluvial sediments. on fluvial morphology. I General views and graphic methods. *Geogrf. Annlr.* **49A**, 333–43.

SWINZOW, G.K., 1962: Investigation of shear zones in the ice sheet margin, Thule area, Greenland. *J. Glaciol.* **4** (32), 215–29.

SWITHINBANK, C.W., 1969: Giant iceberg in the Weddell Sea. *Polar Rec.* **14** (91), 477–8.

SWITHINBANK, C.W.M., McCLAIN, C.P. and LITTLE, P., 1977: Drift tracks of Antarctic icebergs. *Polar Rec.* **18** (116), 495–501.

SYVITSKI, J.P.M., 1980: Flocculation, agglomeration, and zooplankton pelletization of suspended sediment in a fjord receiving glacial meltwater. In Freeland, H.J., Farmer, D.M. and Levings, C.D. (eds.), *Fjord oceanography*, Plenum Press, New York, 615–23.

TAMM, O., 1924: Experimental studies on chemical processes in the formation of glacial clay. *Sveriges Geologiska Undersoknig Arsbok* **18** (5), 1–20.

TANAKA, T., 1962: Preferential grinding mechanism of binary solid mixtures whose components are of different grindability. *Symp. Zerkleinen. Chemie, Weiheim*, VDI Verlag, Dusseldorf, 360–73.

TARR, R.S., 1908: Some phenomena of the glacier margins in the Yakutat Bay region, Alaska. *Zeit. Gletscherk.* **III**, 81–110.

TCHERNIA, P., 1974: Etude de la derive antarctique Est-Ouest au moyen d'icebergs suivis par le satellite E'ole. *C.R. Acad. Sci. Paris* **278** (serb), 667–70.

TCHERNIA, P. and JEANNIN, P.F., 1980: Observations on the Antarctic East Wind Drift using tabular icebergs tracked by satellite Nimbus F. (1975-1977). *Deep-Sea Res.* **27A**, 467–74.

—— 1983: *Quelques aspects de la circulation oceanique Antarctique révélés par l'observation de la dérive d'icebergs (1972-1983).* CNRS, Museum National d'Histoire Naturelle.

TEER, D.G. and ARNELL, R.D., 1975a: Friction theories. In Halling, J. (ed.), *Tribology*, Macmillan, London, 72–93.

—— 1975b: Wear. In Halling J. (ed.), *Tribology*, Macmillan, London, 94–127.

THOMAS, R.H., 1973a: The creep of ice shelves: interpretation of observed behaviour. *J. Glaciol.* **12** (64), 55–70.

—— 1973b: The dynamics of the Brunt Ice Shelf, Coats Land, Antarctica. *Br. Antarct. Surv. Sci. Rep.* **79** (45pp).

—— 1979: The dynamics of marine ice sheets. *J. Glaciol.* **24** (90) 167–8.

THOMAS, R.H. and COSLETT, P.H., 1970: Bottom melting of ice shelves and the mass balance of Antarctica. *Nature* (London) **228** (5266), 47–9.

THOMAS, R.H. and MACAYEAL, D.R., 1982: Derived characteristics of the Ross Ice Shelf. *J. Glaciol.* **28** (100), 397–412.

THORARISSON, S., 1939: Hoffellsjökull, its movements and drainage. *Geogrf. Annlr.* **21**, 189–215.

TILLY, G.P., 1969: Erosion caused by impact of solid particles. In Scott, D. (ed.), *Treatise on materials science and technology* 13, Wear, New York Academic Press, 289–319.

TILLY, G.P. and SAGE, W., 1970: Wear. In Scott, D. (ed.), *Treatise on materials science and technology* 16, New York, Academic Press, 447–65.

TODD, D.K., 1980: *Groundwater hydrology*, Wiley, 535pp.

TSANG, G., 1982: *Frazil and anchor ice: a monograph.* NRC Subcommittee on Hydraulics of Ice Covered Rivers. Ottawa, Ont., Canada.

VINJE, T.E., 1980: Some satellite-tracked iceberg drifts in the Antarctic. *Ann. Glaciol.* **1**, 83–7.

VINOGRADOV, U.N., 1981: Glacier erosion and sedimentation in the volcanic regions of Kamchatka. *Ann. Glaciol.* **2**, 164–9.

VIVIAN, R., 1975: *Les glaciers des alpes occidentales.* Imprimerie Allier, Grenoble (513pp).

VIVIAN, R.A. and ZUMSTEIN, J., 1973: Hydrologie sous-glaciaire au glacier d'Argentière (Mont Blanc, France). *Int. Ass. Sci. Hydrol. Pub.* **95**, 53–64.

VIVIAN, R.A., 1979: Comments in *J. Glaciol.* **23** (89), 386.

VOIGHT, B. (ed.), *Rockslides and avalanches 1.* Elsevier, Amsterdam.

WADHAMS, P., 1980: Ice characteristics in the seasonal sea ice zone. *Cold Reg. Sci. and Tech.* **2**, 37–87.

WADHAMS, P. and HORNE, R.J., 1980: An analysis of ice

profiles obtained by submarine sonar in the Beaufort Sea. *J. Glaciol.* **25** (93), 401–24.

WADHAMS, P., KRISTENSEN, M. and ORHEIM, O., 1983: The response of Antarctic icebergs to ocean waves. *J. Geophys. Res.* **88** (C10), 6053–65.

WAKAHAMA, G., KUROIWA, D., KOBAYASHI, O., TANUMA, K., ENDO, Y., MIZUNO, Y. and KOBAYASHI, S., 1973: Observations of permeating water through a glacier body. *Low Temp. Sci. A.* **31**, 217–19.

WALDER, J.S. and HALLET, B., 1979: Geometry of former subglacial water channels and cavities. *J. Glaciol.* **23** (89), 335–46.

WALDER, J.S., 1982: Stability of sheet flow of water beneath temperate glaciers and implications for glacier surging. *J. Glaciol.* **28** (99), 273–93.

WALKER, H.J., 1974: The Colville River and the Beaufort Sea: Some interactions. In Read, J.C. and Sater, J.E. (eds.), *The coast and shelf of the Beaufort Sea*, Arlington (Va), AINA, 513–40.

WALKER, R.G. and CANT, D.J., 1979: Sandy fluvial systems. In Walker, R.G. (ed.), *Facies models*, Geoscience Reprint Series 1, Geological Assoc. Canada, Ontario, 23–31.

WEEKS, W.F. and MELLOR, 1978: Some elements of iceberg technology. In Husseiny, A.A. (ed.), *Iceberg utilization*, Pergamon Press, Oxford, 45–97.

WEEKS, W.F. and CAMPBELL, W.J., 1973: Icebergs as a freshwater source: an appraisal. *J. Glaciol.* **12** (65), 207–33.

WEEKS, W.F., BARNES, P.W., REARIC, D.M. and REIMNITZ, E., 1983: Statistical aspects of ice gouging on the Alaskan Shelf of the Beaufort Sea. *CRREL Rpt.* **83–21**, (34pp).

WEERTMAN, J., 1957: On the sliding of glaciers. *J. Glaciol.* **3** (21)., 33–8.

—— 1958: Transport of boulder by glaciers and ice sheets. *Int. Ass. Sci. Hydrol. Bull.* **10**, 44.

—— 1961: Mechanism for the formation of inner moraines found near the edge of cold ice caps and ice sheets. *J. Glaciol.* **3** (30), 965–78.

—— 1964: The theory of glacier sliding. *J. Glaciol.* **5** (39), 287–303.

—— 1966: Effect of a basal water layer on the dimensions of ice sheets. *J. Glaciol.* **6** (44), 191–207.

—— 1968: Diffusion law for the dispersion of hard particles in an ice matrix that undergoes simple shear deformation. *J. Glaciol.* **7** (50), 161–5.

—— 1969: Water lubrication mechanism of glacier surges. *Can. J. Earth Sci.* **6** (4/2), 929–42.

—— 1972: General theory of water flow at the base of

a glacier or ice sheet. *Rev. Geophys. Space Phys.* **10** (1), 287–333.

—— 1973: Creep of ice. In Whalley, W., Jones, S.J. and Gold, L.W. (eds.), *Physics and chemistry of ice*, R. Soc. Can., Ottawa, 320–37.

—— 1976: Sliding – no sliding zone effect and age determination of ice cores. *Quat. Res.* **6**, 203–7.

—— 1979: The unsolved general glacier sliding problem. *J. Glaciol.* **23** (89), 97–115.

—— 1983: Creep deformation of ice. *Ann. Rev. Earth Planet. Sci.* **11**, 215–40.

—— (ed.), 1983: *Workshop on the Jakobshavns Glacier (Greenland)*. NSF Washington, DC.

WILLIAMS, P.J., 1982: *The surface of the earth.* Longman, London (212pp).

WILSON, A.T., 1964: Origin of the ice ages: an ice shelf theory for Pleistocene glaciation. *Nature* (London) **201** (4915), 147–9.

WINDOM, A.L., 1969: Atmospheric dust records in permanent snowfields: implications to marine sedimentation. *Geol. Soc. Am. Bull.* **80**, 761–82.

WINTGES, Von TH. and HEUBERGER, H., 1980: Parabelrisse, Sichelbrüche und Sichelwannen im Vereinigungsbereich zweier zillertaler Gletscher (Tirol). *Zeit. Gletscherk. Glazialgeol.* **16** (1), 11–23.

WOLD, B. and ØSTREM, G., 1979: Subglacial constructions and investigations at Bondhusbreen, Norway. *J. Glaciol.* **23** (89), 363–79.

WORDIE, J.M. and KEMP, S., 1933: Observations on certain Antarctic icebergs. *Geogr. J.* **81** (5), 428–34.

WRIGHT, L.D. and COLMAN, J.M., 1973: Variations in morphology of major river deltas as functions of ocean wave and river discharge regimes. *Am. Assoc. Petrol. Geol. Bull.* **57**, 370–98.

WRIGHT, L.D., 1977: Sediment transport and deposition at river mouths: a synthesis. *Geol. Soc. America Bull.* **88**, 857–68.

WRIGHT, R. and ANDERSON, J.B., 1982: The importance of sediment gravity flow to sediment transport and sorting in a glacial marine environment: Eastern Weddell Sea, Antarctica. *Geol. Soc. Am. Bull.* **93** (10), 951–63.

WRIGHT, R., ANDERSON, J.B. and FRISCO, P.P., 1983: Distribution and association of sediment gravity flow deposits and glacial/glacial-marine sediments around the continental margin of Antarctica. In Molnia, B.F. (ed.), *Glacial-marine sedimentation*, Plenum Press, New York, 265–300.

YEVTEYEV, S.A., 1959: Opredeleniye kolichestva morennogo materiala, perenosimogo lednikami vostochnogo poberezh'ya Antarktidy. *Informatsionnyy Byulleten' Sovetskoy Antarkticheskoy Ekspeditsii* **11**, 14–16.

YOUNG, G.J., 1982: Hydrological relationships in a glacierized mountain basin. In Glen, J.W., *Hydrological aspects of alpine and high-mountain areas, Int. Ass. Sci. Hydrol. Publ.* **138**, 51–62.

ZANDI, I., 1971: in, *Hydraulic transport of bulky materials*. Pergamon Press, Oxford, 24–5.

ZOTIKOV, I.A., ZAGORODNOV, V.S. and RAIKOVSKY, J.V., 1980: Core drilling through the Ross Ice Shelf (Antarctica) confirmed basal freezing. *Science* **207** (4438), 1463–5.

ZOTIKOV, I.A., ZAGORODNOV, V.S. and RAIKOVSKY, J.V., 1979: Sea ice on bottom of Ross Ice Shelf. *Ant. Jnl. US* **14** (5), 65–66.

ZOTIKOV, I.A., 1982: *Thermophysics of glaciers*. Leningrad, Hydrometeodat (287pp).

Symbols and notation used in the text

A	surface accumulation	c_i	specific heat capacity of ice
A_a	apparent area of clast	c_w	specific heat capacity of sea water
A_b	abrasion rate	c	constant related to melting- point depression of ice
A_{bb}	ablation rate		
A_c	abrasion rate on bed hummock crests	c_l	length of clast short axis
A_{diff}	difference in mean annual ablation over moraine and ice	c'	parameter related to number of clasts cutting bedrock and ratio of wear particles removed
A_f	surface area of clast created by fracture		
A_i	ice accretion rate (negative melt rate; cf M)	D	constant in basal sliding
		D_f	damping factor
		D_g, D_g'	drag coefficients
A_p	projected area of iceberg normal to current	D_r	sediment quantity from valley side rockfall
A_r	real contact area	$d*$	diffusion coefficient for dissolved matter
A_s	reactive area of surface of solids		
A_t	abrasion rate of bedrock troughs	d	radius of indentation
A_2	area of clast in transverse plane	D	linear rate of dissolution
a	area (usually of glacier surface)		
a_d	adhesion	E	Young's modulus
a_i	surface area of permanent experimental indentation	E_c	parameter related to number of clasts cutting bedrock
a_m	amplitude	E_{com}	energy required for comminution of debris
a'	roughness constant		
a_l	length of clast long axis	E_f	rate of grooving by sea ice keels
$a*$	amplitude of bed irregularities	E_r	erosion rate of fluvial channels
		E_{ri}	erosion index
B	temperature-dependent ice hardness parameter	E_r'	total bed erosion of glacier
		e	base of Naperian logarithms
B_i	Bond index of material hardness	ee	exponent
B', b'	centres of buoyancy of iceberg	E	elastic energy
b	area of glacier bed		
b_l	length of clast intermediate axis	F	force pressing clast to bed
		F_c	coagulation or flocculation rate
C_{av}	rate of cavity formation	F_d	drag
C_d	drag coefficient	F_i	Froude number
C_e	condensation/evaporation ratio	F_p	force exerted on clast by boundaries in debris flow
C_{es}	equilibrium concentration or solubility		
C_o	debris concentration	F_r	force resisting particle motion
C_s	concentration of suspended sediments	F_s	force required to initiate sliding
C_s'	measured value of suspended sediment concentration	F'	shape correction factor
		f	radio frequency
C_{sol}	concentration of solutes in meltwater	ff	reaction order exponent
C_*	sediment concentration-dependent diffusion constant	fn	correction factor for ice flow close to bed
c	cohesion	F	friction force

G	shear modulus		J_d	hydraulic mean depth
G_b	bedload transport rate		J_s	concentration of solutes
G_c	elastic energy absorbed per unit area of crack propagation		j	constant related to interatomic potential
G_d	depth of iceberg plough mark			
G_l	length of iceberg plough mark		$K_{(\ )}$	thermal conductivity (subscripts d, i, p, r refer to debris, ice, particle and rock)
G_w	width of iceberg plough mark			
$G'g'$	centres of gravity of iceberg			
$G*$	sediment transport rate		K_b	bulk modulus
g	gravitational acceleration		K_c	fracture toughness (critical stress intensity factor)
g_w	groundwater flux			
			K_f	stress intensity factor
H	total ice thickness		K_1	constant
H_b	total thickness of iceberg		K_2	constant proportional to thickness of basal debris layer of constant debris thickness
H_d	hardness (H_1 designates clast; H_2 designates bedrock)			
H_r	thickness of rock layer			
H_v	total thickness of varves		k, k_1, ... k_n	constants
H'	total hydraulic head			
H( )	rock hardness coefficient		k_b	wave number
h	ice thickness		k_d	dimensional parameter related to channel size, particle hardness, etc.
h_b	thickness of iceberg			
h_k	draft of sea ice keel		k_g	velocity coefficient for descending turbidity current, related to bedslope
h_d	thickness of debris layer			
h_{di}	depth of overflow layer of influent stream		k_i	thermal diffusivity of ice
			$k*$	wave number of bed undulations
h_f	thickness of sea ice floe		k_t*	transitional wave number
h_i	depth of indentation		k_*, k_*'	material constants
h_l	thickness of frontal lobe of turbidity flow		k', k''	drag coefficients for water flow over bed and fluid interface respectively
h_l'	thickness of uniform segment of turbidity flow			
			k'_*	permeability of flow till
h_p	pressure head		L	latent heat of fusion of ice
h_r	thickness of regelation layer		L_{crit}	critical lodgement index
h_s	vertical thickness of surface sediment layer accumulating on fluvial bedform		L_f	length of leeward face of dune
			L_o	lake level
h_t	elevation of total head		L_t	time lag
h_v	thickness of varve		L_t'	time lag at specified distance
h_w	height of submerged ice wall		L_v	length of glacial lake
h_z	elevation above datum		$L*$	length of flow (in till)
			L'	ice sheet half-width
I	moment of inertia		l	length of side of bed hummock
I_d	proportion of glacier bed covered with debris		l_c	length of crack
			l_e	elongation term
IRD	ice rafted debris		l_g	length of abrasional groove
i	hydraulic gradient		l_r	length of ripples
			l_v	distance of varve thickness sampling point from lake infall
J	mechanical equivalent of heat		l'	height of dunes or ripples

M	ice melt	P_s	precipitation of sediments onto ice surface from atmosphere
M_a	mass of material eroded by abrasion		
M_b	melt rate of iceberg	P_v	vapour pressure or critical pressure at which cavities form
M_c	metacentre of iceberg		
M_d	mass of crushed and fractured material	P_w	water pressure
		P_{wc}	critical water pressure for cavity opening
M_f	mass of specimen after testing		
M_{gm}	ice melt due to geothermal heat	P_z	precipitation on watershed
M_h	Moh's scratch hardness	P_∞	constant external pressure
M_i	mass of specimen before testing	$P(h)$	probability density function of sea ice keel draft
M_m	mass of material removed by mechanical meltwater erosion		
		$P(c)$	probability of particle collisions
M_o	mass of superglacial sediments	P	ploughing force
M_p	mass of particle	p_w	pore water pressure
M_q	mass of chemical substances resulting from meltwater erosion		
		Q	activation energy
M_s	total quantity of matter	Q_g	geothermal heat flux
M_{sm}	ice melt due to shear heating and sliding	Q_h	sensible heat exchange
		Q_i	mass flux or discharge of ice
M_t	melt rate of ice tunnel	Q_i^*	heat flux from sea ice
M'	mass balance of ice sheet or glacier	Q_L	specific latent heat exchange from evaporation or melting of surface lake
m	mean quadratic slope in direction of sliding		
		Q_R	heat input from solar radiation
m'	dry mass of sediments	Q_s	solute discharge
m_*	exponent	Q_s^*	heat flux from the sea
		Q_T	lake heat storage
N_k	constant	Q_w	water discharge
N_p	permeability to liquid water	Q_{wf}	vertical water flux
N_r	reaction rate parameter	q	consolidation factor of melt-out debris
N_t	number of sea ice keels per kilometre	qi	correction for upper, low density firn layer
n	exponent in ice flow law (visco-plastic parameter)		
n_*	porosity	R_e	Reynolds number
n'	Manning roughness coefficient	R_g	gas constant
		R_i	Richardson number
P	pressure	R_t	radius of circular tunnel or pipe
P_a	precipitation over glacier surface	R_t'	total resistance to ice ploughing
P_c	collapse pressure of cavity	R'	amplitude/wavelength ratio of glacier bed
P_{cg}	capillary pressure		
P_g	pressure gradient	R''	hydraulic radius
P_{go}	partial pressure of non-condensible gas at specified bubble radius	R_*	coefficient depending upon ice/water interface roughness and current velocity
P_h	hydrostatic pressure		
P_i	cryostatic pressure (ice overburden pressure)	r	radius or subscript for rock
		r_b	radius of bubble prior to collapse
P_n	normal pressure	r_c	particle or clast radius
P_n'	effective normal pressure	r_d	local sediment deposition rate
P_o	local or reference water pressure	r_f	radius of sea ice floe

r'	radius of asperity tip	t'	time since base of an ice shelf contacted water
r^*	transitional particle radius		
r^{**}	distance from centre of meltwater tunnel	U	total horizontal ice velocity
		U_b	velocity at bed of an ice mass
S_{alb}	salinity of deep ocean water in fjord	U_c	velocity due to creep
S_{alv}	salinity of upper, brackish water layer near a glacier front	U_{cw}	critical velocity for breaking of internal hydrodynamic waves
S_b	storage component of subglacial sediments	U_e	velocity of fluid relative to sphere
S_c	size grade attained by 80% sample particle	U_{en}	entrainment velocity in estuarine circulation
S_d	deposition rate of sediments	U_f	drift velocity of sea ice floe/iceberg
S_{ds}	deposition rate of suspended sediments	U_h	velocity of head of turbidity current
S_{du}	upstream deposition rate of suspended sediments	U_n	viscous drag by ice flow towards the bed
S_{dd}	downstream deposition rate of suspended sediments	$\left.\begin{array}{l}U_o(0), \\ U_o(y1)\end{array}\right\}$	Heaviside step function
S_e	sorting coefficient	U_p	velocity of clast
S_i	storage of sediments held in ice	U_p'	relative velocity of clast
SQ_m	discharge of sediments in ice	U_r	relative velocity decrease of clast from frictional drag
SQ_w	discharge of sediments in meltwater		
S_{rd}	local deposition rate of suspended sediments	U_{ry}, U_{rx}	vertical and horizontal velocity components of ripples
S_t	shear strength of wet sediment	U_{ry}', U_{rx}'	vertical and horizontal velocity components of ripples generated by local, non-uniform or transient flow
S_w	internal glacier water storage		
S_*	tunnel cross-section		
s	parameter related to dimensions of bed hummock	U_s	velocity due to basal sliding
		U_{sc}	velocity due to enhanced ice straining
s'	bulk shortening of specimen	U_{sr}	velocity due to regelation
s_r	ratio of bed roughness to clast size	U_{ud}	upward velocity of debris in ice
		U_t	particle fall velocity
T	temperature	U_{tc}	tidal current velocity
T_a	absolute temperature	U_u	vertical ice velocity
T_b	temperature at glacier bed	U_w	velocity of water
T_c	rate of tunnel closure	U_{w*}	velocity of upper water layer relative to lower layer
T_f	tangential stress		
T_f'	freezing temperature		
T_m	temperature at which maximum density occurs	V	volume
		V_c	volume of clast
T_l	thickness of reactive laminar layer	V_d	volume of solids precipitated onto ice surface
T_s	temperature at glacier surface		
T_{sea}	temperature of sea water	V_f	volume of sea ice floe
T_{ss}	temperature of sea surface	V_g	volume of groove
T_{sd}	temperature of debris surface	V_m	volume of subglacial water transported
TDS	total dissolved solids		
t	time	V_n	coefficient proportional to Froude number

V_o	initial volume of solid		clast to x-direction (θ_x)
v	Poisson's ratio	γ'	coefficient related to controlling obstacle size
W	wear rate	γ	shear strain
W_b	width of iceberg	γ_*	ratio of horizontal to vertical ice
W_c	energy required for comminution		velocities
W_c'	energy required for comminution accounting for energy required to take clast to point of fracture	ϵ	strain
		ϵ_f	volumetric freezing strain
W_i	weight of iceberg	ϵ_s	steady-state (secondary) creep
W_r	abrasional wear	ϵ_n	normal strain
W_x	channel width	ϵ_x	longitudinal strain
$W*$	load	ϵ_{xy}	shear strain
wc	water content	ϵ_z	vertical strain
x	distance in direction of flow	θ	angle related to surface loading and potential fracture plane
x_c	distance from ice sheet centre to point where basal debris crops out on ablation zone surface	θ_a	half-angle of conical asperity tip
		θ_j	junction angle
x_e	distance to equilibrium line from ice sheet centre	θ_t	change in temperature with pressure
		θ_x	angle of clast to x-direction
x_m	distance from ice sheet centre to zone of basal melting and regelation	$\theta*$	angle swept out by pivoting clast
y	direction perpendicular to x-direction	η	viscosity
		η_d	viscosity of debris
Z	height above datum	η_i	viscosity of ice
Z_s	height of glacier surface	η_w	viscosity of water
Z_w	depth below phreatic surface		
z	direction perpendicular to x–y plane	Λ	controlling bed obstacle size
		Λ_g	heat due to geothermal heating
α	an angle	Λ_H	heat due to shear heating
α_d	surface slope of debris	Λ_s	heat due to sliding
α_i	impact angle of suspensoids with meltwater channel	Λ_T	total volume of basal ice melt
		λ_b	spacing of bed hummocks
α_r	angle between two superimposed clasts and direction of motion	λ_p	dominant wavelength
		λ_r	ripple wavelength
α_s	slope of snow or ice surface	λ_t	transitional wavelength of bed undulations
β	bed slope	λ_u	wavelength of bed undulations
β_{min}	minimum slope for debris flowage	$\lambda*$	diffusion distance
β_t	constant related to longitudinal tension over bed hummock	μ_c	friction between bed and clast
		μ_k	coefficient of kinetic friction
β_w	angle at which bed steps are tilted upglacier	μ_s	coefficient of static friction
Γ	difference between angle of principal stress (σ_1) and angle of inclination of	Ξ	balance term between fluvial erosion and deposition

ξ	bed roughness factor (power spectral density term)	σ_t	yield strength (in tension)
		σ_{tan}	tangential stress
ξ_s	radar skin depth	σ_y	yield strength (in compression)
		σ'	effective stress
Π	thickness of basal water film	σ_θ	azimuthal stress
ρ	density	$\tau_{(xy)}$	shear stress
ρ_c	density of particles	τ_b	basal shear stress
ρ_d	dry bulk density of rock	τ_{bc}	shear stress of clast–bed interface
ρ_i	density of ice	τ_{ci}	critical shear stress of solid interface
$\rho_i{}^*$	density of ice with dispersed debris	τ_p	peak shear stress (shear yield strength)
ρ_m	bulk density of debris	τ_{max}	maximum shear stress
ρ_r	density of rock	τ_o	cohesive or residual shear strength
ρ_s	density of flow till	τ_s	shear strength of sediments
$\rho_s{}^*$	density of influent stream water		
ρ_u	density of upper water layer	Φ_f	fluid potential
ρ_w	density of water	ϕ	angle of internal friction
ρ_{wl}	density of lake water	ϕ_r	angle of climb of ripples
		$\phi_1{}^*, \phi_2{}^*$	angle between principal (1) and intermediate (2) stresses and glacier bed
$\sigma_{(\,)}$	effective normal stress (subscripts refer to primary, intermediate and minor stresses)		
σ_c	force required to overcome interatomic bonds	ψ	coefficient related to pressure melting point
σ_{cav}	cavitation condition	ψ'	arbitrary scale factor
σ_{crit}	critical value for cavitation		
σ_n	critical applied stress	Ω	viscous drag
σ_p	unconfined compressive strength (ratio peak stress to initial cross-sectional area)	Ω_b	total moment
		Ω_t	torque
		Ω_w	electrical conductivity of meltwater
σ_r	radial stress	ω_c	angular velocity

Glossary

Ablation Net loss of mass by melting, sublimation, blown snow, iceberg calving. Units in m or kg m^{-2} a^{-1} (also in water equivalent).

Abrasion Mechnaical wear of rock surfaces by rubbing and impact.

Accumulation Net gain of mass by snowfall, avalanching, add-freezing. Units: m or kg m^{-2} a^{-1}.

Activation energy The smallest energy that will raise the free energy of a molecule or atom so as to enable it to rearrange itself into a more stable configuration.

Add-freezing Net gain to an ice mass by freezing-on, usually at the base of an ice shelf. Units m or kg m^{-2} a^{-1}.

Adhesion Act of sticking to a surface produced by forces between molecules.

Adsorption Creation of a layer or zone of a substance (adsorbate) on a surface.

Advection A process of large-scale transfer by horizontal motion in an ice mass (also in atmosphere and oceans).

Aerosol A dispersion of solid or liquid particles in a gas.

Albedo The component of incident light diffusely reflected from an ice or snow surface. Freshly fallen snow posseses a high albedo of between 0.9 and 0.8; old snow 0.7; thawing snow between 0.3 and 0.65; ice 0.3–0.6.

Allochthonous Referring to a sedimentary body composed of fragments (particles or clasts) transported to the site of deposition (i.e. not created *in situ*).

Amber ice North American usage for ice containing a dispersion of fine-gained sediments which impart an amber-like appearance. Usually located in the uppermost sections of basal ice.

Amictic Perennially frozen lake, no seasonal water turnover.

Ankerite Carbonate mineral, usually, $FeCO_3$.

Anion An ion carrying a negative charge.

Anti-dune Small-scale sedimentary cross-bedding of ripple-like character produced by ripple-migration against the predominant current direction (by erosion on lee-face, deposition on stoss-face) usually by a high current velocity when water surface develops standing waves.

Argillaceous Fine-grained detrital sedimentary rocks (e.g. clay, mudstone, shale, marl, siltstone) dominated by clay and silt grades (i.e. particles less than 0.0625 mm).

Asperity Small protuberance on the surface of a rock or clast.

Attrition Process of progressive wear through continuous or semi-continuous contacts; applicable to both bedrock and clasts.

Authigenic Mineralogical change or development *in situ* during or after deposition (e.g. secondary overgrowth).

Autochthonous Referring to a sediment body which has developed *in situ*, without the transport of constituents.

Avalanche Catastrophic failure and transport, usually of unconsolidated debris and/or snow, on steep slopes.

Bar (i) cgs unit of pressure equal to 10^5 N m^{-2} (0.1 MPa). (ii) partly-linear deposit of sand and/or gravel forming in a river.

Barotropic current A current in ocean or estuary induced by longitudinal pressure gradient force due to surface slope.

Basal shear stress Force exerted by an ice mass at its base. It may be equated with 'friction' between ice and the bed.

Bedform Organization of sediment into morphological units (e.g. current ripples, dunes, sandwaves, etc.).

Bernouilli's Theorem At any point in a tube through which a fluid is flowing the sum of pressure energy, potential energy and kinetic energy is constant:

$$\int \frac{dP}{\rho} - \frac{d\phi}{dt} \pm \frac{1}{2} v^2 \pm V = A$$

where P is pressure of fluid of density, ρ; v is the fluid velocity; ϕ the velocity potential; and V the gravitational potential.

Benthic Pertaining to the sea floor (i.e. great depth).

Biological productivity Organic fertility of a water mass (lake, sea, ocean, etc.).

Biomass Amount of living organisms or matter in a given area. Unit: weight or volume of organisms per unit area or volume of the environment.

Bioturbation Disturbance to and destruction of primary sedimentary units (bedding, structures, lamination, etc.) by the activity of animals particularly boring and burrowing.

Bottomsets One of three principal types of bed in a delta and forming on the sea or lake floor in the distal part of the delta (cf. topset, foreset).

Bottom traction current Flow close to the seabed or a

river bed which can winnow sediments and transport fine particles.

Boundary layer A thin layer or narrow zone close to the solid boundary of a flowing substance (e.g. water, ice, mud, etc.) in which fluid viscosity is of major importance. Beyond the boundary layer friction is negligible. The thickness of a layer is often proportional to: $\sqrt{\eta/\rho}$, where η = viscosity and ρ = density.

Braiding In a stream the characteristic of anastomosing (inter-weaving) channels which constantly shift between shallow banks of sediments (bars). Braided streams are typical in areas of steep gradients and high discharge.

Braidplain Extensive region of gently sloping terrain composed of interlapping alluvial fans resulting from the sediment discharge of glacial meltwater streams.

Break-up The disintegration of annual sea-ice (also of lake and river ice) by wave, wind, current or tidal action in spring, as a result of weakening of the ice by surface melting and the development of drainage canals.

Brine Salt–water mixture.

Brittle behaviour Exhibited by rocks which fracture at low strains (i.e. 3–5%).

Brownian motion Random, erratic movement, usually of particles.

Buoyancy Upward thrust experienced by a body submerged in a fluid and equal to the fluid weight that is displaced.

Buoyant weight Apparent and lesser weight of a body when immersed in a fluid due to the upward thrust it experiences.

Carbonation Chemical weathering process in which weak carbonic acid solutions (i.e. carbon dioxide contained in water) converts oxides of calcium, potassium, sodium, magnesium and iron into carbonates or bicarbonates.

Cataclasis Mechanical breakdown of rocks accomplished by fracture and rotation of mineral grains or aggregates without chemical reconstitution and at low temperatures.

Cation Positively charged ion.

Cavitation Growth and collapse of bubbles in a fluid due to the local static pressure in the fluid being less than the fluid vapour pressure. Bubble collapse generates shock waves.

Cavities Space at base of a glacier, usually in the lee of a bed protuberance.

Cenozoic Geological Period from end of Mesozoic to present (65–0 Ma BP)

Ceramic Usually an artificial product of clay or other silicates but may generally refer to non-metallic, crystalline materials such as ice.

Chattermarks Sometimes also referred to as 'friction cracks', which are crescent-shaped grooves on bedrock caused by stick-slip motion of basal debris over the bed. (cf. sichelbruch)

Chezy Formula Derivation for water flow in a wide, open channel in which depth is assumed to be equal to the hydraulic radius (R):

$$V = c\sqrt{Rs}$$

where V = mean velocity of flow; s = water-surface slope, c = constant dependent on gravity and friction forces.

Cirque Semi-circular hollow eroded into a highland terrain by small cirque glacier (syn. corrie, cwm, coire, kar, combe).

Clausius–clapyron equation Relates the variation of freezing (or boiling) point of a liquid to applied pressure.

Cleavage plane Smooth surface of a crystalline substance along which it parts easily or fractures.

Cohesion Force holding a solid or liquid together due to molecular attraction.

Cold welding Forcing together of two substances at ambient temperatures, often in a shearing manner and with consequent adhesion.

Colloid A substance (particle) with properties different from the solution in which it resides, usually large in size (e.g. $10^{-4} – 10^{-6}$ mm).

Comminution Process of size reduction of particles usually resulting in an increase in total number of particles or fragments.

Competence The ability of a current of water to transport debris.

Composite Usually a mixture of a stiff material with a polymer (e.g. carbon-fibre-reinforced polymers). Wood is a natural composite of lignin (an amorphous polymer) stiffened with cellulose fibres.

Conduit A pipe, tube or channel which conducts water (in an ice mass).

Confining stress A stress or pressure applied equally on all sides (e.g. hydrostatic or glacistatic stress).

Continuity equation In fluids the CE states that the rate of flow through one section of a channel equals the rate of flow through another section (of the same channel) with specific additions or subtractions.

Controlling obstacle size In the theory of glacier basal sliding by J. Weertman the size of a bed protuberance, hummock or obstacle for which ice flow by the two mechanisms of enhanced creep and regelation is least favourable, thus determining the overall sliding velocity.

Convective mixing Homogenization of some property in a fluid resulting from vertical movements due to heat or density variations.

Copepods A small crustacean with no carapace or compound eyes.

Coriolis force The inertial force associated with variation in tangential component of the velocity of a particle. It results in an apparent deflection in the centrifugal force generated by the rotation of the earth:

$$k = 2m \, \omega \sin \phi$$

where m = mass; r = radius of body (earth); ω = angular velocity, ϕ = latitude.

Crescentic gouge (see chattermarks and sichelbruch)

Creep Deformation of a material under stress

Crevasse Crack or fissure in an ice mass usually resulting from a tensile stress regime which fractures the brittle upper layers. Crevasses rarely exceed about 30 m depth (depending upon temperature) as they are closed at depth by creep.

Cross-stratification Syn. cross-bedding and cross-lamination. Possesses minor beds or laminae inclined more or less regularly in straight sloping lines or concave forms at a variety of angles to the original depositional surface or principal bedding plane. Generated by rapidly changing local currents.

Crushing Process by which a homogeneous material is destroyed or reduced to finer fragments resulting from the imposition of a high load.

Cryostatic pressure Pressure exerted by superimposed ice load (cf. hydrostatic pressure).

Darcy's Law Describes flow in a permeable medium in which fluid discharge (Q) can be described by:

$$Q = kiA$$

where A = cross sectional area of discharge; i = hydraulic gradient (head loss divided by length of flow); k = constant of proportionality termed coefficient of permeability.

D'arcy-weisbach equation Describes the pressure difference between two points in a pipe in terms of mean flow and a friction coefficient. Used for calculating head loss.

Decollement Plane of decoupling between two units and along which there may be relative motion, typically sliding.

Deflation Transportation by wind of loose surface debris.

Desorption Reverse of adsorption (q.v.).

Detrital Pertaining to particles or minerals and occasionally small rock clasts derived from the breakdown, by weathering or erosion, of a pre-existing rock.

Diachronous Time transgressive, as applied to geological units having the appearance of continuous beds but which exhibit development at different places at different times.

Diamict A non-genetic term covering both diamict*itite* and diamict*on*. The former is used to describe a poorly sorted, non-calcareous, terrigenous sediment possessing a wide range of particle sizes. The latter is used for a non-lithified form of diamictite. A *glacial* diamict would imply an ice-related genesis and would describe till.

Diatom Single-cell marine plant or freshwater plant possessing siliceous frustules.

Diffusion The process by which one substance tends to mix with another. The rate at which diffusion takes place is an important parameter in many physical processes.

Dimictic A lake exhibiting two overturns or periods of circulation each year (i.e. spring and autumn) as exhibited by deep, freshwater lakes in temperate regions.

Dipole Two equal and opposite charges a short distance apart.

Discharge Volume output of a fluid per unit time. Unit: $m^3 \, s^{-1}$.

Dislocation Defect in a crystalline solid, of two principal types: screw and edge. Dislocations are responsible for creep behaviour.

Dissolution Taking up of a substance by a liquid.

Distributary Irregular, divergent streams flowing away from a main channel but not returning to it.

Doping Addition of an impurity into a substance.

Drag structure Bends or distortions formed by deformation of strata often related to faulting.

Drained strength Strength of a sediment in which, during testing, excess pore pressures are dissipated under conditions of full drainage.

Drape A sedimentary layer or layers laid over a harder core or structure.

Driving stress Mean stresses within an ice mass which are in balance with variable basal shear stresses and longitudinal stress gradients. Computed in a similar manner to basal shear stress. Units: kPa.

Drop stone A clast which falls through a water column into soft or partly-unconsolidated sediment disrupting bedding and other structures. Typically released from melting icebergs and sea ice.

Ductile Ability of a rock to sustain strain or deformation (up to 10%) before fracture.

Dye-tracer A chemical added to stream water in order to trace its route and flow characteristics (e.g. through a glacier). Numerous substances can be used such as common salt (NaCl), Rhodamin-B, and radioactive tags.

Eastwind drift A narrow belt of westerly flowing water close to the Antarctic continent and generated by deflected winds from the ice sheet.

Elastic limit Point at which strain produced in a material results in permanent deformation.

Electrical conductivity Relative ability for a conduction current to flow through a substance under an applied electrical field (reciprocal of resistivity). Units: mhos m^{-1}.

Electrolyte A substance that conducts electricity in a solution due to the presence of ions.

Entrainment Process of incorporation or picking up of material by a variety of media (e.g. ice, water) also the mixing of two differently flowing water bodies by interfacial friction.

Epilimnion Uppermost layer of water in a lake possessing a uniform, and typically warmer, temperature, and exhibiting relatively uniform mixing induced by wave and wind action.

Equilibrium constant Number indicating equilibrium of a chemical reaction and given by the product of the activities of the equation divided by the product of the activities of the reactants.

Equipotential surface A surface on which the potential is everywhere equal for the attractive force concerned.

Esker Sinuous, narrow, steep-sided ridge composed of irregular stratified sediment deposited in contact with ice in either an open channel or an enclosed conduit.

Estuarine circulation A pattern of water circulation in an estuary or fjord in which the mass outward (i.e. seaward transport of fresh water as an uppermost, low density prism (resulting from river or stream input at the estuary head) is compensated by the landward flux of more saline and denser sea water, at depth.

Estuary Mouth or seaward part of a river where there is mixing of freshwater and seawater and where there are pronounced tidal influences.

Euler number Inertial force divided by the pressure-gradient force.

Eutectic The minimum freezing point achieved in a eutectic mixture (i.e. a solid solution of two or more substances which have the property of lowering each other's freezing point).

Euxinic A condition of semi-stagnant marine water bodies, or organic-rich brine pools in which reducing conditions are present.

Evacuation Process(es) by which materials weathered-out or loosened are removed from the place of production.

Extrusion flow A now rejected hypothesis by which the lower parts of a glacier are considered to flow more rapidly due to higher pressures.

Fabric Used typically to describe the degree of orientation of features ranging from individual crystals (minerals), through clasts and cleavage to regional joints and induced by stress or flow conditions.

Facies Assemblage of features which characterize a sediment (also snow and ice) and indicate its depositional environment (e.g. lithology, texture, structures, bedding features, fossil content, etc.). Facies pertaining to rock type are termed lithofacies.

Fall velocity Rate at which a suspended particle will settle through a fluid medium.

Fan A roughly triangular and fan-shaped section of a cone describing a sedimentary depositional form resulting from sudden reduction in stream velocity (associated with transition from highland to lowland).

Fast fracture High-speed failure in a material subject to critical or repeated stresses resulting from the propagation of cracks associated with pre-existing defects.

Fast ice Sea ice that forms along and remains attached to the shore and resulting from either *in situ* freezing or impinging sea ice.

Fatigue Failure of a material through continued cycles of a sub-critical stress.

Fecal pellet Organic excrement principally of invertebrates.

Fines Very small particles, usually the clay and silt-size fractions, in a sediment accumulation.

Firn Intermediate state of transformation of snow to ice. Snow becomes firn with the development of coarse crystals and a density in the order of 400 kg m^3. Firn becomes glacier ice at a density of between 800 and 850 kg m^{-1}, when it is impermeable.

Fjord A narrow, steep-sided and often long inlet, bay or estuary formed by the submergence of a glacial trough.

Flocculation Coagulation of fine, usually clay-sized material into particles of greater mass termed floccules.

Floodstage The stage at which a stream overflows its banks.

Flowline Surface expression of the three-dimensional path traced by a packet of ice within a moving ice mass. Flowlines, on a regional scale, tend to be normal to the surface contours.

Flow till A till that undergoes transport and subsequent modification by flow due to a high water content.

Fluid potential Mechanical energy per unit mass of fluid at any given location referenced to some datum. Proportional to fluid head (product of head and gravitational acceleration).

Fluidization Mixing of a liquid (or gas) with unconsolidated, fine-grained sediments resulting in liquid-like flow.

Foliation The parallel orientation of bands in a material such as rock or ice.

Forefield The zone in front of an ice mass from which it may have recently retreated.

Foreset bed The steep, avalanching frontal slope of a delta.

Fracture toughness A term defining when fast fracture will occur through a crack reaching some initial size in a material, subject to a given stress. Also called critical stress intensity factor. (Units: MN m$^{-3/2}$).

Frazil ice A soupy suspension of small ice crystals which develops in the initial stages of sea, lake or river ice growth.

Free energy (Gibbs Function) The energy that would be freed or required during a reversible process and expressed as:

$$G = (E - T)N$$

where E = heat content or enthalpy; T = absolute temperature; N = entropy.

Freeze–thaw Frost action. The mechanical weathering resulting from repeated freezing and thawing of water in pores and cracks in a rock.

Freeze-up The period of formation of a continuous ice cover on lakes, in rivers and on the sea.

Freeze strain Force due to the change of phase of water held in pores in a rock.

Friction Force resisting relative motion between two contacting surfaces or bodies.

Froude number A dimensionless number (F) used in fluid mechanics to distinguish between rapid and tranquil flow conditions by the ratio of inertial to gravitational forces:

$$F = U^2 /dg$$

where U = mean flow velocity; d = water depth and g = acceleration due to gravity.

Frustule Microscopic, siliceous and often ornate skeleton of a diatom, made up of two valves (epivalve and hypovalve).

Geothermal heat Flux of heat energy resulting from radioactive decay within the interior of the earth. Temperatures increase with depth in the earth at a rate dependent upon regional gradient and thermal conductivity of sub-surface rocks. Significant differences in heat flux are found in stable continental shields, ocean basins, volcanic regions etc. Units: milliwatts per square metre (mW m^{-2}).

Glaci-tectonics Study of major structural features found in ice masses.

Glen Flow Law Empirical relationship describing creep behaviour (cf. viscosity) in ice in which deformation is a function of a power of the applied stress. First proposed by J.W. Glen from experimental work undertaken in the Cavendish Laboratory, Cambridge in the early 1950s.

Gliding Translation of one part of a crystal during creep.

Graded beds Sedimentary units in which there is pronounced sorting in which coarsest material is at the bottom and the finest at the top. Typically produced by turbidity currents.

Granulation Process of reduction into grains or small particles, usually by crushing.

Gravity flow Process of sediment transport down steep slopes in which intergranular friction is overcome to give rise to flow. Four principal flow types may be recognized: grain flow, debris flow, liquid flow and turbidity flow.

Grounding line Zone at which an ice mass entering a water body comes afloat. Characterized by a pronounced change in the slope of the ice surface (decrease) as basal shear stress goes to zero. Found along inner parts of ice shelves, ice tongues and many outlet glaciers.

Gyre An essentially closed circulation system occurring in ocean basins (e.g. Beaufort Gyre)

Halocline A pronounced steepening in the vertical salinity gradient of sea water.

Heat pump A system which obtains heat from a material at a low temperature. Based upon second law of thermo-dynamics in which mechanical work is necessary. If a liquid is expanded and evaporates, heat is extracted from the parent material. If there is subsequent compression of the gas the heat is given up.

Heaviside function A unit or step function which represents a step change in a magnitude with an infinite rate of change.

Hertzian contact Contact between two bodies whose geometry is defined by circular arcs.

Hinge The line or zone along which a change in the

amount and/or direction of dip takes place in a fold.

Homopycnal flow Flow in which influent water is of the same density as the main water body and resulting in easy mixing between the two.

Hooke's law For an elastic material or within the elastic limit of a material the strain or deformation is proportional to the applied stress.

Hydraulic geometry The character of the channels of a river system particularly the variations in top width, mean depth and flow velocity at a particular cross-section and between cross-sections.

Hydraulic gradient In a pipe or conduit, the gradient of a line joining the fluid levels in vertical piezometers at intervals along the pipe.

Hydraulic radius (syn. hydraulic mean depth). Cross sectional area of a stream channel divided by the wetted perimeter.

Hydrograph A graphical representation of river or stream discharge as a function of time at a specified point.

Hyperpycnal flow Flow in which influent water is of greater density that the main body of water and resulting in a descending flow or turbidity current.

Hypolimnion Lowermost layer of water in a lake with typically uniform temperature, often relatively stagnant and poorly oxygenated.

Hypopycnal flow Flow in which influent water is of a lesser density than the main water body resulting in a distinctive near-surface lens or prism.

Hysteresis A phenomenon found frequently in elastic materials whereby on release of an applied stress the strain lags behind (i.e. the strain for a given stress is lower when the stress is increasing than when decreasing).

Ice cap A permanent ice mass of at least an order of magnitude smaller size than an ice sheet (i.e. not exceeding 10^4 km^2).

Ice rafting Process whereby sediments are transported on or in a floating ice mass (sea ice or iceberg) and later released by melting-out or overturning.

Ice sheet A large permanent ice mass in excess of 10^6 km^2.

Ice shelf A large floating ice mass composed of terrestrial ice (often due to coalescence of several ice streams and superimposed snowfall and sometimes basally accreted sea ice). Almost exclusively confined at present to Antarctica. Largest example is Ross Ice Shelf (536 070 km^2).

Ice stream Fast-moving stream of ice embedded within an ice sheet or ice cap. Widths vary from a few to tens of kilometres and lengths up to several hundred kilometres. Velocity can be up to an order of magnitude greater than adjacent ice. Margins frequently marked by extensive crevassing. Occur typically within a few hundred kilometres of the coast.

Imbrication Steeply inclined and overlapping arrangement of platy clasts and other structures.

Inertial interaction Solid contacts between particles during relative motion.

Interflow cf. Homopycnal flow.

Interstitial water Small quantities of water contained within the interstices or voids of a soil.

Inverse Square Law A frequently occurring relationship referring the intensity of an effect to the reciprocal of the square of the distance from the cause (e.g. law of gravitation).

Ion Electrically charged atom. A hydrogen proton (i.e. hydrogen atom without its electron) is a hydrogen ion. (cf. cation, anion).

Ion exchange Replacement of some ions by others in a solution containing ions.

Irradiance Radiant energy incident upon a surface. Units: W m^{-2}.

Isochromatics Study of the interference fringes of uniform hue which are observed with a white light source during photoelastic strain analyses. Fringes join points of equal phase retardation.

Isochrons A notional line joining points of equal age within an ice mass.

Isomorphous substitution Characteristic of two or more crystalline substances to possess similar chemical compositions and to be interchangeable.

Isothermal State of maintaining constant temperature.

Isotope Atoms of the same element (i.e. with same atomic number) which differ in mass number. Different isotopes of an element possess different numbers of neutrons in their nuclei.

Jökulhlaup (Icelandic) Catastrophic burst of water from a glacier or ice cap, resulting from the sudden release of super-, en-, or sub-glacial water. Often occur periodically.

Junction growth The increase in real area of contact between two bodies due to deformation associated with increased friction forces.

Katabatic wind Cold, high density and near-surface air flow which drains down the slopes of ice masses generating high wind velocities in coastal regions (e.g. in excess of 100 m s^{-1}). Most developed in Antarctica.

Kettle hole Roughly circular depression in a till sheet resulting from the slow melt-out of buried ice and subsidence. Frequently occupied by small lakes.

Kinematics A branch of mechanics which deals with the precise description and measurment of motion but without reference to the forces or masses concerned.

Krumbein roundness A qualitative scheme for estimating the roundness of clasts by visual comparison with a standard set of shapes: the scale extends from 0.1 (angular), to 0.9 (very well rounded). Devised by W.C. Krumbein.

Lag A remnant surface or deposit of relatively unconsolidated materials produced by the selective removal of finer fractions by water or air.

Latent heat Quantity of heat released or absorbed in an isothermal transformation of phase. Specific latent heat of fusion is the heat required to transform a unit mass of a solid to a liquid at the same temperature. Unit: $J\ kg^{-1}$.

Levée A natural sedimentary bank constructed by peak flow (bankfull) discharge of streams and rivers.

Limb The part of a fold lying between two hinges.

Liquidus Locus of points in a temperature-composition diagram indicating maximum solubility or saturation of a solid constituent or phase in the liquid.

Liquifaction Increase in pore water fluid pressure in a sediment which decreases shearing resistance, and resulting in the material behaving like a fluid.

Lithic fragment Rock fragments occurring in a later formed rock. They comprise intraformational or exotic varieties.

Lithofacies Assemblage of specified attributes that distinguish one rock unit from another.

Lithology A general term usually relating to macroscopic characteristics of a rock such as texture and composition.

Littoral Appertaining to the shore. Definitions vary in rigorousness but often applied to the zone between high and low water. Roughly synonymous with beach.

Load structure Sedimentary structure resulting from density instability in unconsolidated, often water-saturated units and include diapirs, flame structures, sand pillows, sandstone balls and convolute laminations.

Lodgement Process whereby material in traction in basal layers of an ice mass ceases to move, becomes (temporarily) stationary and ostensibly part of the bed.

Loose boundary channel A stream or river channel composed on unconsolidated sediments.

Manning formula An empirical method for deriving the discharge of a river:

$$Q = A \left[\frac{1.5}{n} \right] R^{2/3} s^{1/2}$$

where A = cross-sectional area, R = hydraulic radius, s = bed slope, and n = roughness coefficient (the principal source of error).

Mass balance Algebraic sum of accumulation and ablation. Change in mass per unit area relative to the previous summer surface. Unit: $m^3\ m^{-2}\ a^{-1}$ (water equivalent), or m (of ice) $m^{-2}\ a^{-1}$.

Meander Sinuosity exhibited by a river or stream.

Melting point Temperature at which a substance changes from solid to liquid phase.

Metacentre A term used in stability of a buoyant body. The metacentre is defined for a tilted or listing body as the point of intersection of the line joining the centre of gravity (G') and centre of buoyancy of the body when untilted and the vertical through the new centre of buoyancy. If the metacentre lies *above* G' the body is stable, *below* G' unstable and will roll over.

Metalimnion Zone or layer of thermally stratified lake in which there is a pronounced temperature gradient (decreasing with depth). Syn. thermocline.

Metamorphism Process of alteration of earth materials by the application of pressure, heat and, often in the case of rocks, chemically active fluids.

Miller Index A form of crystallographic notation for forms and faces of crystals.

Moduli Factors used in the conversion of units from one system to another.

Mohs' hardness An arbitary scale of hardness applied to minerals ranging from 1 (talc), 5 (apatite) to 10 (diamond).

Mohr-coulomb circle Graphic representation of the stress state in a rock body. The Mohr envelope is that of a series of Mohr circles – the locus of points whose coordinates represent the failure stress.

Monomictic A lake with only one annual turnover or circulation period.

Moulin A near-vertical, often helical pothole in a glacier through which surface meltwater descends into the body of the ice.

Natural frequency Frequency or period of free oscillation of any system.

Nepheloid layer A layer close to the sea bed in which there is an increase in suspended material. Layer thickness is typically a few hundred metres with concentrations of $0.3–0.01\ mg\ 1^{-1}$. Often caused by frictional effects of thermohaline or weak turbidity currents on sea floor muds.

Nunatak (Inuit) rocky eminence rising above surrounding icefield or glacier. Common features in Greenland and Antarctica.

Nye channel A channel cut into bedrock by subglacial meltwater.

Oligotrophic A state characterized by plant nutrient deficiency.

Open channel flow Water flow in a channel in which there is a free upper boundary with the atmosphere.

Osmotic pressure Pressure which it is necessary to apply to a solution to prevent flow of solvent through a semi-permeable membrane separating the solution and the pure solvent.

Overflow cf. hypopycnal flow.

'P' wave Compressional acoustic wave generated by a seismic explosion.

Palaeo-hydrology Study of former hydraulic processes preserved in sedimentary rock sequences.

Particle path 'Trajectory' or locus traced out by a packet of ice within an ice mass by virtue of its horizontal and vertical flow components (related to surface mass balance, strain, ice thickness and distance from ice divide).

Parting structure A plane along which a sedimentary unit will separate.

Pelagic Referring to the open sea environment.

Permeability Ability of a material or aggregate to allow the passage of water. Unit: mm s^{-1}.

pH Measure of acidity. The number of grammes of hydrogen ions (H$^+$) per litre of solution and defined by:

$$pH = \log_{10}(1/H^+)$$

pH of pure neutral water is 7. Addition of acid increases pH. An increase in alkalinity decreases pH.

Photosynthesis Process by which green plants are able to produce carbohydrates from atmospheric CO_2 and water in the presence of sunlight.

Phreatic zone The region of saturation below the water table where particle interstices are filled with water. (Phreatic surface is identical with water table.)

Plankton Small aquatic organisms which float, drift or swim feebly.

Ploughing The gouging, grooving or furrowing action by a harder substance of a softer one. The product of such a process is termed a ploughmark and is usually applied to the effects of iceberg or sea ice keel scouring of sea floor sediments.

Plucking Removal by ice of rock fragments from bedrock, often previously weakened by fractures and/or weathering activity. Principal process is adhesive action of ice onto rock when ice temperatures are close to the melting point.

Poisson's ratio Ratio of lateral to longitudinal strain.

Polarized light Light in which wave vibrations are only in one plane as a result of passing through a polarizer or Nichol prism. In ordinary light wave vibrations take place in all directions normal to the ray propagation direction.

Polycrystalline ice Ice composed of a number of interlocking crystals.

Polymer A large molecule, formed by joining of two or more molecules of the same compound, possessing an identical empirical formula but greater molecular weight.

Polymictic Lake with no pronounced or persistent thermal stratification.

Portal Exit location of a melt stream at the snout or front of an ice mass.

Potential Magnetostatic, electrostatic or gravitational varieties, referenced to a point. The last mentioned represents the work done in bringing unit mass from an infinitely distant place to the point. Gravitational potentials are always negative.

Potentiometric surface Hypothetical surface equivalent to the static head of groundwater, and defined as the level to which water will rise in a borehole or well. A water table is an example of a potentiometric surface.

Precipitate Insoluble substance created by a chemical reaction in a solution.

Pressure gradient Change in pressure with distance. Unit: N m^{-2} m^{-1}

Progradation Sediment build-up in front of some feature, used typically of deltas.

Prolate clast A clast elongated along its major ('a') axis.

Push moraine A ridge of till constructed by the bulldozing effect of an advancing glacier snout. Often occur annually.

Pycnocline A layer of water characterized by a pronounced vertical density gradient.

'R' channel (Röthlisberger Channel). A channel incised upwards into overlying ice by fast flowing subglacial meltwater.

Radio echo sounding Technique using electromagnetic energy to determine ice thickness by accurately measuring the two-way travel time of a transmitted radio pulse, multiplied by the radio wave propagation velocity.

Radiolaria Marine planktonic unicellular animals (protozoa) in which central capsule and outer spicules are siliceous.

Recessional flow Flow characterized by a hydrograph

exhibiting a decrease in runoff rate (e.g. after a melting event or a rainstorm).

Recrystalization Formation of new crystalline grains in a substance (e.g. rock or ice) in the solid state, resulting from stress and temperature influences.

Reeh calving An explanation for the calving of icebergs from the front of glaciers and ice shelves based upon beam theory, developed by Nils Reeh. Stresses are greatest at a cross-section of a floating glacier at a distance of about the ice thickness from the ice front and stresses are of a magnitude sufficient to cause fracture.

Regelation Solidification or refreezing which is experienced by ice close to its melting point following melting induced by sufficient pressure.

Relaxation Time required for a substance or process to return to its normal state after release of some condition.

Resedimentation Process of mobilization of already deposited sediments, their transport, reworking and redeposition.

Reynolds number A non-dimensional ratio of inertial to viscous forces in a fluid flow. The Reynolds Number (Re) is given by:

$$Re = U\rho l/\eta$$

where U = fluid velocity; ρ = fluid density l = channel depth and η = fluid viscosity. Above a certain Re fluid flow is turbulent and below it laminar, and usually corresponds to a critical U.

Rheology Study of the flow and deformation of matter.

Rhythmite Unit of a rhythmic sedimentary succession independent of thickness, bedding complexity, time or seasonal factors.

Richardson number A number describing the balance between density stabilizing forces and the destabilizing velocity shear at the boundary between two fluids in relative motion. The Richardson number is used to define conditions which determine entrainment and diffusion of water between layers in lakes and estuaries.

Ripples Small-scale bedform or irregularity usually composed of material no greater than 0.7 mm. They are roughly triangular in cross-section and with a gentle upstream (stoss) side and a steeper, downstream (lee) side.

Rock flour Residue of clay- or fine silt-grade particles resulting from glacial wear and comminution.

Saltation Movement of particles transported by water

or wind which are too heavy to remain in suspension. Motion is in a serics jumps.

Sandur (Icelandic) Extensive outwash plain consisting of a sequence of coalescing, gently sloping fans of sandy alluvial material deposited principally by meltwater. Plural: sandar

Sandwave Large dune-like sedimentary bedform or bar.

Schollen (German) Large blocks of country rock found in till sheets, particularly those of North Germany.

Sea ice Any form of ice originating from the freezing of sea water.

Seiche A standing-wave oscillation on the surface of an enclosed or semi-enclosed water body such as a lake, varying in period and generated principally by local changes in atmospheric pressure and surface wind or tidal stress.

Sensible heat Heat which produces a change in a body which is detected by the senses. Measured by the product of specific heat capacity, body mass and change in temperature.

Sepiolite Hydrous magnesium silicate $(Mg_2Si_3O_6(OH)_4)$. In its amorphous form it is given the name meerschaum.

Sessile Pertaining to marine or water creatures which are permanently attached by their base to substrata.

Shear strength Internal resistence of a substance to shear stress (akin to cohesion).

Shear zone Region of brecciated material (ice or rock).

Sichelbrüche (German), literally sickle- or crescentic-shaped fracture in bedrock due to stresses created by impinging clasts.

Silico-flagellate A protozoan belonging to the order Chysomonadina with a skeleton of siliceaous rings and spines.

Sintering Transformation of particles into a coherent solid mass under the effects of heat at a temperature below the melting point.

Skin depth Depth below the surface of a material at which incident radiation has attenuated to l/e (i.e. approx. 37%).

Sliding Process by which a glacier moves over its bed. Basal sliding has two principal mechanisms: enhanced creep and regelation.

Slipface Lee surface of a dune or sandwave characterized by advancing foreset beds where material lies close to the angle of rest.

Slump structure A genetic term for any sedimentary structure produced by subaqueous slumping.

Slurry A mixture of solid and liquid which is able to flow.

Smectite Montmorillonite–saponite group of clay minerals.

Sole Term given to the lowermost part of an ice mass. May be distinct, lying above a definable bed or may merge with increasing content of debris into basal sedimentary horizons (sub-sole drift). Sliding may take place at the sole.

Solute A substance which is dissolved in a solvent and hence forming a solution.

Specific heat capacity The quantity of heat needed to raise the temperature of unit mass by one degree. Unit: $J\ kg^{-1}\ K^{-1}$.

Specific yield Storage capacity of rocks with respect to groundwater.

Sprag mark A ploughmark in sea floor sediments created by irregular motion of an ice keel.

Standing wave A waveform on a water body in which there is vertical oscillation without forward migration. US: rooster tail.

Steady state The condition whereby a physical system maintains, over a specified and finite period, equilibrium or balance between two or more major variables.

Stereology Study of the geometrical arrangement of material constituents from section planes and thin sections.

Stilling basin A bowl-shaped depression used in artificial waterways (associated with dams) in which bedload and some suspended load is allowed to concentrate and settle.

Stoke's Law A small sphere will reach a constant velocity when falling through a viscous medium under gravitational forces.

Stoss face Side of a bed hummock or sedimentary ripple facing direction of flow.

Strain Change in shape or volume of a substance as a result of stress-deformation.

Stratigraphy Sequence in time, related characteristics and correlation of stratified materials.

Streak-out Load or drag structure in sediments giving rise to wave or flame shaped plumes.

Striation (striae) grooves produced during wear processes by sliding action, may affect both clasts and bedrock.

Subglacial lake A body of water dammed beneath an ice mass ranging size from a few m^2 to $10^4\ km^2$.

Sub-polar glacier An ice mass which is polythermal, possessing zones in which, during the summer, ice is below the pressure melting point, but on which there is pronounced surface melting.

Surge Catastrophic advance of an ice mass, often periodic, in which velocities suddenly increase by an order of magnitude. Ice surface may become intensely crevassed. Frontal advance may be up to several tens of kilometres.

Suspensoid Particle carried in suspension.

Tabular iceberg A flat-topped iceberg typically originating from the calving of ice shelves or outlet glaciers.

Tabular set A form of cross-bedding in which the units are bound by planar, quasi-parallel surfaces thus forming a tabular body.

Talus (scree) accumulation of clasts resulting predominantly from weathering and free-fall, usually at the foot of steep slopes and cliffs.

Temperate glacier A body of ice which is at or close to the pressure melting point (pmp) throughout a large proportion of its mass. It should not possess any significant zones of cold ice (i.e. well below the pmp).

Tephra Volcanic fragments resulting from eruption (includes ash, scoria, pumice, bombs, lapilli, cinders, etc.).

Terminal grade A characteristic particle size or texture to which rock lithologies are reduced by comminution/abrasion.

Thermal conductivity Rate of transfer of heat along or through a body by conduction. Unit: $W\ m^{-2}\ K^{-1}$.

Thermocline cf. metalimnion.

Thermodynamics General laws governing the processes concerning conservation of energy and heat exchange.

Thermohaline A vertical circulation of seawater created by density differences which result from variations in both temperature and salinity.

Tidewater glacier (principally North American term) A glacier extending down to sea level and entering the open sea or more commonly a bay or fjord.

Till A deposit resulting from the activity of glaciers. Usually poorly sorted, often with massive structure, little graded bedding or lamination, contains striated and angular clasts, often well compacted.

Topset bed Near horizontal layer of sediments deposited on the top of an advancing delta and grading landwards into alluvial deposits. As it advances it truncates and covers foreset beds.

Torque Force or moment of a force which generates rotation.

Torsion Twisting motion about an axis produced by two opposing couples in parallel planes.

Toughness Critical strain energy release rate of a substance.

Torus 'Doughnut-shaped' solid of a circular or elliptical cross-section.

Transitional wavelength Conceptually similar to

controlling obstacle size (q.v.) in Weertman sliding theory and defined by J.F. Nye and B. Kamb in their theories of glacier sliding as a wavelength for bed undulations below which basal sliding takes place by regelation and above which it takes place by creep.

Transmissivity (of light) The ability of a substance and more particularly a fluid, to allow the passage of light.

Tribology (Greek) The study of rubbing surfaces.

Triple junction Zone of communality between three ice crystals forming a small space in which gas or liquid may reside.

Tsunami (Japanese) Large wave in oceans or seas caused by major underwater earthquake shock. Wavelengths are often 50–200 km and group velocities 500–1000 km hr^{-1}.

Turbidite A sediment body resulting from deposition by a turbidity current characterized by graded bedding, moderate sorting and pronounced primary structures.

Turbulence In fluids, motion at any point characterized by rapid variations in magnitude and direction.

TDS Total Dissolved Solids.

Unconformity A major break in time between rock units of deposition and/or structural style. Surfaces of erosion, non-deposition or denudation may constitute the plane of unconformity.

Underflow cf. Hyperpycnal.

Valency Arrangement and hence combining power of an atom determined by organization of outer electrons.

Vapour pressure In an enclosed space the pressure exerted by molecules given off by a substance (liquids and solids).

Varve (see rhythmite)

Veneer A thin but somewhat extensive layer of sediments.

Viscosity Fluid or material property describing resistance to relative motion within itself (deformation). Constant of proportionality termed coefficient of viscosity. Unit: Ns m^{-2}. Kinematic viscosity is the ratio of coefficient of viscosity to density. Unit: m^{-2} s^{-1}.

Wear Mass loss induced by rubbing together of two surfaces.

Weertman film A thin flim or sheet of water (often only a few μm in thickness) at the base of a glacier resting upon an impermeable bed which discharges water produced by sliding and geothermal heating effects.

Westwind drift (Circum-Antarctic Current) the dominant eastwards flow of water in the Southern Ocean.

White roughness A term used in radio engineering but applied to spectra of spatial irregularities (i.e. glacier beds) in which each part of the bed when referred to its mean datum is identical and isotropic at all scales.

Winnowing Size or mass sorting of particles by a current.

Work hardening Strain hardening. The increase in stress needed to produce further deformation in the plastic region. Each strain increment hardens the material and larger stresses are required to achieve further strain.

X-ray diffraction Study of the phenomenon of apparent bending of X-rays when passing near opaque objects which can be used to investigate substances as diffraction depends upon crystal structure.

Yield strength The stress at which permanent deformation first occurs; cf. yield point.

Young's modulus Modulus of elasticity. Ratio of the stress per unit area to the longitudinal strain.

Znigg form field A method of describing the form of a clast, devised by Th. Znigg, in which four form fields are isolated – spheres, rods, discoids and blades – on a triangular diagram defined by various sphericity ratios. A more flexible and useful trivariant system is the Folk-sphericity-form diagram.

Index